T0199813

Parabolic Equations
on an Infinite Strip

PURE AND APPLIED MATHEMATICS

A Program of Monographs, Textbooks, and Lecture Notes

EXECUTIVE EDITORS

Earl J. Taft
Rutgers University
New Brunswick, New Jersey

Zuhair Nashed
University of Delaware
Newark, Delaware

CHAIRMEN OF THE EDITORIAL BOARD

S. Kobayashi
University of California, Berkeley
Berkeley, California

Edwin Hewitt
University of Washington
Seattle, Washington

EDITORIAL BOARD

M. S. Baouendi
Purdue University

Donald Passman
University of Wisconsin-Madison

Jack K. Hale
Brown University

Fred S. Roberts
Rutgers University

Marvin Marcus
University of California, Santa Barbara

Gian-Carlo Rota
Massachusetts Institute of Technology

W. S. Massey
Yale University

David Russell
University of Wisconsin-Madison

Leopoldo Nachbin
Centro Brasileiro de Pesquisas Físicas
and University of Rochester

Jane Cronin Scanlon
Rutgers University

Anil Nerode
Cornell University

Walter Schempp
Universität Siegen

Mark Teply
University of Wisconsin-Milwaukee

MONOGRAPHS AND TEXTBOOKS IN
PURE AND APPLIED MATHEMATICS

1. *K. Yano,* Integral Formulas in Riemannian Geometry (1970)*(out of print)*
2. *S. Kobayashi,* Hyperbolic Manifolds and Holomorphic Mappings (1970) *(out of print)*
3. *V. S. Vladimirov,* Equations of Mathematical Physics (A. Jeffrey, editor; A. Littlewood, translator) (1970) *(out of print)*
4. *B. N. Pshenichnyi,* Necessary Conditions for an Extremum (L. Neustadt, translation editor; K. Makowski, translator) (1971)
5. *L. Narici, E. Beckenstein, and G. Bachman,* Functional Analysis and Valuation Theory (1971)
6. *D. S. Passman,* Infinite Group Rings (1971)
7. *L. Dornhoff,* Group Representation Theory (in two parts). Part A: Ordinary Representation Theory. Part B: Modular Representation Theory (1971, 1972)
8. *W. Boothby and G. L. Weiss (eds.),* Symmetric Spaces: Short Courses Presented at Washington University (1972)
9. *Y. Matsushima,* Differentiable Manifolds (E. T. Kobayashi, translator) (1972)
10. *L. E. Ward, Jr.,* Topology: An Outline for a First Course (1972) *(out of print)*
11. *A. Babakhanian,* Cohomological Methods in Group Theory (1972)
12. *R. Gilmer,* Multiplicative Ideal Theory (1972)
13. *J. Yeh,* Stochastic Processes and the Wiener Integral (1973) *(out of print)*
14. *J. Barros-Neto,* Introduction to the Theory of Distributions (1973) *(out of print)*
15. *R. Larsen,* Functional Analysis: An Introduction (1973) *(out of print)*
16. *K. Yano and S. Ishihara,* Tangent and Cotangent Bundles: Differential Geometry (1973) *(out of print)*
17. *C. Procesi,* Rings with Polynomial Identities (1973)
18. *R. Hermann,* Geometry, Physics, and Systems (1973)
19. *N. R. Wallach,* Harmonic Analysis on Homogeneous Spaces (1973) *(out of print)*
20. *J. Dieudonné,* Introduction to the Theory of Formal Groups (1973)
21. *I. Vaisman,* Cohomology and Differential Forms (1973)
22. *B. -Y. Chen,* Geometry of Submanifolds (1973)
23. *M. Marcus,* Finite Dimensional Multilinear Algebra (in two parts) (1973, 1975)
24. *R. Larsen,* Banach Algebras: An Introduction (1973)
25. *R. O. Kujala and A. L. Vitter (eds.),* Value Distribution Theory: Part A; Part B: Deficit and Bezout Estimates by Wilhelm Stoll (1973)
26. *K. B. Stolarsky,* Algebraic Numbers and Diophantine Approximation (1974)
27. *A. R. Magid,* The Separable Galois Theory of Commutative Rings (1974)
28. *B. R. McDonald,* Finite Rings with Identity (1974)
29. *J. Satake,* Linear Algebra (S. Koh, T. A. Akiba, and S. Ihara, translators) (1975)

30. *J. S. Golan*, Localization of Noncommutative Rings (1975)
31. *G. Klambauer*, Mathematical Analysis (1975)
32. *M. K. Agoston*, Algebraic Topology: A First Course (1976)
33. *K. R. Goodearl*, Ring Theory: Nonsingular Rings and Modules (1976)
34. *L. E. Mansfield*, Linear Algebra with Geometric Applications: Selected Topics (1976)
35. *N. J. Pullman*, Matrix Theory and Its Applications (1976)
36. *B. R. McDonald*, Geometric Algebra Over Local Rings (1976)
37. *C. W. Groetsch*, Generalized Inverses of Linear Operators: Representation and Approximation (1977)
38. *J. E. Kuczkowski and J. L. Gersting*, Abstract Algebra: A First Look (1977)
39. *C. O. Christenson and W. L. Voxman*, Aspects of Topology (1977)
40. *M. Nagata*, Field Theory (1977)
41. *R. L. Long*, Algebraic Number Theory (1977)
42. *W. F. Pfeffer*, Integrals and Measures (1977)
43. *R. L. Wheeden and A. Zygmund*, Measure and Integral: An Introduction to Real Analysis (1977)
44. *J. H. Curtiss*, Introduction to Functions of a Complex Variable (1978)
45. *K. Hrbacek and T. Jech*, Introduction to Set Theory (1978)
46. *W. S. Massey*, Homology and Cohomology Theory (1978)
47. *M. Marcus*, Introduction to Modern Algebra (1978)
48. *E. C. Young*, Vector and Tensor Analysis (1978)
49. *S. B. Nadler, Jr.*, Hyperspaces of Sets (1978)
50. *S. K. Segal*, Topics in Group Rings (1978)
51. *A. C. M. van Rooij*, Non-Archimedean Functional Analysis (1978)
54. *L. Corwin and R. Szczarba*, Calculus in Vector Spaces (1979)
53. *C. Sadosky*, Interpolation of Operators and Singular Integrals: An Introduction to Harmonic Analysis (1979)
54. *J. Cronin*, Differential Equations: Introduction and Quantitative Theory (1980)
55. *C. W. Groetsch*, Elements of Applicable Functional Analysis (1980)
56. *I. Vaisman*, Foundations of Three-Dimensional Euclidean Geometry (1980)
57. *H. I. Freedman*, Deterministic Mathematical Models in Population Ecology (1980)
58. *S. B. Chae*, Lebesgue Integration (1980)
59. *C. S. Rees, S. M. Shah, and C. V. Stanojevic*, Theory and Applications of Fourier Analysis (1981)
60. *L. Nachbin*, Introduction to Functional Analysis: Banach Spaces and Differential Calculus (R. M. Aron, translator) (1981)
61. *G. Orzech and M. Orzech*, Plane Algebraic Curves: An Introduction Via Valuations (1981)
62. *R. Johnsonbaugh and W. E. Pfaffenberger*, Foundations of Mathematical Analysis (1981)
63. *W. L. Voxman and R. H. Goetschel*, Advanced Calculus: An Introduction to Modern Analysis (1981)
64. *L. J. Corwin and R. H. Szcarba*, Multivariable Calculus (1982)
65. *V. I. Istrătescu*, Introduction to Linear Operator Theory (1981)
66. *R. D. Järvinen*, Finite and Infinite Dimensional Linear Spaces: A Comparative Study in Algebraic and Analytic Settings (1981)

MONOGRAPHS AND TEXTBOOKS IN
PURE AND APPLIED MATHEMATICS

1. *K. Yano*, Integral Formulas in Riemannian Geometry (1970) *(out of print)*
2. *S. Kobayashi*, Hyperbolic Manifolds and Holomorphic Mappings (1970) *(out of print)*
3. *V. S. Vladimirov*, Equations of Mathematical Physics (A. Jeffrey, editor; A. Littlewood, translator) (1970) *(out of print)*
4. *B. N. Pshenichnyi*, Necessary Conditions for an Extremum (L. Neustadt, translation editor; K. Makowski, translator) (1971)
5. *L. Narici, E. Beckenstein, and G. Bachman*, Functional Analysis and Valuation Theory (1971)
6. *D. S. Passman*, Infinite Group Rings (1971)
7. *L. Dornhoff*, Group Representation Theory (in two parts). Part A: Ordinary Representation Theory. Part B: Modular Representation Theory (1971, 1972)
8. *W. Boothby and G. L. Weiss (eds.)*, Symmetric Spaces: Short Courses Presented at Washington University (1972)
9. *Y. Matsushima*, Differentiable Manifolds (E. T. Kobayashi, translator) (1972)
10. *L. E. Ward, Jr.*, Topology: An Outline for a First Course (1972) *(out of print)*
11. *A. Babakhanian*, Cohomological Methods in Group Theory (1972)
12. *R. Gilmer*, Multiplicative Ideal Theory (1972)
13. *J. Yeh*, Stochastic Processes and the Wiener Integral (1973) *(out of print)*
14. *J. Barros-Neto*, Introduction to the Theory of Distributions (1973) *(out of print)*
15. *R. Larsen*, Functional Analysis: An Introduction (1973) *(out of print)*
16. *K. Yano and S. Ishihara*, Tangent and Cotangent Bundles: Differential Geometry (1973) *(out of print)*
17. *C. Procesi*, Rings with Polynomial Identities (1973)
18. *R. Hermann*, Geometry, Physics, and Systems (1973)
19. *N. R. Wallach*, Harmonic Analysis on Homogeneous Spaces (1973) *(out of print)*
20. *J. Dieudonné*, Introduction to the Theory of Formal Groups (1973)
21. *I. Vaisman*, Cohomology and Differential Forms (1973)
22. *B. -Y. Chen*, Geometry of Submanifolds (1973)
23. *M. Marcus*, Finite Dimensional Multilinear Algebra (in two parts) (1973, 1975)
24. *R. Larsen*, Banach Algebras: An Introduction (1973)
25. *R. O. Kujala and A. L. Vitter (eds.)*, Value Distribution Theory: Part A; Part B: Deficit and Bezout Estimates by Wilhelm Stoll (1973)
26. *K. B. Stolarsky*, Algebraic Numbers and Diophantine Approximation (1974)
27. *A. R. Magid*, The Separable Galois Theory of Commutative Rings (1974)
28. *B. R. McDonald*, Finite Rings with Identity (1974)
29. *J. Satake*, Linear Algebra (S. Koh, T. A. Akiba, and S. Ihara, translators) (1975)

30. *J. S. Golan*, Localization of Noncommutative Rings (1975)
31. *G. Klambauer*, Mathematical Analysis (1975)
32. *M. K. Agoston*, Algebraic Topology: A First Course (1976)
33. *K. R. Goodearl*, Ring Theory: Nonsingular Rings and Modules (1976)
34. *L. E. Mansfield*, Linear Algebra with Geometric Applications: Selected Topics (1976)
35. *N. J. Pullman*, Matrix Theory and Its Applications (1976)
36. *B. R. McDonald*, Geometric Algebra Over Local Rings (1976)
37. *C. W. Groetsch*, Generalized Inverses of Linear Operators: Representation and Approximation (1977)
38. *J. E. Kuczkowski and J. L. Gersting*, Abstract Algebra: A First Look (1977)
39. *C. O. Christenson and W. L. Voxman*, Aspects of Topology (1977)
40. *M. Nagata*, Field Theory (1977)
41. *R. L. Long*, Algebraic Number Theory (1977)
42. *W. F. Pfeffer*, Integrals and Measures (1977)
43. *R. L. Wheeden and A. Zygmund*, Measure and Integral: An Introduction to Real Analysis (1977)
44. *J. H. Curtiss*, Introduction to Functions of a Complex Variable (1978)
45. *K. Hrbacek and T. Jech*, Introduction to Set Theory (1978)
46. *W. S. Massey*, Homology and Cohomology Theory (1978)
47. *M. Marcus*, Introduction to Modern Algebra (1978)
48. *E. C. Young*, Vector and Tensor Analysis (1978)
49. *S. B. Nadler, Jr.*, Hyperspaces of Sets (1978)
50. *S. K. Segal*, Topics in Group Rings (1978)
51. *A. C. M. van Rooij*, Non-Archimedean Functional Analysis (1978)
54. *L. Corwin and R. Szczarba*, Calculus in Vector Spaces (1979)
53. *C. Sadosky*, Interpolation of Operators and Singular Integrals: An Introduction to Harmonic Analysis (1979)
54. *J. Cronin*, Differential Equations: Introduction and Quantitative Theory (1980)
55. *C. W. Groetsch*, Elements of Applicable Functional Analysis (1980)
56. *I. Vaisman*, Foundations of Three-Dimensional Euclidean Geometry (1980)
57. *H. I. Freedman*, Deterministic Mathematical Models in Population Ecology (1980)
58. *S. B. Chae*, Lebesgue Integration (1980)
59. *C. S. Rees, S. M. Shah, and C. V. Stanojević*, Theory and Applications of Fourier Analysis (1981)
60. *L. Nachbin*, Introduction to Functional Analysis: Banach Spaces and Differential Calculus (R. M. Aron, translator) (1981)
61. *G. Orzech and M. Orzech*, Plane Algebraic Curves: An Introduction Via Valuations (1981)
62. *R. Johnsonbaugh and W. E. Pfaffenberger*, Foundations of Mathematical Analysis (1981)
63. *W. L. Voxman and R. H. Goetschel*, Advanced Calculus: An Introduction to Modern Analysis (1981)
64. *L. J. Corwin and R. H. Szcarba*, Multivariable Calculus (1982)
65. *V. I. Istrătescu*, Introduction to Linear Operator Theory (1981)
66. *R. D. Järvinen*, Finite and Infinite Dimensional Linear Spaces: A Comparative Study in Algebraic and Analytic Settings (1981)

67. *J. K. Beem and P. E. Ehrlich,* Global Lorentzian Geometry (1981)
68. *D. L. Armacost,* The Structure of Locally Compact Abelian Groups (1981)
69. *J. W. Brewer and M. K. Smith, eds.,* Emmy Noether: A Tribute to Her Life and Work (1981)
70. *K. H. Kim,* Boolean Matrix Theory and Applications (1982)
71. *T. W. Wieting,* The Mathematical Theory of Chromatic Plane Ornaments (1982)
72. *D. B. Gauld,* Differential Topology: An Introduction (1982)
73. *R. L. Faber,* Foundations of Euclidean and Non-Euclidean Geometry (1983)
74. *M. Carmeli,* Statistical Theory and Random Matrices (1983)
75. *J. H. Carruth, J. A. Hildebrant, and R. J. Koch,* The Theory of Topological Semigroups (1983)
76. *R. L. Faber,* Differential Geometry and Relativity Theory: An Introduction (1983)
77. *S. Barnett,* Polynomials and Linear Control Systems (1983)
78. *G. Karpilovsky,* Commutative Group Algebras (1983)
79. *F. Van Oystaeyen and A. Verschoren,* Relative Invariants of Rings: The Commutative Theory (1983)
80. *I. Vaisman,* A First Course in Differential Geometry (1984)
81. *G. W. Swan,* Applications of Optimal Control Theory in Biomedicine (1984)
82. *T. Petrie and J. D. Randall,* Transformation Groups on Manifolds (1984)
83. *K. Goebel and S. Reich,* Uniform Convexity, Hyperbolic Geometry, and Nonexpansive Mappings (1984)
84. *T. Albu and C. Năstăsescu,* Relative Finiteness in Module Theory (1984)
85. *K. Hrbacek and T. Jech,* Introduction to Set Theory, Second Edition, Revised and Expanded (1984)
86. *F. Van Oystaeyen and A. Verschoren,* Relative Invariants of Rings: The Noncommutative Theory (1984)
87. *B. R. McDonald,* Linear Algebra Over Commutative Rings (1984)
88. *M. Namba,* Geometry of Projective Algebraic Curves (1984)
89. *G. F. Webb,* Theory of Nonlinear Age-Dependent Population Dynamics (1985)
90. *M. R. Bremner, R. V. Moody, and J. Patera,* Tables of Dominant Weight Multiplicities for Representations of Simple Lie Algebras (1985)
91. *A. E. Fekete,* Real Linear Algebra (1985)
92. *S. B. Chae,* Holomorphy and Calculus in Normed Spaces (1985)
93. *A. J. Jerri,* Introduction to Integral Equations with Applications (1985)
94. *G. Karpilovsky,* Projective Representations of Finite Groups (1985)
95. *L. Narici and E. Beckenstein,* Topological Vector Spaces (1985)
96. *J. Weeks,* The Shape of Space: How to Visualize Surfaces and Three-Dimensional Manifolds (1985)
97. *P. R. Gribik and K. O. Kortanek,* Extremal Methods of Operations Research (1985)
98. *J.-A. Chao and W. A. Woyczynski, eds.,* Probability Theory and Harmonic Analysis (1986)
99. *G. D. Crown, M. H. Fenrick, and R. J. Valenza,* Abstract Algebra (1986)
100. *J. H. Carruth, J. A. Hildebrant, and R. J. Koch,* The Theory of Topological Semigroups, Volume 2 (1986)

101. *R. S. Doran and V. A. Belfi,* Characterizations of C*-Algebras: The Gelfand-Naimark Theorems (1986)

102. *M. W. Jeter,* Mathematical Programming: An Introduction to Optimization (1986)

103. *M. Altman,* A Unified Theory of Nonlinear Operator and Evolution Equations with Applications: A New Approach to Nonlinear Partial Differential Equations (1986)

104. *A. Verschoren,* Relative Invariants of Sheaves (1987)

105. *R. A. Usmani,* Applied Linear Algebra (1987)

106. *P. Blass and J. Lang,* Zariski Surfaces and Differential Equations in Characteristic p > 0 (1987)

107. *J. A. Reneke, R. E. Fennell, and R. B. Minton.* Structured Hereditary Systems (1987)

108. *H. Busemann and B. B. Phadke,* Spaces with Distinguished Geodesics (1987)

109. *R. Harte,* Invertibility and Singularity for Bounded Linear Operators (1988).

110. *G. S. Ladde, V. Lakshmikantham, and B. G. Zhang,* Oscillation Theory of Differential Equations with Deviating Arguments (1987)

111. *L. Dudkin, I. Rabinovich, and I. Vakhutinsky,* Iterative Aggregation Theory: Mathematical Methods of Coordinating Detailed and Aggregate Problems in Large Control Systems (1987)

112. *T. Okubo,* Differential Geometry (1987)

113. *D. L. Stancl and M. L. Stancl,* Real Analysis with Point-Set Topology (1987)

114. *T. C. Gard,* Introduction to Stochastic Differential Equations (1988)

115. *S. S. Abhyankar,* Enumerative Combinatorics of Young Tableaux (1988)

116. *H. Strade and R. Farnsteiner,* Modular Lie Algebras and Their Representations (1988)

117. *J. A. Huckaba,* Commutative Rings with Zero Divisors (1988)

118. *W. D. Wallis,* Combinatorial Designs (1988)

119. *W. Więsław,* Topological Fields (1988)

120. *G. Karpilovsky,* Field Theory: Classical Foundations and Multiplicative Groups (1988)

121. *S. Caenepeel and F. Van Oystaeyen,* Brauer Groups and the Cohomology of Graded Rings (1989)

122. *W. Kozlowski,* Modular Function Spaces (1988)

123. *E. Lowen-Colebunders,* Function Classes of Cauchy Continuous Maps (1989)

124. *M. Pavel,* Fundamentals of Pattern Recognition (1989)

125. *V. Lakshmikantham, S. Leela, and A. A. Martynyuk,* Stability Analysis of Nonlinear Systems (1989)

126. *R. Sivaramakrishnan,* The Classical Theory of Arithmetic Functions (1989)

127. *N. A. Watson,* Parabolic Equations on an Infinite Strip (1989)

128. *K. J. Hastings,* Introduction to the Mathematics of Operations Research (1989)

Other Volumes in Preparation

Parabolic Equations on an Infinite Strip

N.A. Watson

University of Canterbury
Canterbury, New Zealand

CRC Press
Taylor & Francis Group
Boca Raton London New York

CRC Press is an imprint of the
Taylor & Francis Group, an **informa** business

First published 1989 by Marcel Dekker, Inc.

Published 2019 by CRC Press
Taylor & Francis Group
6000 Broken Sound Parkway NW, Suite 300
Boca Raton, FL 33487-2742

© 1989 by Taylor & Francis Group, LLC
CRC Press is an imprint of Taylor & Francis Group, an Informa business

First issued in paperback 2019

No claim to original U.S. Government works

ISBN 13: 978-0-367-45117-2 (pbk)
ISBN 13: 978-0-8247-7999-3 (hbk)

This book contains information obtained from authentic and highly regarded sources. Reasonable efforts have been made to publish reliable data and information, but the author and publisher cannot assume responsibility for the validity of all materials or the consequences of their use. The authors and publishers have attempted to trace the copyright holders of all material reproduced in this publication and apologize to copyright holders if permission to publish in this form has not been obtained. If any copyright material has not been acknowledged please write and let us know so we may rectify in any future reprint.

Except as permitted under U.S. Copyright Law, no part of this book may be reprinted, reproduced, transmitted, or utilized in any form by any electronic, mechanical, or other means, now known or hereafter invented, including photocopying, microfilming, and recording, or in any information storage or retrieval system, without written permission from the publishers.

For permission to photocopy or use material electronically from this work, please access www. copyright.com (http://www.copyright.com/) or contact the Copyright Clearance Center, Inc. (CCC), 222 Rosewood Drive, Danvers, MA 01923, 978-750-8400. CCC is a not-for-profit organization that provides licenses and registration for a variety of users. For organizations that have been granted a photocopy license by the CCC, a separate system of payment has been arranged.

Trademark Notice: Product or corporate names may be trademarks or registered trademarks, and are used only for identification and explanation without intent to infringe.

**Visit the Taylor & Francis Web site at
http://www.taylorandfrancis.com**

**and the CRC Press Web site at
http://www.crcpress.com**

Library of Congress Cataloging-in-Publication Data

Watson, N. A.
   Parabolic equations on an infinite strip / N. A. Watson
      p.   cm. -- (Monographs and textbooks in pure and applied
  mathematics : 127)
   Bibliography: p.
   Includes index.
   ISBN 0-8247-7999-1
   1. Differential equations, Parabolic. I. Title. II. Series:
  Monographs and textbooks in pure and applied mathematics : v. 127.
  QA371.W34 1988
  515.3'53--dc 19   88-32083
                                          CIP

Preface

This book is concerned with solutions of second order, linear, parabolic partial differential equations on an infinite strip. Particular attention is paid to their integral representation, their initial values in several senses, and the relations between these. The main purpose is to provide a text that takes graduate students rapidly into an area of current research. This is achieved by choosing a narrow field with relatively few prerequisites. Apart from standard undergraduate analysis, the only background knowledge required for reading this book is some general analysis and measure theory which is not always included at the undergraduate level. Some of this material has been included in the present volume; the remainder can be found in Rudin's Real and Complex Analysis. Thus the time spent hunting for the appropriate form of a particular result is minimized, and the contents are made more readily accessible to the student.

In order to keep the main ideas of the arguments as clear as possible, the main body of the text deals only with the heat equation. Each chapter contains a section devoted to the description of the changes necessary to extend its results to more general parabolic

equations. The choice of material on the heat equation in the main part
of the text has been restricted to those theorems and proofs that are
capable of extension to general parabolic equations, although often with
some essential modification. The only exception is Chapter 1, where a
fundamental solution of the heat equation is produced and studied, whereas that of a general parabolic equation must be constructed and many of
its properties are consequences of the construction.

Again in the interests of accessibility, full details of all proofs are
included, and only minimal amounts of notation and terminology are
used. However, a good deal of additional terminology is not only indexed
but also explained in the bibliographical notes at the end of each chapter.
These notes are intended to assist the transition to reading research
papers; to chronicle the development of the topics covered, in conjunction
with the bibliography; and to draw attention to results on the heat equation that have yet to be extended to more general parabolic equations.
The bibliography is intended to be as complete as possible for the
period 1960-85, and only a few papers which contain no new results
have been purposely omitted. Many results on parabolic equations are
analogs of earlier theorems on harmonic functions, but references to
the harmonic case have also been omitted, except in rare instances
where, for one reason or another, they have not already been acknowledged in papers on parabolic equations.

Almost all of the material presented here has not previously
appeared in any book. Indeed, several results have not previously been
published anywhere, so there is something here for the expert as well
as the student. In particular, Sections 2 and 4 of Chapter 4, Sections
2 and 3 of Chapter 6, and Section 4 of Chapter 8 are devoted almost
exclusively to new results.

Without the encouragement of the late Professor Richard Bellman,
I might never have started this book. Without the excellent typing of
Mrs. Ann Tindall, I might never have finished it.

N. A. Watson

Notation and Conventions

We use R^n to denote n-dimensional real euclidean space, $\underline{N}$ the set of natural numbers, $\underline{Q}$ the set of rational numbers, and $\underline{Q}^+$ the set of positive rationals. Most of the material is presented in $\underline{R}^{n+1} = \{(x,t) : x = (x_1, \ldots, x_n) \in \underline{R}^n,\ t \in \underline{R}\}$, the variables $x_1, \ldots, x_n$ being called the space variables, and t the time variable or the exceptional variable. The class of all real-valued functions on $\underline{R}^n$ which possess partial derivatives of all orders is denoted by $C^\infty(\underline{R}^n)$.

The euclidean norm of a point $x \in \underline{R}^n$ is denoted by $\|x\|$. The open ball in $\underline{R}^n$, with center x and radius r, is denoted by $B(x,r)$, and the corresponding closed ball by $\bar{B}(x,r)$. We do not need a special notation for balls in $\underline{R}^{n+1}$. A unit ball is a ball of radius 1, and a unit sphere is its boundary. We use σ_n to denote the surface area of the unit sphere in $\underline{R}^n$, so that

$$\sigma_n = \frac{2\pi^{n/2}}{\Gamma(n/2)}$$

where Γ is Euler's gamma function. The volume of a unit ball in $\underline{R}^n$ is denoted by ω_n, so that

$$\omega_n = \frac{\pi^{n/2}}{(n/2)\Gamma(n/2)} = \sigma_n/n$$

By convention, a/bc means a/(bc) and not (a/b)c. We use m to denote Lebesgue measure in $\underline{R}^n$, so that $\omega_n = m(B(0,1))$. However, in integrals we use the customary dx rather than dm.

If μ is a signed measure on $\underline{R}^n$ with finite total variation, we use μ^+ and μ^- to denote its positive and negative variations, and $|\mu|$ to denote its total variation. If a property holds almost everywhere with respect to a measure ν, we say that it holds ν-a.e. A function is called μ-integrable if its integral with respect to μ exists and is finite, and integrable if it is m-integrable. A function is locally integrable on a set S if it is integrable over every compact subset of S. A measure μ is said to have compact support if there is a compact set K such that $|\mu|(\underline{R}^n \setminus K) = 0$. All measures are assumed to be Borel measures and finite on compact sets. All functions are assumed to be Borel measurable and extended real valued, unless the contrary is explicitly stated. A measure is called singular if it is singular with respect to m.

If f is a function from A into B, we write $f: A \to B$, $f^+(x) = \max\{f(x),0\}$ and $f^-(x) = \max\{-f(x),0\}$ for all $x \in A$. Furthermore, if x_0 is a limit point of A, $g: A \to B$, and $f(x)/g(x) \to 1$ as $x \to x_0$, we write $f(x) \sim g(x)$ as $x \to x_0$. Also, f is said to have compact support if the closure of the set $\{x \in A: f(x) \neq 0\}$ is compact. The set of all continuous real-valued functions on A with compact support is denoted by $C_c(A)$.

For any set E, the boundary of E is written ∂E, the closure of E is $\bar{E}$, and the characteristic function of E is χ_E. If F is another set, then

$$E \setminus F = \{x : x \in E, \ x \notin F\}$$
$$E - F = \{x - y : x \in E, \ y \in F\}$$
$$E + F = \{x + y : x \in E, \ y \in F\}$$

A strip in $\underline{R}^{n+1}$ is usually written as $\underline{R}^n \times \,]0,a[$. This includes the possibility that $a = \infty$, unless otherwise stated.

In $\underline{R}^{n+1}$, partial differentiation with respect to the ith variable is denoted by D_i for $1 \leq i \leq n$, and with respect to the (n+1)th variable by D_t. For a second order derivative, $D_i D_i$ is written as D_i^2.

    If $1 \le p \le \infty$, then L^p denotes the Lebesgue space of all functions f on $\underline{R}^n$ such that

$$\int_{\underline{R}^n} |f(x)|^p \, dx < \infty \qquad (p < \infty)$$

$$\operatorname*{ess\,sup}_{\underline{R}^n} |f| < \infty \qquad (p = \infty)$$

Contents

Preface   iii

Notation and Conventions   v

CHAPTER 1   Fundamental Solutions   1

  1. A fundamental solution of the heat equation   1

  2. Gauss-Weierstrass integrals   6

  3. Initial behavior of Gauss-Weierstrass integrals
      of functions   16

  4. The semigroup property   19

  5. Uniqueness of the Gauss-Weierstrass representation   21

  6. Fundamental solutions of linear parabolic equations   25

  7. Bibliographical notes   30

CHAPTER 2 Non-negative Solutions   33

1. The maximum principle on circular cylinders   33

2. Convergent sequences of integrals   37

3. An extension of the maximum principle to an
   infinite strip   39

4. Some consequences of the extended maximum principle   48

5. Further results on the convergence of sequences
   of integrals   50

6. The Gauss-Weierstrass integral representation
   of non-negative temperatures   57

7. Maximal rates of decay of non-negative temperatures   61

8. Solutions of linear parabolic equations   65

9. Bibliographical notes   67

CHAPTER 3 The Semigroup Property, Cauchy Problem,
           and Gauss-Weierstrass Representation   70

1. The integral means M_b and the classes Σ_b 70

2. The class S_a and the semigroup property   75

3. Behavior of the means M_b 82

4. The class U_a and the Cauchy problem   85

5. Uniqueness of fundamental solutions   95

6. The class R_a and the Gauss-Weierstrass representation   98

7. Linear parabolic equations   103

8. Bibliographical notes   104

CHAPTER 4 Initial Limits of Gauss-Weierstrass Integrals   111

1. Parabolic limits   111

2. Negative results on improving the parabolic
   limit theorem   119

3. Maximal functions   126

4. An improvement of the parabolic limit theorem   137

5. Linear parabolic equations   143

6. Bibliographical notes   143

CHAPTER 5 Normal Limits and Representation Theorems   146

  1. A fundamental relationship   147

  2. A covering theorem   152

  3. Non-negative singular measures and the representation
     of non-negative temperatures   160

  4. Temperatures in the class R_a 165

  5. Linear parabolic equations   177

  6. Bibliographical notes   179

CHAPTER 6 Hyperplane Conditions and
           Representation Theorems   181

  1. Convergent sequences of integrals   182

  2. Temperatures which satisfy L^p conditions   185

  3. Temperatures which satisfy one-sided L^p conditions   194

  4. A result like Egoroff's theorem and a covering lemma   201

  5. Non-negative temperatures and a measure growth condition   205

  6. Linear parabolic equations   215

  7. Bibliographical notes   219

CHAPTER 7 The Initial Measure of a
           Gauss-Weierstrass Integral   221

  1. Preliminary results   221

  2. Determination of the initial measure using
     abundant Vitali coverings   228

  3. Determination of the initial measure under
     milder conditions   231

  4. Linear parabolic equations   240

  5. Bibliographical notes   242

CHAPTER 8 Maximum Principles and Initial Limits   246

  1. The strong maximum principle   247

  2. The weak maximum principle   252

  3. Convergence of sequences of integrals   256

Contents

4. Existence of initial limits   263

5. Linear parabolic equations   271

6. Bibliographical notes   272

References   275

Index   287

Parabolic Equations
on an Infinite Strip

1

Fundamental Solutions

In this chapter, we try to convey the essence of the basic concept of a fundamental solution of the heat equation. This is a solution from which other solutions can be generated by means of an integral transform acting on functions and signed measures, and the results presented here are used repeatedly throughout the book. In section 6, which does not contain any proofs, we outline the corresponding situation for more general second order, linear, parabolic, partial differential equations, and describe how far the results and methods of sections 1 to 5 can be generalized.

1. A FUNDAMENTAL SOLUTION OF THE HEAT EQUATION

Let E be an open set in $\underline{R}^{n+1} = \{(x,t) : x \in \underline{R}^n, t \in \underline{R}\}$. We say that a function $u : E \to \underline{R}$ is a <u>solution of the heat equation</u> on E if $D_1^2 u$, ..., $D_n^2 u$, and $D_t u$ exist, are continuous functions, and satisfy

$$\sum_{i=1}^{n} D_i^2 u - D_t u = 0$$

throughout E. A solution of the heat equation will usually be called a
temperature.

Let S denote the strip $\underline{R}^n \times [0, a[$, where $0 < a \leq \infty$. A fundamental
solution of the heat equation is a function Γ on $S \times S$ which has the following properties:

(i) For each fixed point (y_0, s_0) of S, the function

$$(x, t) \longmapsto \Gamma(x, t; y_0, s_0)$$

is a solution of the heat equation on the substrip $\underline{R}^n \times]s_0, a[$.

(ii) For each continuous, real-valued function f on $\underline{R}^n$ which has
compact support, and each $s \in [0, a[$,

$$\int_{\underline{R}^n} \Gamma(x, t; y, s) f(y) \, dy \to f(\xi)$$

as $(x, t) \to (\xi, s^+)$, for all $\xi \in \underline{R}^n$.

A fundamental solution of the heat equation can be given explicitly.
For all $(x, t) \in \underline{R}^n \times]0, \infty[$, we put

$$W(x, t) = (4\pi t)^{-n/2} \exp(-\|x\|^2/4t)$$

Throughout this book, the symbol W will be reserved for this function.
Then Γ_0, defined by

$$\Gamma_0(x, t; y, s) = \begin{cases} W(x - y, t - s) & \text{if} \quad t > s \\ 0 & \text{if} \quad t \leq s \end{cases}$$

is a fundamental solution of the heat equation on $\underline{R}^n \times [0, a[$ for any
$a > 0$, provided that we restrict it to those values of s and t which lie
in $[0, a[$. The first property (i) of a fundamental solution is easily veri-
fied using elementary calculus; in fact, for any fixed point $(y_0, s_0) \in$
$\underline{R}^{n+1}$, the function $(x, t) \longmapsto \Gamma_0(x, t; y_0, s_0)$ is a temperature on $\underline{R}^{n+1} \setminus$
$\{(y_0, s_0)\}$. To show that Γ_0 possesses the second property (ii) is more
difficult, and we break up the proof into several steps.

Lemma 1.1

Whenever $x \in \underline{R}^n$ and $s, t \in \underline{R}$ with $t > s$, we have

$$\int_{\underline{R}^n} W(x - y, t - s) \, dy = 1$$

Proof. Put $r = \|x - y\|$ and $\tau = 4(t - s)$. Then, if σ_n denotes the surface area of a unit sphere in $\underline{R}^n$, a change to polar coordinates gives

$$\int_{\underline{R}^n} W(x - y, t - s) \, dy = (\pi \tau)^{-n/2} \int_0^\infty \exp(-r^2/\tau) \sigma_n r^{n-1} \, dr$$

Putting $\rho = r^2/\tau$, we deduce that

$$\int_{\underline{R}^n} W(x - y, t - s) \, dy = (\sigma_n/2) \pi^{-n/2} \int_0^\infty e^{-\rho} \rho^{(n/2)-1} \, d\rho = 1$$

Lemma 1.2

If δ is an arbitrary positive constant, $\xi \in \underline{R}^n$ and $s \in \underline{R}$, then

$$\int_{\|\xi - y\| > \delta} W(x - y, t - s) \, dy \to 0$$

as $(x, t) \to (\xi, s^+)$.

Proof. Suppose that $\|x - \xi\| \le \delta/2$. Then, for all y such that $\|\xi - y\| > \delta$, we have

$$\|x - y\| \ge \|y - \xi\| - \|\xi - x\| > \delta/2 \ge \|x - \xi\|$$

so that

$$\|\xi - y\| \le \|\xi - x\| + \|x - y\| \le 2\|x - y\|$$

and hence

$$\exp\left(-\frac{\|x - y\|^2}{4(t - s)}\right) \le \exp\left(-\frac{\|\xi - y\|^2}{16(t - s)}\right)$$

Therefore, writing $\tau = 4(t - s)$, we obtain

$$\int_{\|\xi-y\|>\delta} W(x - y, t - s)\, dy \le \int_{\|\xi-y\|>\delta} (\pi\tau)^{-n/2} \exp(-\|\xi-y\|^2/4\tau)\, dy$$

so that a change to polar coordinates, with $r = \|\xi - y\|$, gives

$$\int_{\|\xi-y\|>\delta} W(x-y, t-s)\, dy \le (\pi\tau)^{-n/2} \sigma_n \int_\delta^\infty \exp(-r^2/4\tau) r^{n-1}\, dr$$

Putting $\rho = r^2/\tau$, we deduce that

$$\int_{\|\xi-y\|>\delta} W(x-y, t-s)\, dy \le (\sigma_n/2)\pi^{-n/2} \int_{\delta^2/\tau}^\infty e^{-\rho/4}\rho^{(n/2)-1}\, d\rho$$

As $t \to s^+$ we have $\tau \to 0^+$, and therefore $\delta^2/\tau \to \infty$, so that the last integral tends to zero. Hence, given $\epsilon > 0$, we can find $\eta > 0$ such that

$$\int_{\|\xi-y\|>\delta} W(x - y, t - s)\, dy < \epsilon$$

whenever $0 < t - s < \eta$ and $\|x - \xi\| \le \delta/2$.

Theorem 1.1

If $\xi \in \underline{R}^n$, $s \in \underline{R}$, and f is a bounded function on $\underline{R}^n$ which is continuous at ξ, then

$$\int_{\underline{R}^n} W(x - y, t - s)f(y)\, dy \to f(\xi) \tag{1}$$

as $(x, t) \to (\xi, s^+)$. Therefore, in particular, the function Γ_0 defined above is a fundamental solution of the heat equation on any strip $\underline{R}^n \times [0, a[$.

Proof. By Lemma 1.1,

$$\int_{\underline{R}^n} W(x-y, t-s)f(y)\, dy - f(\xi) = \int_{\underline{R}^n} W(x-y, t-s)(f(y) - f(\xi))\, dy$$

so that

$$\left| \int_{\underline{R}^n} W(x-y, t-s)f(y) \; dy - f(\xi) \right| \leq \int_{\underline{R}^n} W(x-y, t-s)|\, f(y) - f(\xi)|\; dy \tag{2}$$

Since f is continuous at ξ, given $\epsilon > 0$ we can find $\delta > 0$ such that

$$|\, f(y) - f(\xi)\,| \; < \; \epsilon \quad \text{whenever} \quad \|\, y - \xi\,\| \leq \delta$$

Furthermore, since f is bounded on $\underline{R}^n$, there is a constant M such that $|\, f(y)\,| \leq M$ for all y, and hence, in particular

$$|\, f(y) - f(\xi)\,| \leq |\, f(y)\,| + |\, f(\xi)\,| \leq 2M$$

whenever $\|\, y - \xi\,\| > \delta$. It therefore follows from (2) that

$$\left| \int_{\underline{R}^n} W(x - y, t - s)f(y) \; dy - f(\xi) \right|$$

$$\leq \int_{\|y-\xi\|\leq\delta} W(x - y, t - s)|\, f(y) - f(\xi)|\; dy$$

$$+ \int_{\|y-\xi\|>\delta} W(x - y, t - s)\; |f(y) - f(\xi)|\; dy$$

$$< \epsilon \int_{\|y-\xi\|\leq\delta} W(x - y, t - s)\; dy + 2M \int_{\|y-\xi\|>\delta} W(x - y, t - s)\; dy$$

$$< \epsilon + 2M \int_{\|y-\xi\|>\delta} W(x - y, t - s)\; dy$$

by Lemma 1.1. As $(x, t) \rightarrow (\xi, s^+)$, the last integral tends to zero, by Lemma 1.2. We can therefore find $\eta > 0$ such that

$$\left| \int_{\underline{R}^n} W(x - y, t - s)f(y) \; dy - f(\xi) \right| < 2\epsilon$$

whenever $\|x - \xi\|^2 + (t - s)^2 < \eta^2$ and $t > s$.

The delta-function property. The Dirac δ-measure concentrated at a fixed point ξ of $\underline{R}^n$, is the measure δ_ξ defined by

$$\delta_\xi(A) = \begin{cases} 1 & \text{if} \quad \xi \in A \\ 0 & \text{if} \quad \xi \notin A \end{cases}$$

for every subset A of $\underline{R}^n$. If f is an arbitrary function on $\underline{R}^n$, then

$$\int_{\underline{R}^n} f(y) \, d\delta_\xi(y) = f(\xi)$$

Therefore the conclusion (1) of Theorem 1.1 can be written as

$$\int_{\underline{R}^n} f(y) W(x - y, t - s) \, dy \rightarrow \int_{\underline{R}^n} f(y) \, d\delta_\xi(y)$$

Because of this relation, and the tradition of referring to δ_ξ as a function rather than a measure, it is often said that W has the δ-function property. Thus a fundamental solution is loosely described as a solution with the δ-function property.

A fundamental solution Γ is useful primarily because, for any continuous function $f : \underline{R}^n \rightarrow \underline{R}$ which has compact support, and any $s \in [0, a[$, the function u defined by

$$u(x, t) = \int_{\underline{R}^n} \Gamma(x, t; y, s) f(y) \, dy$$

is also a solution. This will be shown by differentiation under the integral sign in the next section. Furthermore, the second property (ii) of a fundamental solution shows that $u(x, t) \rightarrow f(\xi)$ as $(x, t) \rightarrow (\xi, s^+)$. Also since $\Gamma_0 \geq 0$, if Γ_0 is used in place of Γ then $u \geq 0$ if and only if $f \geq 0$. All these properties are extremely important.

2. GAUSS-WEIERSTRASS INTEGRALS

We shall study integrals of the form

$$\int_{\underline{R}^n} W(x - y, t - s) f(y) \, dy$$

which appeared in Theorem 1.1 with f bounded. Since $W(\xi, \tau)$ tends rapidly to zero as $\| \xi \| \rightarrow \infty$, this integral can exist for unbounded functions f, and it is desirable to study it for as general a function f as possible. In fact, we shall replace $f(y) \, dy$ with $d\mu(y)$ for a suitable "signed measure" μ on $\underline{R}^n$. Usually a signed measure has, by definition, a finite total variation, but this is too restrictive for the present situation, so we adopt the convention which we now describe. Let ν be a signed measure on $\underline{R}^n$ such that, for some $x \in \underline{R}^n$, $\alpha > 0$, and $t > 0$, the function $y \longmapsto W(x - y, t) \exp(\alpha \| y \|^2)$ is ν-integrable over $\underline{R}^n$. Then we write

$$\int_{\underline{R}^n} W(x - y, t) \, d\mu(y) = \int_{\underline{R}^n} W(x - y, t) \exp(\alpha \| y \|^2) \, d\nu(y) \qquad (3)$$

and thus identify $d\mu(y)$ with $\exp(\alpha\|y\|^2)\,d\nu(y)$. But this does not define
a true signed measure, since if ν^+ and ν^- denote the positive and negative variations of ν, there may be a set A such that

$$\int_A \exp(\alpha\|y\|^2)\,d\nu^+(y) = \int_A \exp(\alpha\|y\|^2)\,d\nu^-(y) = \infty$$

so that the ν-integral over A of $\exp(\alpha\|y\|^2)$ is undefined. However, for
every <u>bounded</u> Borel set E, we have $|\nu|(E) < \infty$, so that we can define

$$\mu(E) = \int_E \exp(\alpha\|y\|^2)\,d\nu(y) \tag{4}$$

Thus, restricted to a fixed bounded set, μ is a well-defined measure.
We shall follow the usual practice and speak of <u>a signed measure μ such
that</u>

$$\int_{\underline{R}^n} W(x - y, t)\,d|\mu|(y) < \infty \tag{5}$$

when to be strictly correct we should start with a signed measure ν
such that

$$\int_{\underline{R}^n} W(x - y, t)\exp(\alpha\|y\|^2)\,d|\nu|(y) < \infty$$

and relate μ to ν by (3) and (4). Of course, we can also define the <u>posi-
tive and negative variations</u>, μ^+ and μ^-, of μ by putting

$$\mu^+(E) = \int_E \exp(\alpha\|y\|^2)\,d\nu^+(y), \qquad \mu^-(E) = \int_E \exp(\alpha\|y\|^2)\,d\nu^-(y)$$

for <u>every</u> Borel set E. Then μ^+ and μ^- are non-negative measures and,
if $|\mu| = \mu^+ + \mu^-$, so is $|\mu|$, and

$$|\mu|(E) = \int_E \exp(\alpha\|y\|^2)\,d|\nu|(y)$$

for <u>every</u> Borel set E.

Definition. Given a signed measure μ such that (5) holds, the <u>Gauss-
Weierstrass integral</u> u of μ is defined by

$$u(x, t) = \int_{\underline{R}^n} W(x - y, t) \, d\mu(y) \tag{6}$$

If μ is absolutely continuous with respect to Lebesgue measure in $\underline{R}^n$, so that $d\mu(y) = f(y) \, dy$, then u may be called the Gauss-Weierstrass integral of f, rather than of μ.

In (6), it is implicitly assumed that $t > 0$. Similar results hold if t is replaced by $t - s$ in the integrand, provided that $t > s$. We have taken $s = 0$ for simplicity.

For each fixed y, the function $(x, t) \longmapsto W(x - y, t)$ is a temperature on $\underline{R}^n \times]0, \infty[$. It is therefore reasonable to expect that differentiation under the integral sign in (6) will, under appropriate conditions, show that u is also a temperature. We shall prove that this is the case under the mildest of conditions, and thus exhibit a large class of temperatures.

Lemma 1.3

Let f be defined on $]\alpha, \beta[\times A \times B$, where A and B are Borel sub-sets of $\underline{R}^n$, and let μ be a signed measure on B such that

$$\int_B |f(\xi, \eta, \zeta)| \, d|\mu|(\zeta) < \infty \tag{7}$$

for all $(\xi, \eta) \in]\alpha, \beta[\times A$. Define ϕ on $]\alpha, \beta[\times A$ by

$$\phi(\xi, \eta) = \int_B f(\xi, \eta, \zeta) \, d\mu(\zeta)$$

and let (ξ_0, η_0) be a fixed point of $]\alpha, \beta[\times A$. If $D_1 f(\xi, \eta_0, \zeta)$ exists for all $(\xi, \zeta) \in]\alpha, \beta[\times B$, and there is a positive function g on B such that

$$\int_B g(\zeta) \, d|\mu|(\zeta) < \infty \tag{8}$$

and

$$|D_1 f(\xi, \eta_0, \zeta)| \leq g(\zeta) \tag{9}$$

for all $(\xi, \zeta) \in]\alpha, \beta[\times B$, then $D_1 \phi(\xi_0, \eta_0)$ exists, the function $\zeta \longmapsto D_1 f(\xi_0, \eta_0, \zeta)$ is $|\mu|$-integrable over B, and

$$D_1 \phi(\xi_0, \eta_0) = \int_B D_1 f(\xi_0, \eta_0, \zeta) \, d\mu(\zeta)$$

Proof. Since (7) implies that f is both μ^+-integrable and μ^--integrable over B, the result as stated will follow if we prove it for the special case of a non-negative measure, because then we can argue that

$$D_1\phi(\xi_0, \eta_0) = D_1 \int_B f(\xi_0, \eta_0, \zeta)\, d\mu^+(\zeta) - D_1 \int_B f(\xi_0, \eta_0, \zeta)\, d\mu^-(\zeta)$$

$$= \int_B D_1 f(\xi_0, \eta_0, \zeta)\, d\mu^+(\zeta) - \int_B D_1 f(\xi_0, \eta_0, \zeta)\, d\mu^-(\zeta)$$

$$= \int_B D_1 f(\xi_0, \eta_0, \zeta)\, d\mu(\zeta)$$

Suppose, therefore, that μ is non-negative. By the mean value theorem, for each $\zeta \in B$ we have

$$\frac{f(\xi, \eta_0, \zeta) - f(\xi_0, \eta_0, \zeta)}{\xi - \xi_0} = D_1 f(\xi', \eta_0, \zeta)$$

for some ξ' between ξ and ξ_0 (which depends on ζ). Therefore, by (9),

$$\left| \frac{f(\xi, \eta_0, \zeta) - f(\xi_0, \eta_0, \zeta)}{\xi - \xi_0} \right| \leq g(\zeta)$$

for all $\zeta \in B$. Since (8) holds and μ is non-negative, it follows from Lebesgue's dominated convergence theorem that the function $\zeta \longmapsto D_1 f(\xi_0, \eta_0, \zeta)$ is μ-integrable over B and

$$\lim_{\xi \to \xi_0} \int_B \frac{f(\xi, \eta_0, \zeta) - f(\xi_0, \eta_0, \zeta)}{\xi - \xi_0}\, d\mu(\zeta) = \int_B D_1 f(\xi_0, \eta_0, \zeta)\, d\mu(\zeta)$$

Hence

$$D_1\phi(\xi_0, \eta_0) = \lim_{\xi \to \xi_0} \left\{ (\xi - \xi_0)^{-1} \left(\int_B f(\xi, \eta_0, \zeta)\, d\mu(\zeta) - \int_B f(\xi_0, \eta_0, \zeta)\, d\mu(\zeta) \right) \right\}$$

$$= \int_B D_1 f(\xi_0, \eta_0, \zeta)\, d\mu(\zeta)$$

Notation. In the following lemma, and subsequently, we use the symbol C to denote a positive constant which may change in value from one occurrence to the next. This avoids the use of many subscripts in situations where the value of the constant is unimportant.

Lemma 1.4 (Standard estimates for W.)

Given any constant $c \in]0, 1/4[$, we can find a positive constant C such that

$$W(x - y, t - s) \leq C(t - s)^{-n/2} \exp(-c \|x - y\|^2 / (t - s)) \tag{10}$$

$$|D_i W(x - y, t - s)| \leq C(t - s)^{-(n+1)/2} \exp(-c \|x - y\|^2 / (t - s)) \tag{11}$$

for $i = 1, \ldots, n$,

$$|D_i^2 W(x - y, t - s)| \leq C(t - s)^{-(n+2)/2} \exp(-c \|x - y\|^2 / (t - s)) \tag{12}$$

for $i = 1, \ldots, n$, and

$$|D_t W(x - y, t - s)| \leq C(t - s)^{-(n+2)/2} \exp(-c \|x - y\|^2 / (t - s)) \tag{13}$$

whenever $x, y \in \underline{R}^n$ and $s < t$.

Proof. Put $\xi = x - y$ and $\tau = t - s$.
Obviously (10) holds with $C = (4\pi)^{-n/2}$.
For (11), we first write $D_i W$ in the form

$$D_i W(\xi, \tau) = -2^{-n-1} \pi^{-n/2} \tau^{-(n+1)/2} (\xi_i / \tau^{1/2}) \exp(-\|\xi\|^2 / 4\tau)$$

Now note that, if α is a positive constant and $f(r) = r \exp(-\alpha r^2)$ for all $r \geq 0$, then f is bounded by some constant $M(\alpha)$. Therefore, if $c < 1/4$, there is a constant $M = M((1/4) - c)$ such that

$$r \exp(-r^2 / 4) \leq M \exp(-cr^2)$$

for all $r \geq 0$. It follows that, putting $\rho = \|\xi\| / \tau^{1/2}$,

$$|D_i W(\xi, \tau)| = C \tau^{-(n+1)/2} (|\xi_i| / \tau^{1/2}) \exp(-\rho^2 / 4)$$

$$\leq C \tau^{-(n+1)/2} \rho \exp(-\rho^2 / 4)$$

$$\leq C \tau^{-(n+1)/2} \exp(-c\rho^2)$$

For (12), the argument is basically similar. We first write $D_i^2 W$ in the form

$$D_i^2 W(\xi, \tau) = 2^{-n-2} \pi^{-n/2} \tau^{-(n+2)/2} \{2 \exp(-\|\xi\|^2/4\tau) - (\xi_i^2/\tau) \exp(-\|\xi\|^2/4\tau)\}$$

If β is a positive constant and $g(r) = re^{-\beta r}$ for all $r \geq 0$, then g is bounded by some constant $N(\beta)$. Therefore, if $c < 1/4$ and $N = N((1/4) - c)$, we have

$$re^{-r/4} \leq Ne^{-cr}$$

for all $r \geq 0$. Hence, writing $\sigma = \|\xi\|^2/\tau$, we have

$$|D_i^2 W(\xi, \tau)| \leq C\tau^{-(n+2)/2}\{2e^{-\sigma/4} + \sigma e^{-\sigma/4}\}$$

$$\leq C\tau^{-(n+2)/2}e^{-c\sigma}$$

Finally, (13) follows from (12) because

$$|D_t W(\xi, \tau)| \leq \sum_{i=1}^{n} |D_i^2 W(\xi, \tau)|$$

since W is a temperature.

Theorem 1.2

Let μ be a signed measure on R^n such that the Gauss-Weierstrass integral u of μ is defined at a point (x_0, a) of $R^n \times]0, \infty[$. Then u is defined and is a temperature throughout $R^n \times]0, a[$.

Proof. Since $u(x_0, a)$ is defined, we have

$$\int_{R^n} W(x_0 - y, a) \, d|\mu|(y) < \infty \qquad (14)$$

Let $(x, t) \in R^n \times]0, a[$, and put

$$D = \{y \in R^n : \|x_0 - y\|(a^{1/2} - t^{1/2}) \geq \|x_0 - x\|a^{1/2}\}$$

If $y \in D$, then

$$\|y - x\| \geq \|y - x_0\| - \|x_0 - x\| \geq \|y - x_0\| - \|x_0 - y\|(a^{\frac{1}{2}} - t^{\frac{1}{2}})a^{-\frac{1}{2}}$$

$$= \|x_0 - y\|(t/a)^{\frac{1}{2}}$$

so that

$$W(x - y, t) \leq (4\pi t)^{-n/2} \exp(-\|x_0 - y\|^2/4a) = (a/t)^{n/2}W(x_0 - y, a) \tag{15}$$

Therefore, in view of (14),

$$\int_D W(x - y, t)\, d|\mu|(y) \leq (a/t)^{n/2} \int_D W(x_0 - y, a)\, d|\mu|(y) < \infty \tag{16}$$

On the other hand, if $y \notin D$ we have $\|x_0 - y\| < \|x_0 - x\| a^{\frac{1}{2}}(a^{\frac{1}{2}} - t^{\frac{1}{2}})^{-1}$, so that

$$W(x_0 - y, a) \geq (4\pi a)^{-n/2} \exp(-\|x_0 - x\|^2(a^{1/2} - t^{1/2})^{-2}/4)$$

and hence

$$W(x - y, t) \leq (4\pi t)^{-n/2}$$

$$\leq (a/t)^{n/2} \exp(\|x_0 - x\|^2(a^{1/2} - t^{1/2})^{-2}/4)W(x_0 - y, a) \tag{17}$$

Therefore, in view of (14),

$$\int_{\underline{R}^n\backslash D} W(x - y, t)\, d|\mu|(y)$$

$$\leq (a/t)^{n/2} \exp(\|x_0 - x\|^2(a^{1/2} - t^{1/2})^{-2}/4) \int_{\underline{R}^n\backslash D} W(x_0 - y, a)\, d|\mu|(y)$$

$$< \infty$$

This, combined with (16), shows that

$$\int_{\underline{R}^n} W(x - y, t)\, d|\mu|(y) < \infty$$

for an arbitrary point $(x,t) \in \underline{R}^n \times]0, a[$, so that u is defined throughout this strip.

We now prove that u is a temperature on $\underline{R}^n \times]0, a[$. Given any numbers h, k and κ such that $0 < h < k < \kappa < a$, let (x,t) be an arbitrary point of $\underline{R}^n \times]h, k[$, and let $y \in \underline{R}^n$. Put $c = k/4\kappa$, so that $c < 1/4$ and $c/t > 1/4\kappa$. Then, by Lemma 1.4, there is a positive constant C such that

$$| D_i W(x - y, t)| \le C t^{-(n+1)/2} \exp(-c \|x - y\|^2/t)$$

$$\le C h^{-(n+1)/2} \exp(-\|x - y\|^2/4\kappa)$$

$$= C h^{-(n+1)/2} \kappa^{n/2} W(x - y, \kappa) \qquad (18)$$

(recall that C may vary from one line to the next),

$$| D_i^2 W(x - y, t)| \le C t^{-(n+2)/2} \exp(-c \|x - y\|^2/t)$$

$$\le C h^{-(n+2)/2} \kappa^{n/2} W(x - y, \kappa) \qquad (19)$$

and similarly

$$| D_t W(x - y, t)| \le C h^{-(n+2)/2} \kappa^{n/2} W(x - y, \kappa) \qquad (20)$$

If D is defined as above, relative to the present (x, κ) rather than (x,t), then (15) (with t replaced by κ) holds if $y \in D$, and (17) (with t replaced by κ) holds if $y \notin D$. Therefore, whatever the value of y, we have

$$W(x-y, \kappa) \le (a/\kappa)^{n/2} \exp(\|x_0 - x\|^2 (a^{1/2} - \kappa^{1/2})^{-2}/4) W(x_0 - y, a)$$

Hence, given any $\rho > 0$, whenever $\|x_0 - x\| < \rho$ we have

$$W(x - y, \kappa) \le (a/\kappa)^{n/2} \exp(\rho^2/4(a^{1/2} - \kappa^{1/2})^2) W(x_0 - y, a)$$

It therefore follows from (18), (19) and (20) that

$$| D_i W(x - y, t)| \le C W(x_0 - y, a) \qquad (21)$$

$$| D_i^2 W(x - y, t)| \le C W(x_0 - y, a) \qquad (22)$$

and

$$|D_t W(x - y, t)| \leq C W(x_0 - y, a) \tag{23}$$

for all x, y, t such that $\|x - x_0\| < \rho$, $y \in \underline{R}^n$ and $t \in]h, k[$. We shall apply Lemma 1.3. First we choose $i \in \underline{N}$ with $i \leq n$, and take $\xi = x_i$, η the point obtained from x by replacing x_i with t, $]\alpha, \beta[$ any bounded interval containing x_i, A any bounded Borel set containing η, $B = \underline{R}^n$ and $\zeta = y$. We then choose ρ such that $\|x_0 - x\| < \rho$, and with the corresponding C we take $g(y) = C W(x_0 - y, a)$, so that (8) and (14) coincide. With $f(\xi, \eta, \zeta) = W(x - y, t)$, the first part of this proof shows that (7) is satisfied, and (9) follows from (21). Hence Lemma 1.3 is applicable, and implies that

$$D_i u(x, t) = \int_{\underline{R}^n} D_i W(x - y, t) \, d\mu(y) \tag{24}$$

Furthermore, the function $y \to D_i W(x - y, t)$ is $|\mu|$-integrable over $\underline{R}^n$, either by Lemma 1.3 or by (21) and (14). It follows that we can again apply that lemma, this time with $f(\xi, \eta, \zeta) = D_i W(x - y, t)$, but with everything else as before, (9) now following from (22). We thus obtain from (24) that

$$D_i^2 u(x, t) = D_i \int_{\underline{R}^n} D_i W(x - y, t) \, d\mu(y) = \int_{\underline{R}^n} D_i^2 W(x - y, t) \, d\mu(y)$$

The proof that

$$D_t u(x, t) = \int_{\underline{R}^n} D_t W(x - y, t) \, d\mu(y)$$

is similar to that of (24). Hence, because W is a temperature

$$\sum_{i=1}^{n} D_i^2 u(x, t) = \int_{\underline{R}^n} \sum_{i=1}^{n} D_i^2 W(x - y, t) \, d\mu(y)$$

$$= \int_{\underline{R}^n} D_t W(x - y, t) \, d\mu(y)$$

$$= D_t u(x, t)$$

Thus u is a temperature on $\underline{R}^n \times]h, k[$ whenever $0 < h < k < a$, and hence on $\underline{R}^n \times]0, a[$.

Remarks. In view of Theorem 1.2, there is never any loss of generality in assuming that any Gauss-Weierstrass integral is a temperature on $\underline{R}^n \times \,]0,a[$ for some positive a.

Theorem 1.2 is analogous to the result that, if a complex power series in z converges at z_0, then it is absolutely convergent to an analytic function on $\{z : |z| < |z_0|\}$.

Corollary

If f is a function on $\underline{R}^n$ such that

$$\int_{\underline{R}^n} \exp(-\alpha\|y\|^2)\,|\,f(y)\,|\;dy < \infty$$

for some positive constant α, then the Gauss-Weierstrass integral u of f is defined and is a temperature throughout $\underline{R}^n \times \,]0,1/4\alpha[$.

Proof. Since

$$\int_{\underline{R}^n} W(y,1/4\alpha)\,|\,f(y)|\;dy = (\alpha/\pi)^{n/2} \int_{\underline{R}^n} \exp(-\alpha\|y\|^2)\,|\,f(y)\,|\;dy$$

$$< \infty$$

we see that $u(0,1/4\alpha)$ is defined. The result now follows from Theorem 1.2.

Example. Let $f(y) = \exp(\beta\|y\|^2)$. The above corollary implies that the Gauss-Weierstrass integral u of f is defined on $\underline{R}^n \times \,]0,1/4\alpha[$ for every $\alpha > \beta$, and is therefore defined on $\underline{R}^n \times \,]0,1/4\beta[$. Since

$$\int_{\underline{R}^n} W(y,1/4\beta)f(y)\;dy = (\beta/\pi)^{n/2} \int_{\underline{R}^n} dy = \infty$$

u is not defined at $(0,1/4\beta)$, and so the result is sharp. Furthermore, we can evaluate $u(x,t)$ explicitly, for all $(x,t) \in \underline{R}^n \times \,]0,1/4\beta[$. In this strip we have $4\beta t < 1$, and it is convenient to write $s = (1 - 4\beta t)^{1/2}$. Now

$$\beta \|y\|^2 - \|x - y\|^2 (4t)^{-1} = (4t)^{-1} \sum_{i=1}^{n} ((4\beta t - 1)y_i^2 + 2x_i y_i - x_i^2)$$

$$= -(4t)^{-1} \sum_{i=1}^{n} (s^2 y_i^2 - 2x_i y_i + x_i^2)$$

$$= -(4t)^{-1} \sum_{i=1}^{n} \{(sy_i - s^{-1}x_i)^2 - (s^{-2} - 1)x_i^2\}$$

Putting $z = sy - s^{-1}x$, we obtain

$$\beta \|y\|^2 - \|x - y\|^2 (4t)^{-1} = -\|z\|^2 (4t)^{-1} + (s^{-2} - 1)(4t)^{-1}\|x\|^2$$

$$= -\|z\|^2 (4t)^{-1} + \beta s^{-2} \|x\|^2$$

Therefore

$$u(x, t) = (4\pi t)^{-n/2} \int_{\underline{R}^n} \exp\left(-\frac{\|z\|^2}{4t} + \frac{\beta \|x\|^2}{s^2}\right) s^{-n} \, dz$$

$$= (1 - 4\beta t)^{-n/2} \exp\left(\frac{\beta \|x\|^2}{1 - 4\beta t}\right) \int_{\underline{R}^n} W(z, t) \, dz$$

$$= (1 - 4\beta t)^{-n/2} \exp\left(\frac{\beta \|x\|^2}{1 - 4\beta t}\right)$$

by Lemma 1.1.

3. INITIAL BEHAVIOR OF GAUSS-WEIERSTRASS INTEGRALS OF FUNCTIONS

We shall prove a generalization of property (ii) of the definition of a fundamental solution. This involves, in particular, Gauss-Weierstrass integrals of discontinuous functions. Of course, the corollary in the previous section shows that we need not restrict ourselves to bounded functions, but this is incidental.

We begin with a lemma whose full generality will not be needed until later.

Lemma 1.5

Let μ be a signed measure on $\underline{R}^n$ whose Gauss-Weierstrass integral u is defined at $(0, a)$, let $x_0 \in \underline{R}^n$, let $\delta \in]0, 1[$, and let $B = B(x_0, r)$. Then

$$\int_{\underline{R}^n \setminus B} W(x - y, t) \, d\mu(y) \to 0 \quad \text{as} \quad t \to 0^+$$

uniformly in $B(x_0, \delta r)$.

Proof. We first observe that, for any $\xi, \eta \in \underline{R}^n$,

$$\|\xi\|^2 \leq (\|\xi - \eta\| + \|\eta\|)^2 \leq 2(\|\xi - \eta\|^2 + \|\eta\|^2) \tag{25}$$

Suppose that $t < a/2$, $\|x - x_0\| < \delta r$, and $\|y - x_0\| > r$. Two applications of (25) give

$$2\|x - y\|^2 \geq \|y\|^2 - 2\|x\|^2 \geq \|y\|^2 - 4(\|x - x_0\|^2 + \|x_0\|^2)$$

which, together with the easy inequality

$$\|x - y\| \geq \|y - x_0\| - \|x_0 - x\| > r - \delta r$$

gives

$$\frac{\|x - y\|^2}{4t} = \frac{(a - 2t)\|x - y\|^2}{4at} + \frac{2\|x - y\|^2}{4a}$$

$$\geq \frac{(a - 2t)(1 - \delta)^2 r^2}{4at} + \frac{\|y\|^2 - 4\delta^2 r^2 - 4\|x_0\|^2}{4a}$$

It follows that

$$\left| \int_{\underline{R}^n \setminus B} W(x - y, t) \, d\mu(y) \right|$$

$$\leq \int_{\underline{R}^n \setminus B} W(x - y, t) \, d|\mu|(y)$$

$$\leq C t^{-n/2} \exp\left(- \frac{(a - 2t)(1 - \delta)^2 r^2}{4at} + \frac{\delta^2 r^2 + \|x_0\|^2}{a}\right) \int_{\underline{R}^n} \exp(-\|y\|^2/4a) \, d|\mu|(y)$$

$$= C t^{-n/2} \exp(-(1 - \delta)^2 r^2/4t) \int_{\underline{R}^n} W(y, a) \, d|\mu|(y)$$

Since u is defined at $(0, a)$, the integral is finite. Therefore the last expression, which is independent of x, tends to zero with t.

Theorem 1.3

Let f be a function on $\underline{R}^n$ whose Gauss-Weierstrass integral u is defined at $(0, a)$, and let $\xi \in \underline{R}^n$. Then

$$\limsup_{(x,t) \to (\xi, 0^+)} u(x, t) \leq \limsup_{\eta \to \xi} f(\eta) \qquad (26)$$

and

$$\liminf_{(x,t) \to (\xi, 0^+)} u(x, t) \geq \liminf_{\eta \to \xi} f(\eta) \qquad (27)$$

Proof. It is sufficient to prove (26), since (27) will then follow from (26) applied to -f and its Gauss-Weierstrass integral -u.

Since the result is otherwise trivial, we suppose that the right hand side of (26) is not $+\infty$. Let λ be an arbitrary real number which satisfies

$$\limsup_{\eta \to \xi} f(\eta) < \lambda \qquad (28)$$

Then there is a ball $B = B(\xi, r)$ in $\underline{R}^n$ such that $f(\eta) < \lambda$ for all $\eta \in B$.

For each $(x, t) \in \underline{R}^n \times]0, a[$, Theorem 1.2 shows that $u(x, t)$ is defined. We write

$$u(x, t) = \int_B W(x - y, t) f(y) \, dy + \int_{\underline{R}^n \backslash B} W(x - y, t) f(y) \, dy$$

and note that, by Lemma 1.5, the second integral tends to zero as $(x, t) \to (\xi, 0^+)$. Since $f(y) < \lambda$ for all $y \in B$, and $W > 0$, the first integral is less than

$$\lambda \int_B W(x - y, t) \, dy$$

By Lemmas 1.1 and 1.2,

$$\lambda \int_B W(x - y, t) \, dy = \lambda \left(1 - \int_{\underline{R}^n \backslash B} W(x - y, t) \, dy \right)$$

$$\to \lambda \quad \text{as} \quad (x, t) \to (\xi, 0^+)$$

It follows that, as $(x, t) \to (\xi, 0+)$,

$$\lim \sup u(x, t) \leq \lim \sup \lambda \int_B W(x - y, t) \, dy = \lambda$$

Since this is true for every λ which satisfies (28), we have proved (26).

Corollary

Let f be a function on $\underline{R}^n$ whose Gauss-Weierstrass integral u is defined at $(0, a)$, and suppose that f is continuous at ξ. Then $u(x, t) \to f(\xi)$ as $(x, t) \to (\xi, 0^+)$.

Proof. If f is continuous at ξ, then the right-hand sides of (26) and (27) are both equal to $f(\xi)$. The same is therefore true of the left hand sides.

Remark. Neither Theorem 1.3, nor its corollary, assumes that f is finite-valued. In particular, if $f(\eta) \to \infty$ as $\eta \to \xi$, the corollary implies that $u(x, t) \to \infty$ as $(x, t) \to (\xi, 0^+)$.

4. THE SEMIGROUP PROPERTY

If u is a temperature on $\underline{R}^n \times \,]0, a[$, and the formula

$$u(x, t) = \int_{\underline{R}^n} W(x - y, t - s) u(y, s) \, dy$$

holds whenever $x \in \underline{R}^n$ and $0 < s < t < a$, we say that u has the semi-group property. We shall show that all Gauss-Weierstrass integrals have this property, after first showing that W itself does.

Theorem 1.4

Suppose that $x \in \underline{R}^n$ and $0 < s < t$. Then

$$W(x, t) = \int_{\underline{R}^n} W(x - y, t - s) W(y, s) \, dy \qquad (29)$$

Furthermore, if u is a Gauss-Weierstrass integral on $\underline{R}^n \times \,]0, a[$, then

$$u(x, t) = \int_{\underline{R}^n} W(x - y, t - s)u(y, s) \, dy \qquad (30)$$

if $t < a$.

Proof. The formula (29) is proved by very elementary techniques. If $x, y \in \underline{R}^n$ and $0 < s < t$, we have

$$s\|x - y\|^2 + (t - s)\|y\|^2 = \sum_{i=1}^{n} (sx_i^2 + ty_i^2 - 2sx_iy_i)$$

$$= \sum_{i=1}^{n} (t(y_i - st^{-1}x_i)^2 + x_i^2 st^{-1}(t - s))$$

$$= t\|y - st^{-1}x\|^2 + st^{-1}(t - s)\|x\|^2$$

so that

$$\exp\left(-\frac{\|x-y\|^2}{4(t-s)}\right)\exp\left(-\frac{\|y\|^2}{4s}\right) = \exp\left(-\frac{s\|x-y\|^2 + (t-s)\|y\|^2}{4s(t-s)}\right)$$

$$= \exp\left(-\frac{\|y - st^{-1}x\|^2}{4st^{-1}(t-s)}\right)\exp\left(-\frac{\|x\|^2}{4t}\right)$$

Hence, putting $z = y - st^{-1}x$, we obtain

$$\int_{\underline{R}^n} W(x - y, t - s)W(y, s) \, dy$$

$$= (4\pi)^{-n}(t - s)^{-n/2}s^{-n/2}\exp\left(-\frac{\|x\|^2}{4t}\right)\int_{\underline{R}^n}\exp\left(-\frac{\|z\|^2}{4st^{-1}(t-s)}\right)dz$$

By Lemma 1.1,

$$\int_{\underline{R}^n}\exp\left(-\frac{\|z\|^2}{4st^{-1}(t-s)}\right)dz = (4\pi st^{-1}(t-s))^{n/2}$$

so that

$$\int_{\underline{R}^n} W(x - y, t - s)W(y, s) \, dy = (4\pi)^{-n/2}t^{-n/2}\exp(-\|x\|^2/4t)$$

and (29) is proved.

To obtain (30) from (29), we use Fubini's theorem on interchanging the order of integration in a repeated integral. Since, by (29),

$$\int_{\underline{R}^n} d|\mu|(z) \int_{\underline{R}^n} W(x-y, t-s)W(y-z, s)\, dy = \int_{\underline{R}^n} W(x-z, t)\, d|\mu|(z) < \infty$$

whenever $0 < s < t < a$ and $x \in \underline{R}^n$, Fubini's theorem is applicable. Thus

$$\begin{aligned}
u(x, t) &= \int_{\underline{R}^n} W(x - z, t)\, d\mu(z) \\
&= \int_{\underline{R}^n} d\mu(z) \int_{\underline{R}^n} W(x - y, t - s)W(y - z, s)\, dy \\
&= \int_{\underline{R}^n} W(x - y, t - s)\, dy \int_{\underline{R}^n} W(y - z, s)\, d\mu(z) \\
&= \int_{\underline{R}^n} W(x - y, t - s)u(y, s)\, dy
\end{aligned}$$

Remark. Formula (29) is actually a special case of (30), since W can be considered as the Gauss-Weierstrass integral of the Dirac δ-measure concentrated at 0, that is

$$W(x, t) = \int_{\underline{R}^n} W(x - y, t)\, d\delta_0(y)$$

The semigroup property is extremely important, particularly in Chapters II and III.

5. UNIQUENESS OF THE GAUSS-WEIERSTRASS REPRESENTATION

In Section 2, we saw that a Gauss-Weierstrass integral is always a temperature. If u is such a temperature, the question of whether it can be the Gauss-Weierstrass integral of two distinct measures naturally arises. We now show that this cannot happen.

Theorem 1.5

Suppose that u is a temperature on $\underline{R}^n \times \,]0, a[$, and that u is the Gauss-Weierstrass integral of two signed measures μ_1 and μ_2. Then $\mu_1 = \mu_2$.

Proof. Since the Gauss-Weierstrass integrals of μ_1 and μ_2 are defined at $(0, a/2)$, putting $\gamma = 1/2a$, we have

$$\int_{\underline{R}^n} \exp(-\gamma\|y\|^2)\, d|\mu_i|(y) = (2\pi a)^{n/2} \int_{\underline{R}^n} W(y, a/2)\, d|\mu_i|(y) < \infty$$

for each $i \in \{1, 2\}$. Therefore, if we put $d\nu_i(y) = \exp(-\gamma\|y\|^2)\, d\mu_i(y)$ for each i, then each ν_i is a signed measure of finite total variation. Hence, if λ is defined to be $\nu_1 - \nu_2$, then λ is a well-defined signed measure with $|\lambda|(\underline{R}^n) < \infty$. The result of the theorem will follow if we show that λ is the zero measure. Note that

$$\int_{\underline{R}^n} W(x-y, t) \exp(\gamma\|y\|^2)\, d\lambda(y) = \int_{\underline{R}^n} W(x-y, t)\, d\mu_1(y) - \int_{\underline{R}^n} W(x-y, t)\, d\mu_2(y)$$

$$= 0 \tag{31}$$

for all $(x, t) \in \underline{R}^n \times]0, a[$.

By the Hahn decomposition theorem, there are disjoint Borel sets A and B, with $A \cup B = \underline{R}^n$, such that

$$\lambda^+(E) = \lambda(A \cap E) \quad \text{and} \quad \lambda^-(E) = -\lambda(B \cap E)$$

for every Borel set E. Furthermore, λ^+ and λ^- are regular measures since they are non-negative and have finite total variation. Suppose that λ is not null, so that either $\lambda^+(A) > 0$ or $\lambda^-(B) > 0$. It is sufficient to consider the former case, since the latter follows from this applied to $-\lambda$. If $\lambda^+(A) = \alpha > 0$, then because of the regularity of λ^+ and λ^-, there exist a compact set $K \subseteq A$ such that

$$\lambda(K) = \lambda^+(K) > 3\alpha/4 \tag{32}$$

and a bounded open set $G \supseteq K$ such that

$$\lambda^-(G) = \lambda^-(G \setminus K) < \alpha/4 \tag{33}$$

By Urysohn's lemma, there is a continuous function ϕ on $\underline{R}^n$ such that $\phi(y) = 1$ for all $y \in K$, $\phi(y) = 0$ for all $y \in \underline{R}^n \setminus G$, and $0 \leq \phi(y) \leq 1$ for all $y \in \underline{R}^n$. Given such a function, we put

$$v(x, t) = \int_{\underline{R}^n} W(x - y, t)\phi(y) \exp(-\gamma\|y\|^2)\, dy$$

so that, since

$$\int_{R^n} \exp(-\beta\|y\|^2)\phi(y) \exp(-\gamma\|y\|^2) \, dy < \infty$$

for every $\beta > 0$, v is defined and is a temperature on $R^n \times]0, \infty[$, by Theorem 1.2, Corollary. By Theorem 1.1,

$$\lim_{t \to 0^+} v(x, t) = \phi(x) \exp(-\gamma\|x\|^2) \tag{34}$$

for all $x \in \underline{R}^n$.

We now choose r such that $G \subseteq B(0, r)$, so that $\phi(y) = 0$ whenever $\|y\| \geq r$, and hence, since $0 \leq \phi \leq 1$,

$$0 \leq \exp(\gamma\|x\|^2)v(x, t)$$

$$\leq (4\pi t)^{-n/2} \int_{\|y\|<r} \exp\left(\gamma\|x\|^2 - \frac{\|x-y\|^2}{4t} - \gamma\|y\|^2\right) dy \tag{35}$$

If $0 < t < 1/16\gamma$, then for any $x, y \in \underline{R}^n$ we have

$$\gamma(\|x\|^2 - 2\|y\|^2) \leq 2\gamma\|x-y\|^2 \leq \|x-y\|^2/8t$$

so that

$$\gamma\|x\|^2 \leq \frac{\|x-y\|^2}{8t} + 2\gamma\|y\|^2$$

and hence

$$\gamma\|x\|^2 - \frac{\|x-y\|^2}{4t} - \gamma\|y\|^2 \leq -\frac{\|x-y\|^2}{8t} + \gamma\|y\|^2$$

It therefore follows from (35) and Lemma 1.1 that

$$0 \leq \exp(\gamma \|x\|^2) v(x,t)$$

$$\leq (4\pi t)^{-n/2} \int_{\|y\|<r} \exp\left(-\frac{\|x-y\|^2}{8t} + \gamma\|y\|^2\right) dy$$

$$\leq \exp(\gamma r^2) 2^{n/2} \int_{\|y\|<r} W(x-y, 2t) \, dy$$

$$\leq \exp(\gamma r^2) 2^{n/2} \tag{36}$$

and hence

$$\int_{\underline{R}^n} \exp(\gamma\|x\|^2) v(x,t) \, d|\lambda|(x) \leq \exp(\gamma r^2) 2^{n/2} |\lambda|(\underline{R}^n) < \infty \tag{37}$$

whenever $0 < t < 1/16\gamma$. Since (36) shows that the function $(x,t) \longmapsto \exp(\gamma\|x\|^2) v(x,t)$ is bounded on $\underline{R}^n$, and $|\lambda|(\underline{R}^n) < \infty$, it follows from Lebesgue's dominated convergence theorem and (34) that

$$\lim_{t\to0^+} \int_{\underline{R}^n} \exp(\gamma\|x\|^2) v(x,t) \, d\lambda(x) = \int_{\underline{R}^n} \phi(x) \, d\lambda(x)$$

The properties of ϕ, together with (32) and (33), show that

$$\int_{\underline{R}^n} \phi(x) \, d\lambda(x) = \int_K d\lambda(x) + \int_{G\backslash K} \phi(x) \, d\lambda^+(x) - \int_{G\backslash K} \phi(x) \, d\lambda^-(x)$$

$$\geq \lambda(K) + 0 - \lambda^-(G\backslash K)$$

$$> \alpha/2$$

$$> 0$$

so that

$$\lim_{t\to0^+} \int_{\underline{R}^n} \exp(\gamma\|x\|^2) v(x,t) \, d\lambda(x) > 0 \tag{38}$$

However, if $0 < t < 1/16\gamma$, it follows from Fubini's theorem and (31) that

$$\int_{\underline{R}^n} \exp(\gamma\|x\|^2) v(x,t) \, d\lambda(x)$$

$$= \int_{\underline{R}^n} \exp(\gamma\|x\|^2) \, d\lambda(x) \int_{\underline{R}^n} W(x-y, t) \phi(y) \exp(-\gamma\|y\|^2) \, dy$$

$$= \int_{\underline{R}^n} \phi(y) \exp(-\gamma \|y\|^2) \, dy \int_{\underline{R}^n} W(x - y, t) \exp(\gamma \|x\|^2) \, d\lambda(x)$$

$$= 0 \tag{39}$$

the application of Fubini's theorem being justified by (37). Since (39) contradicts (38), our assumption that λ is not null must be false. It follows that $\mu_1 = \mu_2$.

6. FUNDAMENTAL SOLUTIONS OF LINEAR PARABOLIC EQUATIONS

Consider the linear partial differential equation

$$\sum_{i,j=1}^{n} a_{ij}(x,t) D_i D_j u(x,t) + \sum_{j=1}^{n} b_j(x,t) D_j u(x,t) + c(x,t) u(x,t) - D_t u(x,t) = 0 \tag{40}$$

whose coefficients a_{ij}, b_j, and c, are functions defined on the strip $S = R^n \times [0, a[$. Such an equation is called <u>uniformly parabolic</u> if there is a positive constant γ such that, whenever $\xi = (\xi_1, \ldots, \xi_n) \in R^n$,

$$\sum_{i,j=1}^{n} a_{ij}(x,t) \xi_i \xi_j \geq \gamma \|\xi\|^2 \tag{41}$$

for all $(x,t) \in S$. The word <u>uniformly</u> refers to the fact that γ is the same for all points (x,t). If this were not the case, then the left hand side of (41) would merely be positive, and the equation would be called <u>parabolic</u>. For the heat equation, there is equality in (41) with $\gamma = 1$.

A <u>solution</u> of (40), on an open subset E of S, is a function u for which all the derivatives that appear in (40), with coefficients that are not identically zero, exist, are continuous and real-valued, and satisfy (40) on E.

A <u>fundamental solution</u> of (40) is defined analogously to that of the heat equation, that is, it is a function Γ on $S \times S$ which has the following properties:

(i)   For each fixed point (y_0, s_0) of S, the function

$$(x,t) \longmapsto \Gamma(x, t; y_0, s_0)$$

is a solution of (40) on the substrip $\underline{R}^n \times]s_0, a[$.

(ii) For each continuous real-valued, function f on $\underline{R}^n$ which has compact support, and each $s \in [0, a[$,

$$\int_{\underline{R}^n} \Gamma(x, t; y, s) f(y) \; dy \rightarrow f(\xi)$$

as $(x, t) \rightarrow (\xi, s^+)$, for all $\xi \in \underline{R}^n$.

Although all the results in the previous sections can be extended to a wide class of parabolic equations, the overall treatment and some of the methods of proof cannot be similarly extended. In particular, we cannot simply write down a fundamental solution, and first have to establish its existence. This is not just for pedantic reasons, but because parabolic equations have been found which do not possess fundamental solutions. In many cases, a fundamental solution can be constructed by what is known as the parametrix method. This is a good method because it produces a nonnegative fundamental solution Γ, as well as good upper estimates for Γ and the moduli of those of its derivatives which appear in (40). These estimates are analogous to those in Lemma 1.4, with $W(x - y, t - s)$ replaced by $\Gamma(x, t; y, s)$ on the left-hand sides only, and with the range of values of c reduced; similar estimates for the mixed second derivatives are also obtained. These are sometimes called Eidel'man's estimates. The applicability of the parametrix method depends on the coefficients of (40) satisfying certain growth and smoothness conditions; a smoothness condition is one which demands continuity or differentiability of some order.

A function g on a subset E of $\underline{R}^n$ is said to satisfy a Hölder condition of exponent α, or to be Hölder continuous of exponent α, on E if there is a constant M such that

$$| f(\xi) - f(\eta) | \leq M \| \xi - \eta \|^{\alpha}$$

for all $\xi, \eta \in E$. Here $0 < \alpha \leq 1$.

The parametrix method can be used to construct a fundamental solution of (40), provided that the equation is uniformly parabolic, the coefficients a_{ij}, b_j and c are uniformly continuous, bounded, satisfy a Hölder condition in x of exponent α, for some $\alpha \in {]}0, 1]$, uniformly in t, and also satisfy a Hölder condition in t, uniformly in x. The method also provides extra information, apart from Eidel'man's estimates. Corresponding to Lemma 1.1, we have the pair of results

$$\int_{\underline{R}^n} \Gamma(x, t; y, s) \; dy \leq C$$

for some constant C, and

$$\int_{\underline{R}^n} \Gamma(x,t;y,s)\, dy \to 1 \qquad (42)$$

as $(x,t) \to (\xi, 0^+)$ for any $\xi \in \underline{R}^n$, the first of which follows from the upper estimate for Γ and Lemma 1.1, and the second of which follows from the definition of a fundamental solution by taking the f in (ii) to be equal to 1 in $B(\xi, 1)$ and to 0 outside $B(\xi, 2)$. The upper estimate for Γ and Lemma 1.2 show that

$$\int_{\|\xi - y\| > \delta} \Gamma(x,t;y,s)\, dy \to 0$$

as $(x,t) \to (\xi, s+)$, for any positive constant δ. The generalization of Theorem 1.1, namely that

$$\int_{\underline{R}^n} \Gamma(x,t;y,s) f(y)\, dy \to f(\xi) \qquad (43)$$

as $(x,t) \to (\xi, s^+)$ for any bounded function f which is continuous at ξ, is proved during the construction. The upper estimate for Γ can be used to extend the proof of Lemma 1.5 to the general case.

The parametrix method provides a fundamental solution Γ with the following property: Given any positive constant δ, there is a constant K such that $\Gamma(x,t;y,s) \leq K$ whenever $\|x - y\| > \delta$ and $t - s > \delta$. This can easily be deduced from the upper estimate for Γ. The method also shows that there is only one fundamental solution with this property, and also provides details that can be used to prove basic results on uniqueness of solution of the Cauchy problem, which is a closely related question. Let $s \in [0, a[$. Given a continuous function f on $\underline{R}^n$, the problem of finding a continuous function u on $\underline{R}^n \times [s, a[$ which satisfies (40) on $\underline{R}^n \times]s, a[$ and satisfies $u(x,s) = f(x)$ for all $x \in \underline{R}^n$, is called the Cauchy problem for (40) on the strip $\underline{R}^n \times [s, a[$. Details from the parametrix construction of Γ imply that, if f is bounded and continuous on $\underline{R}^n$ and

$$u(x,t) = \int_{\underline{R}^n} \Gamma(x,t;y,s) f(y)\, dy$$

then u is a bounded solution of the Cauchy problem for (40) on $\underline{R}^n \times [s, a[$, and is the only such solution. Thus, bounded solutions of the Cauchy problem are unique. It follows that

$$\Gamma(x, t; z, r) = \int_{\underline{R}^n} \Gamma(x, t; y, s) \Gamma(y, s; z, r) \, dy \qquad (44)$$

whenever $0 \le r < s < t < a$ and $x, z \in \underline{R}^n$, which generalizes the semi-group property for W of Theorem 1.4. To see this, fix (z, r) and note first that the left-hand side of (44), as a function of (x, t), is a bounded solution of the Cauchy problem for (40) on $\underline{R}^n \times [s, a[$, given the continuous function $\Gamma(\cdot, s; z, r)$ on $\underline{R}^n \times \{s\}$. Next, note that, since the function $\Gamma(\cdot, s; z, r)$ is continuous and bounded on $\underline{R}^n$, the right-hand side of (44), as a function of (x, t), is the only bounded solution of the Cauchy problem for (40) on $\underline{R}^n \times [s, a[$, given $\Gamma(\cdot, s; z, r)$ on $\underline{R}^n \times \{s\}$, in view of (43). Hence (44) holds. It now follows, using the upper estimate for Γ and the method of proof of the last part of Theorem 1.4, that

$$u(x, t) = \int_{\underline{R}^n} \Gamma(x, t; y, s) u(y, s) \, dy$$

whenever $x \in \underline{R}^n$ and $0 < s < t < a$, if u is defined by

$$u(x, t) = \int_{\underline{R}^n} \Gamma(x, t; y, 0) \, d\mu(y)$$

We now consider generalizing Theorems 1.2, 1.3 and 1.5. In order to carry over the proofs of these results to the general case, we require not only the upper estimate for Γ, but also a similar lower estimate which is not a consequence of the parametrix construction. More precisely, we require that there exist positive constants C_1, C_2, c_1 and c_2 such that

$$C_1 (t-s)^{-n/2} \exp(-c_1 \|x-y\|^2 / (t-s)) \le \Gamma(x, t; y, s) \qquad (45)$$

$$\le C_2 (t-s)^{-n/2} \exp(-c_2 \|x-y\|^2 / (t-s)) \qquad (46)$$

whenever $x, y \in \underline{R}^n$ and $0 \le s < t < a$. Indeed, many of the results proved throughout this book depend on these inequalities for their extension from the heat equation to more general parabolic equations. The inequality (45) will be referred to as the sharp lower estimate for Γ. It is not difficult to imagine that many calculations which require upper and lower estimates for $W(x - y, t - s)$ will be capable of generalization to parabolic equations with fundamental solutions which satisfy (45) and (46).

The conditions on the coefficients which ensure that all the above results hold are quite strong, particularly the requirement that the coefficients be bounded. If we begin to relax these conditions, then we start to obtain less information about the fundamental solution, until eventually we cannot prove even its existence. In this situation, it is sometimes still possible to obtain related results, along the following lines. Let f be a continuous, real-valued function on $\underline{R}^n$, with compact support. Then one can show that there is a unique bounded solution u_f of the Cauchy problem on $\underline{R}^n \times [0, a[$ with f as the given function on $\underline{R}^n$, and that $u_f \geq 0$ if $f \geq 0$. This can be done under mild conditions on the coefficients without using the concept of a fundamental solution. Given this result, we proceed as follows. We fix $(x, t) \in \underline{R}^n \times]0, a[$, and define a mapping Λ, on the class $C_c(\underline{R}^n)$ of all continuous real-valued functions on $\underline{R}^n$ with compact support, by

$$\Lambda(f) = u_f(x, t)$$

Then, since $\Lambda(f) \geq 0$ whenever $f \geq 0$, Λ is a positive linear functional on $C_c(\underline{R}^n)$. Therefore, by the Riesz representation theorem, there is a unique positive measure $\mu_{x, t}$ on $\underline{R}^n$ such that

$$\Lambda(f) = \int_{\underline{R}^n} f(y) \, d\mu_{x, t}(y)$$

so that

$$u_f(x, t) = \int_{\underline{R}^n} f(y) \, d\mu_{x, t}(y) \tag{47}$$

Since (x, t) is arbitrary, we have obtained a representation for u_f on $\underline{R}^n \times]0, a[$. The measure $\mu_{x, t}$ is called the <u>parabolic measure</u> at (x, t) for the given parabolic equation. Comparing (47) with the representation

$$u_f(x, t) = \int_{\underline{R}^n} f(y) \Gamma(x, t; y, 0) dy$$

for the case where there is a fundamental solution with all the properties described above, we see that the existence of that fundamental solution includes an assertion that $\mu_{x, t}$ is, for every $(x, t) \in R^n \times]0, a[$, absolutely continuous with respect to Lebesgue measure; for then $\Gamma(x, t; y, 0) dy = d\mu_{x, t}(y)$, by the uniqueness part of the Riesz representation theorem. It has been shown that $\mu_{x, t}$ can be singular for every (x, t).

Fundamental solutions are not unique, but there is usually some
extra condition which only one of them will satisfy, such as the one
given above in the discussion of the parametrix method. We shall return
to this question in Chapter III when we have proved some theorems
about uniqueness of solution of the Cauchy problem, as these matters
are intimately related.

7. BIBLIOGRAPHICAL NOTES

Details of the construction of a fundamental solution by the parametrix
method, under the conditions given in section 6, are given in several
texts, that of Il'in, Kalashnikov and Oleinik [1] being the clearest and
most complete. The Hölder condition on t can be dropped, at the ex-
pense of losing some useful information, as has been demonstrated by
Eidel'man [1, 3], Aronson [1] and Robinson [1]. However, the Hölder
condition on x is necessary to get a fundamental solution Γ such that
$0 < \Gamma(x, t; y, s) < K$ whenever $t - s > \delta$. This condition, together with
the representation

$$u(x, t) = \int_{\underline{R}^n} \Gamma(x, t; y, s) f(y) \, dy$$

for a solution of the Cauchy problem on $\underline{R}^n \times [s, a[$ with $u(\cdot, s) = f$,
implies that, whenever $t - s > \delta$,

$$|u(x, t)| \leq K \int_{\underline{R}^n} |f(y)| \, dy \qquad\qquad (48)$$

and Il'in [1] has produced an equation

$$D_t u(x, t) = a(x, t) D_1^2 u(x, t)$$

on $\underline{R} \times]0, 1[$, with a continuous coefficient whose range lies in $[1/2,
3/2]$, such that (48) fails. See also the notes on parabolic measure
below.

Constructions of fundamental solutions for equations with unbounded
coefficients have been given, under various conditions, by Krzyżański
and Szybiak [1, 2], Eidel'man [2, 3], Krzyżański [2, 3], and Eidel'man
and Porper [1]. The existence of fundamental solutions under still
weaker growth conditions was established by Aronson and Besala [2],
whose methods were extended to a wider class of equations by
Chabrowski [4], and subsequently improved to cover still more

equations by Besala [6], who was also able to weaken the smoothness conditions. Lower estimates for various fundamental solutions with unbounded coefficients were obtained by Besala [3], Aronson and Besala [3], and Guenther [2]. These are not as sharp as estimate (45), which was established for equations with bounded coefficients by Aronson [4, 5], and is also proved in the survey by Eidel'man and Porper [2].

Having proved the necessary results on the Cauchy problem in his paper [1], Krzyżański proved the existence of parabolic measure for certain equations with unbounded, not necessarily Hölder continuous coefficients in his paper [7]. These results were generalized still further by Chabrowski [5]. The existence of a uniformly parabolic equation, with bounded and uniformly continuous coefficients, such that the parabolic measure $\mu_{x,t}$ is singular with respect to Lebesgue measure for every point (x,t), was demonstrated by Fabes and Kenig [1].

For the heat equation, special fundamental solutions, which have properties not possessed by the Γ_0 of section 1, have been constructed and utilized by Cakała [1] and Jones [1].

Theorem 1.2 was proved for the heat equation in the case $n = 1$ by Tychonoff [1], using a different method. The proof in the text is due to Flett [1] for the heat equation, and its extension to more general equations was given by Watson [3]. Prior to Watson's paper, results of a similar kind were only able to assert that the integral represents a solution on some unspecified substrip $\underline{R}^n \times \,]0, b[$ with $b < a$, as in Guenther's paper [1].

The proof of Theorem 1.3 is similar to that of a slightly weaker result, which was given by Watson [2]. Only minor changes are necessary to extend it to more general equations: the estimates (45) and (46) are needed, and (42) must be used instead of Lemma 1.1.

Theorem 1.4, and its extension to the general case, are classical.

The proof of Theorem 1.5 is taken from Aronson's paper [5].

The terminology of this chapter is not all standard terminology, as there is sometimes no standard. The Gauss-Weierstrass integral may be referred to as the Poisson, Fourier-Poisson, Poisson-Weierstrass, or Weierstrass integral, depending on the author. The semigroup property may similarly be referred to as the Huygens property, or the Poisson (or Fourier-Poisson, etc.) representation. Its generalization (44) is sometimes called Kolmogorov's equation, or the Chapman-Kolmogorov relation. The term 'semigroup property' refers to the fact that the relation implies that methods of the theory of semigroups of operators can be used to study temperatures, as in the book by Butzer and Berens [1]. The definition of a fundamental solution varies, but ours is the least restrictive. A Hölder condition is often called a Lipschitz condition.

   Many authors refer to the fundamental solution of a parabolic equation, despite the lack of uniqueness which we shall demonstrate in
Chapter III. In Theorem 3.12, Corollary, we shall prove that there is
only one non-negative fundamental solution Γ on $\underline{R}^n \times [0, a[$ with the
property that the integral

$$\int_{\underline{R}^n} \Gamma(x, t; y, s) f(y) \, dy$$

defines a solution of the parabolic equation on $\underline{R}^n \times]s, a[$, for each continuous, real-valued function f with compact support. This is the fundamental solution to which the authors refer. Except when considering the
uniqueness of fundamental solutions, we shall always deal, in our sections on general parabolic equations, only with the fundamental solution
which has these extra properties, and we shall use the definite article
when referring to it.

2

Non-negative Solutions

In this chapter, we establish the fundamental result that every non-negative temperature on $\underline{R}^n \times]0, a[$ can be represented as the Gauss-Weierstrass integral of a non-negative measure on $\underline{R}^n$. This involves a notion of convergence of a sequence of measures, as well as the maximum principle on a circular cylinder and an extension of this to $\underline{R}^n \times]0, a[$. This extension, Theorem 2.4, is also essential for the results of Chapter III. To conclude the present chapter, we use the representation theorem to characterize Gauss-Weierstrass integrals and find the greatest rate at which a positive temperature on $\underline{R}^n \times]0, a[$ can decay as $t \to 0^+$ or $\|x\| \to \infty$.

1. THE MAXIMUM PRINCIPLE ON CIRCULAR CYLINDERS

Essentially, a maximum principle asserts that the functions of a specified class cannot attain a strict maximum at any interior point of their domains of definition. In the case of temperatures, a strict maximum cannot even be attained on certain parts of the boundary.

Theorem 2.1

Let B be an open ball in R^n, and let u be a temperature on the cylinder $C = B \times]a, b[$. If

$$\lim \sup u(x, t) \le 0 \qquad (1)$$

whenever (x, t) approaches any point of $\partial C \setminus (B \times \{b\})$ from within C, then $u \le 0$ throughout C.

Proof. Given any $\tau \in]a, b[$, consider the restriction of u to the cylinder $C_\tau = B \times]a, \tau]$. Suppose that there is a point $(x_1, t_1) \in C_\tau$ such that

$$u(x_1, t_1) = \ell > 0$$

Choose k such that $0 < k < \ell/(t_1 - a)$, and define v on C by

$$v(x, t) = u(x, t) - k(t - t_1)$$

Given any $\epsilon > 0$ and any point $(y, s) \in \partial C \setminus (B \times \{b\})$, we can find an open set $N_{y,s}$ containing (y, s), such that $u(x, t) < \epsilon$ for all $(x, t) \in N_{y,s} \cap C$, because of (1). Since $\partial C \setminus (B \times \{b\})$ is compact and is covered by the family of open sets $\{N_{y,s} : (y, s) \in \partial C \setminus (B \times \{b\})\}$, there is a finite subfamily which still covers $\partial C \setminus (B \times \{b\})$. It follows that there exists $\delta > 0$ such that $u(x, t) < \epsilon$ for all $(x, t) \in C$ whose distance from $\partial C \setminus (B \times \{b\})$ is less than δ. For such points which also lie in C_τ, if $0 < \epsilon < \ell - k(t_1 - a)$ we have

$$v(x, t) \le u(x, t) + k(t_1 - a) < \epsilon + k(t_1 - a) < \ell$$

Since $v(x_1, t_1) = u(x_1, t_1) = \ell$, it follows that v attains a maximum over C_τ at some point (x_2, t_2), whose distance from $\partial C \setminus (B \times \{b\})$ is not less than δ. Because u is a temperature

$$\sum_{i=1}^n D_i^2 v = \sum_{i=1}^n D_i^2 u = D_t u = D_t v + k$$

throughout C. However, at (x_2, t_2) we have $D_i^2 v \le 0$ for all i, and

$D_t v \geq 0$, which is a contradiction since $k > 0$. The only assumption that we have made is that there is a point $(x_1, t_1) \in C_\tau$ such that $u(x_1, t_1) > 0$, and so this must be false. Therefore $u \leq 0$ on C_τ. Since τ is an arbitrary element of $]a, b[$, it follows that $u \leq 0$ on C.

Remark. The maximum principle will be improved or extended in Theorem 2.4 and Chapter VIII, as well as in Theorem 2.7 where it takes the form of a minimum principle.

As an application of Theorem 2.1, we shall prove that every non-negative temperature on $R^n \times]0, a[$ has a property closely related to the semigroup property. This will later be used to establish the semigroup property itself (Theorem 2.6), which in turn is used to prove the Gauss-Weierstrass integral representation. This contrasts with the proof of Theorem 1.4, where the semigroup property was derived from the Gauss-Weierstrass representation. However, in Chapter III we shall show that $D_1 W$ has the semigroup property but is not the Gauss-Weierstrass integral of any measure.

We require the following result, which is given in more generality than is needed in this chapter.

Lemma 2.1

If μ is a non-negative measure on R^n with compact support, then its Gauss-Weierstrass integral u is defined on $R^n \times]0, \infty[$, and $u(x, t) \to 0$ as $\|x\| \to \infty$, uniformly for t in $]0, \infty[$.

Proof. Since

$$\int_{R^n} W(x - y, t) \, d\mu(y) \leq (4\pi t)^{-n/2} \mu(R^n) < \infty$$

for all $x \in R^n$ and $t > 0$, u is defined on $R^n \times]0, \infty[$. Choose r such that $B(0, r)$ contains the support of μ. Whenever $\|y\| < r < \|x\|$, we have $\|x - y\| \geq \|x\| - \|y\| \geq \|x\| - r$. Furthermore, if $\alpha = n/2$ and $\beta = (\|x\| - r)^2/4$, then elementary calculus can show that

$$t^{-n/2} \exp(-(\|x\| - r)^2/4t) = t^{-\alpha} e^{-\beta/t} \leq (\alpha/\beta e)^\alpha = (2n/(\|x\| - r)^2 e)^{n/2}$$

Therefore, whenever $\|x\| > r$,

$$u(x, t) = \int_{\|y\| < r} W(x - y, t) \, d\mu(y)$$

$$\leq (4\pi t)^{-n/2} \int\limits_{\|y\|<r} \exp(-(\|x\| - r)^2/4t)\, d\mu(y)$$

$$\leq \mu(\underline{R}^n)(n/2\pi e)^{n/2}(\|x\| - r)^{-n}$$

from which it follows immediately that $u(x, t) \to 0$ as $\|x\| \to \infty$, uniformly for t in $]0, \infty[$.

Theorem 2.2

If u is a non-negative temperature on $\underline{R}^n \times]0, a[$, and $0 < s < t < a$, then

$$u(x, t) \geq \int\limits_{\underline{R}^n} W(x - y, t - s)u(y, s)\, dy \tag{2}$$

for all $x \in \underline{R}^n$.

Proof. Given any $r > 0$, we define a function u_r on $\underline{R}^n \times]s, \infty[$ by putting

$$u_r(x, t) = \int\limits_{\|y\|\leq r} W(x - y, t - s)u(y, s)\, dy$$

and consider the function v_r, given by

$$v_r(x, t) = u(x, t) - u_r(x, t)$$

for all $(x, t) \in \underline{R}^n \times]s, a[$. We shall show that $v_r \geq 0$. By Theorem 1.2, Corollary, u_r is a temperature, and hence v_r is also a temperature. Putting $f(y) = u(y, s)$ whenever $\|y\| \leq r$, $f(y) = 0$ elsewhere, and applying Theorem 1.3, we obtain

$$\limsup_{(x, t)\to(\xi, s^+)} u_r(x, t) \leq \limsup_{\eta\to\xi} f(\eta) \leq u(\xi, s)$$

for all $\xi \in \underline{R}^n$. Therefore

$$\liminf_{(x, t)\to(\xi, s^+)} v_r(x, t) = u(\xi, s) - \limsup_{(x, t)\to(\xi, s^+)} u_r(x, t) \geq 0 \tag{3}$$

for all ξ. By Lemma 2.1, given any $\epsilon > 0$ we can find R such that $u_r(x, t) < \epsilon$ whenever $\|x\| = R$ and $s < t < a$. Therefore

$$v_r(x, t) \geq -u_r(x, t) > -\epsilon$$

whenever $(x, t) \in \partial B(0, R) \times]s, a[$. This, together with (3), implies that

$$\lim \inf(v_r(x, t) + \epsilon) \geq 0$$

whenever (x, t) approaches any point of $\partial(B(0, R) \times]s, a[) \setminus (B(0, R) \times \{a\})$. It now follows from Theorem 2.1 that $v_r \geq -\epsilon$ throughout $B(0, R) \times]s, a[$. Since ϵ is arbitrary, $v_r \geq 0$ on $B(0, R) \times]s, a[$. Since R can be arbitrarily large, $v_r \geq 0$ on $\underline{R}^n \times]s, a[$. Therefore $u_r \leq u$, and (2) follows by making $r \to \infty$.

Remark. Since both sides of (2) are non-negative and u is real-valued, the integral is finite for all $(x, t) \in \underline{R}^n \times]s, a[$, and therefore defines a temperature, by Theorem 1.2.

2. CONVERGENT SEQUENCES OF INTEGRALS

Let $C_c(\underline{R}^n)$ denote the set of all continuous real-valued functions on $\underline{R}^n$ which have compact support. If the distance between two elements f and g of $C_c(\underline{R}^n)$ is defined to be

$$\|f - g\|_\infty = \sup\{|f(x) - g(x)| : x \in \underline{R}^n\}$$

then $C_c(\underline{R}^n)$ is a metric space.

The only result on convergent sequences of integrals that we need at this stage is a corollary of the following general result.

Theorem 2.3

Let κ be a fixed real number, and let $\{\mu_i\}$ be a sequence of non-negative measures on $\underline{R}^n$ such that $\mu_i(\underline{R}^n) \leq \kappa$ for all i. Suppose that $\{\mu_i\}$ converges to a non-negative measure μ in the sense that

$$\lim_{i \to \infty} \int_{\underline{R}^n} g \, d\mu_i = \int_{\underline{R}^n} g \, d\mu \tag{4}$$

for all $g \in C_0(\underline{R}^n)$. If $f_i \to f$ in the metric space $C_0(\underline{R}^n)$, then

$$\lim_{i \to \infty} \int_{\underline{R}^n} f_i \, d\mu_i = \int_{\underline{R}^n} f \, d\mu$$

<u>Proof.</u> Since $f \in C_0(\underline{R}^n)$, (4) implies that

$$\lim_{i \to \infty} \int_{\underline{R}^n} f \, d\mu_i = \int_{\underline{R}^n} f \, d\mu$$

Since $\|f_i - f\|_\infty \to 0$ as $i \to \infty$, and $\mu_i(\underline{R}^n) \leq \kappa$ for all i, we have

$$\left| \int_{\underline{R}^n} (f_i - f) \, d\mu_i \right| \leq \kappa \|f_i - f\|_\infty \to 0$$

as $i \to \infty$. Therefore

$$\lim_{i \to \infty} \int_{\underline{R}^n} f_i \, d\mu_i = \lim_{i \to \infty} \int_{\underline{R}^n} (f_i - f) \, d\mu_i + \lim_{i \to \infty} \int_{\underline{R}^n} f \, d\mu_i$$

$$= \int_{\underline{R}^n} f \, d\mu$$

Corollary

If $f_i \to f$ in the metric space $C_0(\underline{R}^n)$, $\{t_i\}$ is a positive null sequence of real numbers, and $x \in \underline{R}^n$, then

$$\lim_{i \to \infty} \int_{\underline{R}^n} W(x - y, t_i) f_i(y) \, dy = f(x)$$

<u>Proof.</u> By Lemma 1.1,

$$\int_{\underline{R}^n} W(x - y, t_i) \, dy = 1$$

for all i, so that, if μ_i is the non-negative measure on $\underline{R}^n$ given by

$$\mu_i(E) = \int_E W(x - y, t_i) \, dy$$

for every Borel set $E \subseteq \underline{R}^n$, then $\mu_i(\underline{R}^n) = 1$ for all i. Furthermore, for any $g \in C_C(\underline{R}^n)$, Theorem 1.1 shows that

$$\lim_{i \to \infty} \int_{\underline{R}^n} g \, d\mu_i = \lim_{i \to \infty} \int_{\underline{R}^n} W(x - y, t_i) g(y) \, dy$$

$$= g(x)$$

$$= \int_{\underline{R}^n} g(y) \, d\delta_x(y)$$

where δ_x is the Dirac δ-measure concentrated at x. We can therefore apply Theorem 2.3, with $\kappa = 1$ and $\mu = \delta_x$. Thus we obtain

$$\lim_{i \to \infty} \int_{\underline{R}^n} W(x - y, t_i) f_i(y) \, dy = \int_{\underline{R}^n} f(y) \, d\delta_x(y) = f(x)$$

3. AN EXTENSION OF THE MAXIMUM PRINCIPLE TO AN INFINITE STRIP

We shall establish a result, analogous to Theorem 2.1, for temperatures on $\underline{R}^n \times \,]0, a[$. The initial conditions will be strengthened slightly, in that continuity will be assumed on $\underline{R}^n \times [0, a[$. A condition on $u(x, t)$ as $\|x\| \to \infty$ must be imposed, to replace the condition on the curved part $\partial B \times \,]a, b[$ of the boundary of the cylinder $B \times \,]a, b[$ in Theorem 2.1. As might be expected, the condition that $u(x, t) \to 0$ as $\|x\| \to \infty$ would suffice, but in fact a much weaker condition, which allows $u(x, t)$ to grow like $\exp(\alpha \|x\|^2)$, leads to the same conclusion.

We need some preliminary results. The first is a technical lemma which ensures the existence of functions, with derivatives of all orders, which have certain useful properties. The class of functions on $\underline{R}^n$ which have derivatives of all orders will be denoted by $C^\infty(\underline{R}^n)$.

Lemma 2.2

Suppose that $0 < \rho < \sigma$ and $\kappa(\sigma - \rho) < \rho$ for some $\kappa \in \,]0, 1[$. There is a function $f \in C^\infty(\underline{R}^n)$ with the following properties:

(i) $f(x) = 1$ whenever $\|x\| \leq \rho$;

(ii) $f(x) = 0$ whenever $\|x\| \geq \sigma$;

(iii) $0 \leq f(x) \leq 1$ for all $x \in \underline{R}^n$;

(iv) there is a constant M, which depends on κ but not on σ or ρ, such that $|D_if| \leq M(\sigma - \rho)^{-1}$ and $|D_jD_if| \leq M(\sigma - \rho)^{-2}$ for all i, j;

(v) $f(x) = f(y)$ whenever $\|x\| = \|y\|$.

Proof. We first look at functions on $\underline{R}$. Let

$$F(r) = \begin{cases} \exp\left(\dfrac{-1}{1-r^2}\right) & \text{if} \quad |r| < 1, \\ 0 & \text{if} \quad |r| \geq 1 \end{cases}$$

It is an elementary exercise to prove, by induction on k, that $F^{(k)}$ is given by

$$F^{(k)}(r) = p(r)(1 - r^2)^{-q} \exp(-(1-r^2)^{-1}) \qquad (k \geq 1)$$

whenever $|r| < 1$, for some polynomial p and natural number q, both depending on k. It follows that $F^{(k)}(r)$ exists for all $r \in \underline{R}$. Therefore, for each $k \geq 1$, there is a constant M_k such that $|F^{(k)}| \leq M_k$ on $\underline{R}$.
Choose ϵ such that $\kappa(\sigma - \rho) < \epsilon < \min\{\sigma - \rho, \rho\}$, put

$$\gamma = \int_{-1}^{1} F(t)\, dt$$

and define ψ on $\underline{R}$ by

$$\psi(r) = \frac{1}{\gamma\epsilon} \int_{-2\epsilon}^{2\epsilon} F\left(\frac{r-s}{\epsilon}\right) ds$$

The integrand is non-zero if and only if $|r - s| < \epsilon$, that is, $s \in]r - \epsilon,, r + \epsilon[$. Since the range of integration is $]-2\epsilon, 2\epsilon[$, we see that $\psi(r) = 0$ if $r - \epsilon \geq 2\epsilon$ or $r + \epsilon \leq -2\epsilon$, that is, if $|r| \geq 3\epsilon$. Next, if $]r - \epsilon, r + \epsilon[\subseteq]-2\epsilon, 2\epsilon[$, we have

$$\psi(r) = \frac{1}{\gamma\epsilon} \int_{r-\epsilon}^{r+\epsilon} F\left(\frac{r-s}{\epsilon}\right) ds = \frac{1}{\gamma} \int_{-1}^{1} F(t)\, dt = 1$$

Hence $\psi(r) = 1$ whenever $|r| \leq \epsilon$. Finally, if $\epsilon < |r| < 3\epsilon$, then $]r - \epsilon, r + \epsilon[$ meets $]-2\epsilon, 2\epsilon[$ but is not contained in it. Therefore

$$0 \leq \psi(r) < \frac{1}{\gamma \epsilon} \int_{r-\epsilon}^{r+\epsilon} F\left(\frac{r-s}{\epsilon}\right) ds = 1$$

Since the derivatives of F are all bounded, and the range of integration is finite, we can use Lemma 1.3 to prove inductively that $\psi^{(k)}$ exists and can be calculated by differentiation under the integral sign, for all k. For example, to show that this is true with $k = 1$, we apply the lemma with $]\alpha, \beta[$ any bounded superinterval of $]-3\epsilon, 3\epsilon[$, $A = \phi$, $B =]-2\epsilon, 2\epsilon[$, μ Lebesgue measure, $\xi = r$, $\zeta = s$, $f(\xi, \eta, \zeta) = F((r-s)/\epsilon)$, $\phi = \gamma \epsilon \psi$ and $g = M_1$. Hence $\psi \in C^{\infty}(\underline{R}^n)$ and

$$|\psi^{(k)}(r)| = \left| \frac{1}{\gamma \epsilon} \int_{-2\epsilon}^{2\epsilon} \frac{1}{\epsilon^k} F^{(k)}\left(\frac{r-s}{\epsilon}\right) ds \right| \leq 4\gamma^{-1} M_k \epsilon^{-k}$$

for all $k \in \underline{N}$.

Now define f on $\underline{R}^n$ by putting

$$f(x) = \begin{cases} 1 & \text{if} \quad \|x\| < \rho, \\ \psi(2(\|x\| - \rho) + \epsilon) & \text{if} \quad \|x\| \geq \rho \end{cases}$$

Then (i), (iii) and (v) obviously hold. If $\|x\| \geq \sigma$, then

$$2(\|x\| - \rho) + \epsilon \geq 2(\sigma - \rho) + \epsilon > 3\epsilon$$

so that $f(x) = 0$, which proves that (ii) holds. Finally, for each i and j,

$$|D_i f(x)| = |\psi'(2\|x\| - 2\rho + \epsilon) 2x_i / \|x\| | \leq 8\gamma^{-1} M_1 \epsilon^{-1}$$

and

$$|D_j D_i f(x)| \leq 4|\psi''(2\|x\| - 2\rho + \epsilon)| + 4|\psi'(2\|x\| - 2\rho + \epsilon)/\|x\| |$$
$$\leq 16\gamma^{-1} M_2 \epsilon^{-2} + 16\gamma^{-1} M_1 \epsilon^{-2}$$

since $\|x\| \geq \rho > \epsilon$ whenever $D_j D_i f(x) \neq 0$. It now follows from the inequality $\epsilon > \kappa(\sigma - \rho)$ that (iv) holds.

Lemma 2.3

Suppose that $\alpha > 0$, $0 < \delta < 1/4(n+1)$, $1/8 < c < 1/4$, $1 < R \leq \|x\| \leq R + 1$, and $0 < t \leq \delta$. There exists a constant C, which depends only on c, δ, and n, such that

$$W(x,t) + \sum_{i=1}^{n} |D_i W(x,t)| \le C \exp(-(cR^2/\delta) + 2\alpha(R + 1)^2 - 2\alpha\|x\|^2) \tag{5}$$

Proof. By Lemma 1.4, there exists a positive constant C, which depends only on c and n, such that

$$W(x,t) \le Ct^{-n/2} \exp(-c\|x\|^2/t)$$

and

$$\sum_{i=1}^{n} |D_i W(x,t)| \le Ct^{-(n+1)/2} \exp(-c\|x\|^2/t)$$

for all $(x,t) \in \underline{R}^n \times]0, \infty[$. Therefore, whenever $\|x\| \ge R$,

$$W(x,t) \le Ct^{-n/2} \exp(-cR^2/t) \tag{6}$$

and

$$\sum_{i=1}^{n} |D_i W(x,t)| \le Ct^{-(n+1)/2} \exp(-cR^2/t) \tag{7}$$

If β, γ are positive constants, and $\phi(r) = r^{-\gamma}e^{-\beta/r}$ for all $r > 0$, then ϕ is increasing on $]0, \beta/\gamma]$. We use this fact, first with $\gamma = n/2$ and $\beta = cR^2$, then with $\gamma = (n + 1)/2$ and $\beta = cR^2$. Note that, in either case, since $c > 1/8$ and $R > 1$,

$$\delta < \frac{1}{4(n+1)} \le \frac{1}{8\gamma} < \frac{c}{\gamma} < \frac{cR^2}{\gamma} = \frac{\beta}{\gamma}$$

and therefore ϕ is increasing on $]0, \delta]$. It therefore follows from (6) and (7) that, whenever $0 < t \le \delta$ and $\|x\| \ge R$,

$$W(x,t) \le C\delta^{-n/2} \exp(-cR^2/\delta)$$

and

$$\sum_{i=1}^{n} |D_i W(x,t)| \le C\delta^{-(n+1)/2} \exp(-cR^2/\delta)$$

Hence

$$W(x, t) + \sum_{i=1}^{n} |D_i W(x, t)| \leq C \exp(-cR^2/\delta) \tag{8}$$

Finally, whenever $\|x\| \leq R + 1$,

$$1 \leq \exp(2\alpha(R + 1)^2 - 2\alpha\|x\|^2)$$

since $\alpha > 0$. Hence (5) follows from (8).

Green's formula for the heat equation. Let θ be the heat operator and let $\theta*$ denote its adjoint, defined by

$$\theta = \sum_{i=1}^{n} D_i^2 - D_t \quad \text{and} \quad \theta* = \sum_{i=1}^{n} D_i^2 + D_t$$

Then, if v and w are such that all their derivatives which occur in θv and θw exist and are continuous functions on an open set E, we have

$$v\theta w - w\theta*v = \sum_{i=1}^{n} (vD_i^2 w - wD_i^2 v) - (vD_t w + wD_t v)$$

$$= \sum_{i=1}^{n} D_i(vD_i w - wD_i v) - D_t(vw)$$

The last line is the divergence of the continuously differentiable function $\phi : E \rightarrow R^{n+1}$ whose i-th component is $vD_i w - wD_i v$ if $i \leq n$, $-vw$ if $i = n + 1$. It therefore follows from Gauss's divergence theorem that, for any bounded open set Ω whose boundary is piecewise smooth,

$$\iint_{\Omega} (v\theta w - w\theta*v) \; dx \; dt = \int_{\partial\Omega} \left(\sum_{i=1}^{n} (vD_i w - wD_i v)\nu_i - vw\nu_{n+1} \right) d\sigma \tag{9}$$

where $\nu = (\nu_1, \ldots, \nu_n, \nu_{n+1})$ is the outward unit normal to $\partial\Omega$ and σ denotes surface area on $\partial\Omega$.

We can now prove our extension of the maximum principle to $R^n \times]0, a[$. Here, and subsequently, ϕ^+ and ϕ^- denote the positive and negative parts of a function ϕ, defined by

$$\phi^+(\zeta) = \max\{\phi(\zeta), 0\}, \quad \phi^-(\zeta) = \max\{-\phi(\zeta), 0\}$$

for all ζ in the domain of ϕ.

Theorem 2.4

Let u be a continuous, real-valued function on $\underline{R}^n \times [0, a[$ and a temperature on $\underline{R}^n \times]0, a[$. If $u(\cdot, 0) \leq 0$ and

$$\int_0^a dt \int_{\underline{R}^n} u^+(x, t) \exp(-\alpha \|x\|^2)\, dx < \infty$$

for some $\alpha > 0$, then $u \leq 0$ on $\underline{R}^n \times [0, a[$.

Proof. We would like to apply Green's formula for the heat equation (9) with $w = u^+$, $v(x, t) = W(y - x, s - t)$ for some fixed $(y, s) \in \underline{R}^n \times]0, a[$, and $\Omega = \underline{R}^n \times]0, b[$ for some $b \in]0, s[$. However, u^+ may not have partial derivatives at zeros of u, so we replace it by $u + (u^2 + \eta)^{1/2}$ for some $\eta > 0$, and later make $\eta \to 0$ so that $u + (u^2 + \eta)^{1/2} \to u + |u| = 2u^+$. Furthermore, we have Green's formula only on bounded domains, and lack sufficient conditions on u to ensure the finiteness of the appropriate integrals in any case, so we take $v(x, t) = h(x) W(y - x, s - t)$ for a suitable $h \in C_0(\underline{R}^n)$, and later make $h(x) \to 1$ for all x. These are standard ways of overcoming such technical difficulties.

Choose δ such that $0 < \delta < \min\{1/8\alpha, 1/4(n + 1), a\}$. Let (y, s) be any point of $\underline{R}^n \times]0, \delta[$, let $b \in]0, s[$, let $R > 1$, and for each $\epsilon \in [0, b[$ let $\Omega_\epsilon = B(y, R + 1) \times]\epsilon, b[$. Choose $h \in C^\infty(\underline{R}^n)$ with the following properties:

$$h(x) = 1 \quad \text{whenever} \quad \|x - y\| \leq R \tag{10}$$

$$h(x) = 0 \quad \text{whenever} \quad \|x - y\| \geq R + 1 \tag{11}$$

$$0 \leq h(x) \leq 1 \quad \text{for all} \quad x \in \underline{R}^n \tag{12}$$

there exists a constant M, independent of R, such that

$$\sum_{i=1}^n (|D_i h| + |D_i^2 h|) \leq M \quad \text{on} \quad \underline{R}^n \tag{13}$$

The existence of h is assured by Lemma 2.2. We now put

$$v(x, t) = h(x) W(y - x, s - t) \tag{14}$$

into Green's formula (9) on Ω_ϵ for $\epsilon > 0$, noting that $v, D_1 v, \ldots, D_n v$ are all zero on $\partial B(y, R + 1) \times [0, b]$. Thus we obtain

$$\iint_{\Omega_\epsilon} (v\theta w - w\theta^* v) \ dx \ dt$$

$$= \int_{B(y,R+1)} v(x,\epsilon)w(x,\epsilon) \ dx - \int_{B(y,R+1)} v(x,b)w(x,b) \ dx \qquad (15)$$

If w is continuous and real-valued on $\underline{R}^n \times [0,b]$, then vw is uniformly continuous on $\bar{B}(y, R+1) \times [0,b]$, so that

$$\lim_{\epsilon \to 0} \int_{B(y,R+1)} v(x,\epsilon)w(x,\epsilon) \ dx = \int_{B(y,R+1)} v(x,0)w(x,0) \ dx$$

Therefore, making $\epsilon \to 0$ in (15), we obtain

$$\iint_{\Omega_0} (v\theta w - w\theta^* v) \ dx \ dt$$

$$= \int_{B(y,R+1)} v(x,0)w(x,0) \ dx - \int_{B(y,R+1)} v(x,b)w(x,b) \ dx \qquad (16)$$

If we take $w = u + w_\eta$, where $w_\eta = (u^2 + \eta)^{1/2}$ and $\eta > 0$, then w is continuous and real-valued on $\underline{R}^n \times [0,a[$, and on $\underline{R}^n \times]0,a[$ we have

$$\theta w = \theta w_\eta = \left(\frac{u}{w_\eta}\right)\theta u + \frac{w_\eta^2 - u^2}{w_\eta^3} \sum_{i=1}^n (D_i u)^2 = \frac{\eta}{w_\eta^3} \sum_{i=1}^n (D_i u)^2 \geq 0$$

Since (12) holds $v \geq 0$, and so (16) implies that

$$\iint_{\Omega_0} w(x,t)\theta^*(h(x)W(y-x,s-t)) \ dx \ dt$$

$$\geq \int_{B(y,R+1)} W(y-x, s-b)h(x)w(x,b) \ dx$$

$$- \int_{B(y,R+1)} W(y-x,s)h(x)w(x,0) \ dx \qquad (17)$$

Now $w_\eta - |u| = (u^2 + \eta)^{1/2} - (u^2)^{1/2} \leq ((u^2 + \eta) - u^2)^{1/2} = \eta^{1/2}$, so that $w_\eta \to |u|$ as $\eta \to 0$, uniformly on Ω_0. Therefore $w \to u + |u| = 2u^+$ uniformly on Ω_0. Since the ranges of integration in (17) all have finite measure, it follows that

$$\iint_{\Omega_0} u^+(x, t)\, \theta*(h(x)W(y - x, s - t))\, dx\, dt$$

$$\geq \int_{B(y, R+1)} W(y - x, s - b)h(x)u^+(x, b)\, dx$$

$$- \int_{B(y, R+1)} W(y - x, s)h(x)u^+(x, 0)\, dx$$

By hypothesis $u(\cdot, 0) \leq 0$, and so the last integral is zero. Because of (11), it follows that

$$\int_0^b dt \int_{\underline{R}^n} u^+(x, t)\, \theta*(h(x)W(y - x, s - t))\, dx$$

$$\geq \int_{\underline{R}^n} W(y - x, s - b)h(x)u^+(x, b)\, dx \tag{18}$$

Since u^+ is uniformly continuous on $\bar{B}(y, R + 1) \times [b, s]$, $u^+(\cdot, b)$ converges uniformly on $B(y, R + 1)$ to $u^+(\cdot, s)$ as $b \to s^-$. Therefore $hu^+(\cdot, b)$ converges to $hu^+(\cdot, s)$ in the metric space $C_c(\underline{R}^n)$, in view of (11). Theorem 2.3, Corollary, applied to every positive null sequence $\{t_j\} = \{s - b_j\}$, with $f_i = hu^+(\cdot, b_i)$, now shows that

$$\lim_{b \to s^-} \int_{\underline{R}^n} W(y - x, s - b)h(x)u^+(x, b)\, dx = h(y)u^+(y, s) = u^+(y, s)$$

because of (10). Making $b \to s^-$ in (18), we deduce that if the integral is finite

$$u^+(y, s) \leq \int_0^s dt \int_{\underline{R}^n} u^+(x, t)\, \theta*(h(x)W(y - x, s - t))\, dx \tag{19}$$

Since $\theta*W(y - x, s - t) = 0$, a routine calculation shows that

$$\theta*(h(x)W(y - x, s - t))$$

$$= \sum_{i=1}^n (D_i^2 h(x))W(y - x, s - t) + 2 \sum_{i=1}^n (D_i h(x))D_i W(y - x, s - t)$$

This, together with (13) and Lemma 2.3, implies that, if $1/8 < c < 1/4$,

$| \theta*(h(x)W(y - x, s - t))|$

$$\leq \sum_{i=1}^{n} |D_i^2 h(x)| W(y - x, s - t) + 2 \sum_{i=1}^{n} |D_i h(x)| |D_i W(y - x, s - t)|$$

$$\leq 2M\left(W(y - x, s - t) + \sum_{i=1}^{n} |D_i W(y - x, s - t)| \right)$$

$$\leq C \exp(-(cR^2/\delta) + 2\alpha(R + 1)^2) \exp(-2\alpha\|y - x\|^2)$$

whenever $R \leq \|y - x\| \leq R + 1$ and $0 < s - t \leq \min\{\delta, s\}$, where C depends only on c, δ, and n. Furthermore, since $\|x\|^2 \leq 2(\|y\|^2 + \|y - x\|^2)$, we have

$$\exp(-2\alpha\|y - x\|^2) \leq \exp(2\alpha\|y\|^2) \exp(-\alpha\|x\|^2)$$

Also $\theta*(h(x)W(y - x, s - t)) = 0$ if $\|y - x\| < R$ or $\|y - x\| > R + 1$, in view of (10), (11), and the fact that $\theta*W(y - x, s - t) = 0$. It follows that

$$\int_0^s dt \int_{R^n} u^+(x, t) |\theta*(h(x)W(y - x, s - t))| \, dx$$

$$= \int_0^s dt \int_{R \leq \|y-x\| \leq R+1} u^+(x, t) |\theta*(h(x)W(y - x, s - t))| \, dx$$

$$\leq C \exp(-(cR^2/\delta) + 2\alpha(R + 1)^2 + 2\alpha\|y\|^2) \int_0^s dt \int_{R^n} u^+(x, t) \exp(-\alpha\|x\|^2) \, dx$$

The last integral is finite by hypothesis, so (19) holds. Since $\delta < 1/8\alpha$, we can choose $c \in \,]1/8, 1/4[$ such that $\delta < c/2\alpha < 1/8\alpha$. With this c we have $c/\delta > 2\alpha$, so that if we make $R \to \infty$ we obtain

$$\lim_{R \to \infty} \int_0^s dt \int_{R^n} u^+(x, t) \theta*(h(x)W(y - x, s - t)) \, dx = 0$$

It therefore follows from (19) that $u(y, s) \leq u^+(y, s) \leq 0$. Since (y, s) is an arbitrary point of $R^n \times \,]0, \delta[$, we see that $u \leq 0$ on $R^n \times \,]0, \delta[$.

Now let $S = \{\tau : 0 \leq \tau \leq a, \ u \leq 0 \text{ on } R^n \times \,]0, \tau[\}$. If $\sigma = \sup S$, then $\sigma \in S$ and $\sigma \geq \delta > 0$. If $\sigma < a$, then since $u \leq 0$ on $R^n \times \,]0, \sigma[$ and u is continuous, $u(\cdot, \sigma) \leq 0$. Applying the result just proved to the function

$(x, t) \longmapsto u(x, t + \sigma)$ on $\underline{R}^n \times [0, a - \sigma[$, we see that there is $\delta' > 0$ such that this function is non-positive on $\underline{R}^n \times]0, \delta'[$. Hence $u \leq 0$ on $\underline{R}^n \times]0, \sigma + \delta'[$, which contradicts the definition of σ. If follows that $\sigma = a$, and the theorem is proved.

4. SOME CONSEQUENCES OF THE EXTENDED MAXIMUM PRINCIPLE

Theorem 2.4 has several important consequences, some of which will be discussed in later chapters. Here we use it to prove that if u is a non-negative temperature on $\underline{R}^n \times]0, a[$, continuous on $\underline{R}^n \times [0, a[$, and zero initially, then $u = 0$. From this we deduce that u has the semigroup property, which in turn implies that, if u has a zero then u is identically zero.

Theorem 2.5

If u is continuous on $\underline{R}^n \times [0, a[$, a non-negative temperature on $\underline{R}^n \times]0, a[$, and zero on $\underline{R}^n \times \{0\}$, then $u = 0$ on $\underline{R}^n \times [0, a[$.

Proof. Suppose that $0 < s < b < t < a$. By Theorem 2.2,

$$\int_{\underline{R}^n} W(y, t - s)u(y, s) \, dy \leq u(0, t)$$

so that

$$\int_0^b ds \int_{\underline{R}^n} W(y, t - s)u(y, s) \, dy \leq bu(0, t)$$

Furthermore,

$$W(y, t - s) \geq (4\pi t)^{-n/2} \exp(-\|y\|^2/4(t - b))$$

and it therefore follows that, if $\alpha = 1/4(t - b)$,

$$\int_0^b ds \int_{\underline{R}^n} \exp(-\alpha\|y\|^2)u(y, s) \, dy \leq bu(0, t)(4\pi t)^{n/2} < \infty$$

Since $u = u^+$, it now follows from Theorem 2.4 that $u \leq 0$, hence $u = 0$, on $R^n \times [0, b[$. Because b can be arbitrarily close to a, u = 0 on $\underline{R}^n \times [0, a[$.

Theorem 2.6

If u is a non-negative temperature on $\underline{R}^n \times]0, a[$, then

$$u(x, t) = \int_{\underline{R}^n} W(x - y, t - s)u(y, s) \, dy \qquad (20)$$

whenever $x \in \underline{R}^n$ and $0 < s < t < a$.

Proof. Given any $s \in]0, a[$, let v(x, t) denote the right-hand side of (20). By Theorem 2.2, v(x, t) is finite for all $(x, t) \in \underline{R}^n \times]s, a[$, and is therefore a temperature because of Theorem 1.2. By Theorem 1.3, Corollary, v can be extended to a continuous function on $\underline{R}^n \times [s, a[$ by putting $v(\cdot, s) = u(\cdot, s)$. If $w = u - v$ on $\underline{R}^n \times [s, a[$, then w is a temperature on $\underline{R}^n \times]s, a[$, $w \geq 0$ by Theorem 2.2, and $w(\cdot, s) = 0$. Theorem 2.5 now shows that $w = 0$, which is the same as (20).

Theorem 2.7

If u is a non-negative temperature on $\underline{R}^n \times]0, a[$, and there is a point (x_0, t_0) such that $u(x_0, t_0) = 0$, then $u = 0$ throughout $\underline{R}^n \times]0, a[$.

Proof. Let $s \in]0, t_0[$. By Theorem 2.6,

$$\int_{\underline{R}^n} W(x_0 - y, t_0 - s)u(y, s) \, dy = u(x_0, t_0) = 0$$

Since $W(x_0 - y, t_0 - s) > 0$ and $u \geq 0$, it follows that $u(\cdot, s) = 0$. Therefore $u = 0$ on $\underline{R}^n \times]0, t_0[$, since s is arbitrary. It now follows from Theorem 2.6 that, whenever $x \in \underline{R}^n$ and $t_0 \leq t < a$,

$$u(x, t) = \int_{\underline{R}^n} W(x - y, t - (t_0/2))u(y, t_0/2) \, dy = 0$$

Hence $u = 0$ on $\underline{R}^n \times]0, a[$.

5. FURTHER RESULTS ON THE CONVERGENCE
OF SEQUENCES OF INTEGRALS

The theorems presented here we will be using in the following section
to establish the Gauss-Weierstrass integral representation of non-
negative temperatures. To prove the first of them, we need the fact
that $C_c(\underline{R}^n)$ has a countable dense subset.

Lemma 2.4

The metric space $C_c(\underline{R}^n)$ has a countable dense subset.

<u>Proof.</u> Let $N, k \in \underline{N}$, and let q be a polynomial on $\underline{R}^n$, with rational
coefficients, such that $\sup\{|q(x)| : \|x\| = N\} \le 1/k$. Since $\partial B(0, N)$ is
compact and q is continuous there, Tietze's extension theorem shows
that there exists $\phi \in C_c(\underline{R}^n)$ such that $\phi(x) = q(x)$ whenever $\|x\| = N$
and

$$\|\phi\|_\infty = \sup\{|\phi(x)| : x \in \underline{R}^n\} = \sup\{|q(x)| : \|x\| = N\} \le 1/k$$

To each q we associate a single such ϕ, then we put

$$\psi(x) = \begin{cases} q(x) & \text{if} \quad x \in B(0, N), \\ \phi(x) & \text{if} \quad x \notin B(0, N) \end{cases}$$

so that $\psi \in C_c(\underline{R}^n)$ and $\sup\{|\psi(x)| : \|x\| \ge N\} \le 1/k$. Thus, given N, k,
and q, we have a single $\psi = \psi_{N,k,q}$. Since there are only countably
many N, k, and q, the set Ψ of all such functions $\psi_{N,k,q}$ is countable.
   We show that Ψ is dense in $C_c(\underline{R}^n)$. Given any $f \in C_c(\underline{R}^n)$, choose
N such that $B(0, N)$ contains the support of f, and let $B = \bar{B}(0, N)$. The
Stone-Weierstrass theorem, applied to the restriction of f to B, shows
that, given $k \in \underline{N}$, there is a polynomial p on B such that

$$\sup\{|f(x) - p(x)| : x \in B\} < 1/2k \tag{21}$$

Let

$$p(x) = \sum_{|j|=0}^{J} \alpha_j x^j$$

for all $x \in B$, where $j = (j_1, \ldots, j_n)$ is an n-tuple of non-negative integers, x^j denotes the product $x_1^{j_1} \cdots x_n^{j_n}$, $|j| = j_1 + \cdots + j_n$, J is the degree of p, and each $\alpha_j \in \underline{R}$. Since the rational numbers are dense in $\underline{R}$, we can find rationals β_j such that

$$|\alpha_j - \beta_j| < (2k)^{-1} N^{-J} (J + 1)^{-n}$$

for all j such that $0 \le |j| \le J$. Then, putting

$$q(x) = \sum_{|j|=0}^{J} \beta_j x^j$$

for all $x \in \underline{R}^n$, whenever $x \in B$ we have

$$|p(x) - q(x)| \le \sum_{|j|=0}^{J} |\alpha_j - \beta_j| \, |x^j|$$

$$< (2k)^{-1} N^{-J} (J + 1)^{-n} \sum_{|j|=0}^{J} N^{|j|}$$

$$\le 1/2k$$

since the sum contains no more than $(J + 1)^n$ terms, each of which does not exceed N^J. It now follows from (21) that

$$\sup\{|f(x) - q(x)| : x \in B\} \le 1/k \qquad (22)$$

Since $f(x) = 0$ whenever $x \in \partial B$, this implies that $\sup\{|q(x)| : \|x\| = N\}$ $\le 1/k$. With the present N, k, and q, let ψ be the $\psi_{N,k,q}$ of the subset Ψ of $C_0(\underline{R}^n)$. Then $\sup\{|\psi(x)| : \|x\| \ge N\} \le 1/k$ and, since $\psi(x) = q(x)$ whenever $\|x\| < N$, (22) shows that $\sup\{|f(x) - \psi(x)| : \|x\| < N\} \le 1/k$. Since $f(x) = 0$ whenever $\|x\| \ge N$, we have $\|f - \psi\|_\infty \le 1/k$. Therefore Ψ is dense in $C_0(\underline{R}^n)$.

Theorem 2.8

Let κ be a fixed real number, and let $\{\mu_i\}$ be a sequence of non-negative measures on $\underline{R}^n$ such that $\mu_i(\underline{R}^n) \le \kappa$ for all i. Then there is a non-negative measure μ such that $\mu(\underline{R}^n) \le \kappa$, and a subsequence $\{\mu_{i(j)}\}$ such that

$$\lim_{j \to \infty} \int_{\underline{R}^n} f \, d\mu_{i(j)} = \int_{\underline{R}^n} f \, d\mu$$

for all $f \in C_c(\underline{R}^n)$.

Proof. By Lemma 2.4, $C_c(\underline{R}^n)$ has a countable dense subset C_0. Arrange the members of C_0 as a sequence $\{f_j\}$. The sequence of real numbers

$$\left\{ \int_{\underline{R}^n} f_1 \, d\mu_i \right\}$$

is bounded by $\kappa \|f_1\|_\infty$, and therefore has a convergent subsequence. Thus there is a subsequence $\{\mu_i^{(1)}\}$ of $\{\mu_i\}$ such that the sequence

$$\left\{ \int_{\underline{R}^n} f_1 \, d\mu_i^{(1)} \right\}$$

is convergent. Next, the sequence of real numbers

$$\left\{ \int_{\underline{R}^n} f_2 \, d\mu_i^{(1)} \right\}$$

is bounded by $\kappa \|f_2\|_\infty$, so that it too has a convergent subsequence. Therefore, there is a subsequence $\{\mu_i^{(2)}\}$ of $\{\mu_i^{(1)}\}$ such that

$$\left\{ \int_{\underline{R}^n} f_2 \, d\mu_i^{(2)} \right\}$$

is convergent. We successively choose subsequences in this way, so that for each fixed j, $\{\mu_i^{(j)}\}$ is a subsequence of $\{\mu_i^{(j-1)}\}$ and

$$\left\{ \int_{\underline{R}^n} f_j \, d\mu_i^{(j)} \right\}$$

is convergent. The sequence $\{\mu_i^{(i)}\}$ is a subsequence of $\{\mu_i^{(j)}\}$ for each fixed j, and therefore

$$\left\{ \int_{\underline{R}^n} f_j \, d\mu_i^{(i)} \right\}$$

converges for each fixed j.

Given any $f \in C_c(\underline{R}^n)$ and $\epsilon > 0$, choose $f_\ell \in C_0$ such that

$$\|f - f_\ell\|_\infty < \epsilon/3\kappa$$

Then, for each i, k,

$$\left| \int_{\underline{R}^n} f \, d\mu_i^{(i)} - \int_{\underline{R}^n} f \, d\mu_k^{(k)} \right|$$

$$\leq \left| \int_{\underline{R}^n} (f - f_\ell) \, d\mu_i^{(i)} \right| + \left| \int_{\underline{R}^n} f_\ell \, d\mu_i^{(i)} - \int_{\underline{R}^n} f_\ell \, d\mu_k^{(k)} \right| + \left| \int_{\underline{R}^n} (f_\ell - f) \, d\mu_k^{(k)} \right|$$

Now,

$$\left| \int_{\underline{R}^n} (f - f_\ell) \, d\mu_i^{(i)} \right| \leq \kappa \|f - f_\ell\|_\infty < \epsilon/3$$

and similarly for k. Furthermore, since $f_\ell \in C_0$, the sequence of real numbers

$$\left\{ \int_{\underline{R}^n} f_\ell \, d\mu_i^{(i)} \right\}$$

is convergent, and therefore we can find M such that

$$\left| \int_{\underline{R}^n} f_\ell \, d\mu_i^{(i)} - \int_{\underline{R}^n} f_\ell \, d\mu_k^{(k)} \right| < \epsilon/3$$

whenever $i, k \geq M$. It follows that

$$\left| \int_{\underline{R}^n} f \, d\mu_i^{(i)} - \int_{\underline{R}^n} f \, d\mu_k^{(k)} \right| < \epsilon$$

whenever $i, k \geq M$. Therefore the sequence of real numbers

$$\left\{ \int_{\underline{R}^n} f \, d\mu_i^{(i)} \right\}$$

is convergent.
    Put

$$\Lambda(f) = \lim_{i \to \infty} \int_{\underline{R}^n} f \, d\mu_i^{(i)}$$

for all $f \in C_c(\underline{R}^n)$. If $f \geq 0$ then $\Lambda(f) \geq 0$, so that Λ is a positive linear functional on $C_c(\underline{R}^n)$. By the Riesz representation theorem, there is a non-negative measure μ on $\underline{R}^n$ such that

$$\Lambda(f) = \int_{\underline{R}^n} f \, d\mu$$

for all $f \in C_c(\underline{R}^n)$. Hence the subsequence $\{\mu_i^{(i)}\}$ of $\{\mu_i\}$ satisfies

$$\lim_{i \to \infty} \int_{\underline{R}^n} f \, d\mu_i^{(i)} = \int_{\underline{R}^n} f \, d\mu$$

for all $f \in C_c(\underline{R}^n)$.

If we had $\mu(\underline{R}^n) > \kappa$, we could find a ball $B = B(0, r)$ such that $\mu(B) > \kappa$. Then, choosing $\phi \in C_c(\underline{R}^n)$ such that $\phi(x) = 1$ for all $x \in B$ and $0 \leq \phi(x) \leq 1$ for all $x \in \underline{R}^n$, we would have

$$\int_{\underline{R}^n} \phi \, d\mu \geq \mu(B) > \kappa$$

as well as the contradictory estimate

$$\int_{\underline{R}^n} \phi \, d\mu = \lim_{i \to \infty} \int_{\underline{R}^n} \phi \, d\mu_i^{(i)} \leq \limsup_{i \to \infty} \mu_i^{(i)}(R^n) \leq \kappa$$

Hence $\mu(R^n) \leq \kappa$, and the result is proved.

Definition. We say that a sequence of functions $\{f_i\}$ on $\underline{R}^n$ converges locally uniformly on $\underline{R}^n$ to a function f if, for every compact set $K \subseteq \underline{R}^n$, $f_i \to f$ uniformly on K.

Theorem 2.9

Let $\{f_i\}$ be a sequence of continuous functions which converges locally uniformly on $\underline{R}^n$ to a function f, and suppose that there is a positive constant M such that $|f_i| \leq M$ for all i. Let κ be a fixed real number, and let $\{\mu_i\}$ be a sequence of non-negative measures on $\underline{R}^n$ such that:

(i) $\mu_i(\underline{R}^n) \leq \kappa$ for all i;

(ii) $\{\mu_i\}$ converges to a non-negative measure μ, in the sense that

$$\lim_{i\to\infty} \int_{\underline{R}^n} g \, d\mu_i = \int_{\underline{R}^n} g \, d\mu$$

for all $g \in C_c(\underline{R}^n)$;

(iii) to each $\eta > 0$ there corresponds $\rho_\eta > 0$ such that $\mu_i(\underline{R}^n \setminus B(0,\rho_\eta)) < \eta$ for all i.

Then

$$\int_{\underline{R}^n} |f| \, d\mu < \infty \tag{23}$$

and

$$\lim_{i\to\infty} \int_{\underline{R}^n} f_i \, d\mu_i = \int_{\underline{R}^n} f \, d\mu$$

<u>Proof.</u> Given $\epsilon > 0$, choose r_ϵ such that $\mu_i(\underline{R}^n \setminus B(0,r_\epsilon)) < \epsilon/3M$ for all i. Let r_1, r_2, r_3, r_4 be numbers such that $r_\epsilon < r_1 < r_2 < r_3 < r_4$, and for each $j \in \{1,3\}$ let ω_j be a function in $C_c(\underline{R}^n)$ such that $\omega_j(x) = 1$ if $\|x\| < r_j$, $\omega_j(x) = 0$ if $\|x\| > r_{j+1}$, and $0 \leq \omega_j(x) \leq 1$ for all $x \in \underline{R}^n$. Put

$$\psi(x) = \omega_3(x)(1 - \omega_1(x))$$

for all $x \in \underline{R}^n$, so that $\psi \in C_c(\underline{R}^n)$, $\psi(x) = 0$ if $\|x\| < r_1$ or $\|x\| > r_4$, $\psi(x) = 1$ if $r_2 < \|x\| < r_3$, and $0 \leq \psi(x) \leq 1$ for all $x \in \underline{R}^n$.

Since $|f_i| \leq M$ for all i, $\psi(x) = 0$ whenever $\|x\| \leq r_\epsilon$, and $0 \leq \psi \leq 1$ on $\underline{R}^n$,

$$\int_{\underline{R}^n} |f_i| \psi \, d\mu_i \leq M \mu_i(\underline{R}^n \setminus \bar{B}(0,r_\epsilon)) < \frac{\epsilon}{3} \tag{24}$$

for all i. Because $f_i \to f$ uniformly on $\bar{B}(0,r_4)$, we have that $|f_i| \psi \to |f| \psi$ in the metric space $C_c(\underline{R}^n)$. Therefore, by Theorem 2.3 and (24),

$$\int_{\underline{R}^n} |f|\,\psi\,d\mu = \lim_{i\to\infty} \int_{\underline{R}^n} |f_i|\,\psi\,d\mu_i \le \frac{\epsilon}{3}$$

Since $\psi \ge 0$ on $\underline{R}^n$ and $\psi(x) = 1$ whenever $r_2 < \|x\| < r_3$, it follows that

$$\int_{r_2<\|x\|<r_3} |f(x)|\,d\mu(x) \le \frac{\epsilon}{3}$$

Because this holds for r_2 arbitrarily close to r_ϵ, and r_3 arbitrarily large, we deduce that

$$\int_{\|x\|>r_\epsilon} |f(x)|\,d\mu(x) \le \frac{\epsilon}{3} \tag{25}$$

Now, $f_i \to f$ locally uniformly on $\underline{R}^n$, and $|f_i| \le M$ for all i. Therefore $|f| \le M$ and

$$\int_{\|x\|\le r_\epsilon} |f(x)|\,d\mu(x) < \infty$$

This, together with (25), establishes (23).

Let $r > r_\epsilon$, and let h be a function in $C_c(\underline{R}^n)$ such that $h(x) = 1$ whenever $\|x\| \le r_\epsilon$, $h(x) = 0$ whenever $\|x\| > r$, and $0 \le h(x) \le 1$ for all $x \in \underline{R}^n$. Put $\phi = fh$ and $\phi_i = f_ih$ for all i. Since $f_i \to f$ uniformly on $\bar{B}(0,r)$, $\phi_i \to \phi$ in the metric space $C_c(\underline{R}^n)$. Therefore, by Theorem 2.3, there is an integer i_0 such that

$$\left| \int_{\underline{R}^n} \phi_i\,d\mu_i - \int_{\underline{R}^n} \phi\,d\mu \right| < \frac{\epsilon}{3} \tag{26}$$

for all $i > i_0$. Furthermore, for every i, since $h = 1$ on $\bar{B}(0,r_\epsilon)$ and $0 \le h \le 1$ on $\underline{R}^n$, we have

$$\left| \int_{\underline{R}^n} (f_i - \phi_i)\,d\mu_i \right| = \left| \int_{\|x\|>r_\epsilon} (1 - h(x))f_i(x)\,d\mu_i(x) \right|$$

$$\le \int_{\|x\|>r_\epsilon} |f_i(x)|\,d\mu_i(x)$$

$$\le M\mu_i(\underline{R}^n \setminus B(0,r_\epsilon))$$

$$< \frac{\epsilon}{3} \tag{27}$$

by our choice of r_ϵ. Using (25), we can similarly show that

$$\left| \int_{\underline{R}^n} (\phi - f)\, d\mu \right| \leq \int_{\|x\| > r_3} |f(x)|\, d\mu(x) \leq \frac{\epsilon}{3} \tag{28}$$

It follows from (26), (27), and (28) that, for all $i > i_0$,

$$\left| \int_{\underline{R}^n} f_i\, d\mu_i - \int_{\underline{R}^n} f\, d\mu \right|$$

$$\leq \left| \int_{\underline{R}^n} (f_i - \phi_i)\, d\mu_i \right| + \left| \int_{\underline{R}^n} \phi_i\, d\mu_i - \int_{\underline{R}^n} \phi\, d\mu \right| + \left| \int_{\underline{R}^n} (\phi - f)\, d\mu \right|$$

$$< \epsilon$$

6. THE GAUSS-WEIERSTRASS INTEGRAL REPRESENTATION OF NON-NEGATIVE TEMPERATURES

We are now in a position to prove the main result of this chapter. The last part of the theorem implies the result of Theorem 2.5, and the proof depends on that result.

Theorem 2.10

Let u be a non-negative temperature on $\underline{R}^n \times\,]0, a[$. Then there exists a non-negative measure μ on $\underline{R}^n$ such that u is the Gauss-Weierstrass integral of μ on $\underline{R}^n \times\,]0, a[$. If u is also continuous and real-valued on $\underline{R}^n \times\, [0, a[$, then $d\mu(y) = u(y, 0)\, dy$.

Proof. Let $b \in\,]0, a[$. Then, by Theorem 2.6,

$$\int_{\underline{R}^n} W(y, b - s)u(y, s)\, dy = u(0, b)$$

whenever $0 < s < b$. Whenever $(y, s) \in \underline{R}^n \times\,]0, b/2[$,

$$W(y, b - s) \geq (4\pi b)^{-n/2} \exp(-\|y\|^2/2b)$$

so it follows that

$$\int_{\underline{R}^n} \exp(-\|y\|^2/2b)u(y, s)\, dy \leq (4\pi b)^{n/2} u(0, b) \tag{29}$$

For each $s \in \,]0, b/2[$ and every Borel set $E \subseteq \underline{R}^n$, we put

$$\mu_s(E) = \int_E \exp(-\|y\|^2/b)u(y, s) \, dy$$

It then follows from (29) that $\mu_s(\underline{R}^n) \leq (4\pi b)^{n/2}u(0, b)$ for all s, so that we can apply Theorem 2.8. We thus deduce that there is a non-negative measure ν on $\underline{R}^n$ such that $\nu(\underline{R}^n) \leq (4\pi b)^{n/2}u(0, b)$, and a decreasing null sequence $\{s(i)\}$ in $]0, b/2[$ such that

$$\lim_{i \to \infty} \int_{\underline{R}^n} g \, d\mu_{s(i)} = \int_{\underline{R}^n} g \, d\nu \qquad (30)$$

for all $g \in C_c(\underline{R}^n)$.

We aim to apply Theorem 2.9 to the sequence of measures $\{\mu_{s(i)}\}$, so we need to establish that the hypothesis (iii) of that result holds in this case. Using (29) we see that, for any $r > 0$ and all $s \in \,]0, b/2[$,

$$\int_{\|y\| \geq r} \exp(-\|y\|^2/b)u(y, s) \, dy$$

$$\leq \exp(-r^2/2b) \int_{\|y\| \geq r} \exp(-\|y\|^2/2b)u(y, s) \, dy$$

$$\leq (4\pi b)^{n/2} u(0, b) \exp\left(-\frac{r^2}{2b}\right)$$

Because $\exp(-r^2/2b) \to 0$ as $r \to \infty$, it follows that to each $\eta > 0$ there corresponds $r_\eta > 0$ such that

$$\mu_s(\underline{R}^n \setminus B(0, r_\eta)) = \int_{\|y\| \geq r_\eta} \exp(-\|y\|^2/b)u(y, s) \, dy < \eta$$

uniformly for $s \in \,]0, b/2[$; in particular, $\mu_{s(i)}(\underline{R}^n \setminus B(0, r_\eta)) < \eta$ for all i. Hence $\{\mu_{s(i)}\}$ satisfies the hypotheses (i), (ii), and (iii) of Theorem 2.9.

Now fix $(x, t) \in \underline{R}^n \times \,]0, b/8[$, and define f_i on $\underline{R}^n$ by

$$f_i(y) = W(x - y, t - s(i)) \exp(\|y\|^2/b)$$

for every $i \in \underline{N}$ such that $s(i) < t/2$. For such i,

$$t/2 < t - s(i) < t < b/8$$

so that, since $-2\|x - y\|^2 \le 2\|x\|^2 - \|y\|^2$ for all $y \in \underline{R}^n$, we have

$$f_i(y) \le (2\pi t)^{-n/2} \exp\left(-\frac{\|x - y\|^2}{4t} + \frac{\|y\|^2}{b}\right)$$

$$\le (2\pi t)^{-n/2} \exp\left(\left(\frac{\|x\|^2}{4t} - \frac{\|y\|^2}{8t}\right) + \frac{\|y\|^2}{b}\right)$$

$$\le (2\pi t)^{-n/2} \exp(\|x\|^2/4t)$$

There is thus a constant M, which depends on the fixed point (x, t), such that $0 \le f_i \le M$ for all i. Furthermore, $f_i(y) \to W(x - y, t) \exp(\|y\|^2/b)$ locally uniformly on $\underline{R}^n$, since the function $(y, s) \longmapsto W(x - y, t - s) \cdot \exp(\|y\|^2/b)$ is uniformly continuous on $K \times [0, t/2]$ for every compact $K \subseteq \underline{R}^n$. Hence $\{f_i\}$ satisfies the hypotheses of Theorem 2.9.
By Theorem 2.9,

$$\int_{\underline{R}^n} W(x - y, t) \exp(\|y\|^2/b) \, d\nu(y) < \infty$$

and

$$\lim_{i \to \infty} \int_{\underline{R}^n} W(x - y, t - s(i)) u(y, s(i)) \, dy = \int_{\underline{R}^n} W(x - y, t) \exp(\|y\|^2/b) \, d\nu(y)$$

It follows from the semigroup property (Theorem 2.6) that the left-hand side is just $u(x, t)$. Therefore, putting $d\mu(y) = \exp(\|y\|^2/b) \, d\nu(y)$, we obtain

$$u(x, t) = \int_{\underline{R}^n} W(x - y, t) \, d\mu(y) \tag{31}$$

for any $(x, t) \in \underline{R}^n \times]0, b/8[$.
To extend (31) to all $(x, t) \in \underline{R}^n \times]0, a[$, we choose $\tau \in]0, b/8[$ and use the semigroup property. Thus, whenever $(x, t) \in \underline{R}^n \times [b/8, a[$, we have

$$u(x, t) = \int_{\underline{R}^n} W(x - y, t - \tau) u(y, \tau) \, dy$$

$$= \int_{\underline{R}^n} W(x - y, t - \tau) \, dy \int_{\underline{R}^n} W(y - z, \tau) \, d\mu(z)$$

$$= \int_{\underline{R}^n} d\mu(z) \int_{\underline{R}^n} W(x - y, \ t - \tau) W(y - z, \tau) \ dy$$

$$= \int_{\underline{R}^n} W(x - z, t) \ d\mu(z)$$

The changing of the order of integration is justified because everything is non-negative, and the last line follows from Theorem 1.4.

We now consider the case where u is continuous on $\underline{R}^n \times [0, a[$. It follows from Fatou's lemma and (29) that

$$\int_{\underline{R}^n} \exp(-\|y\|^2/2b) u(y, 0) \ dy \leq \liminf_{s \to 0^+} \int_{\underline{R}^n} \exp(-\|y\|^2/2b) u(y, s) \ dy$$

$$\leq (4\pi b)^{n/2} u(0, b)$$

$$< \infty$$

Therefore, by Theorem 1.2, Corollary, the Gauss-Weierstrass integral v of $u(\cdot, 0)$ is defined and is a non-negative temperature on $\underline{R}^n \times]0, b/2[$. If we define $v(\cdot, 0)$ to be equal to $u(\cdot, 0)$, then v is continuous on $\underline{R}^n \times [0, b/2[$, by Theorem 1.3, Corollary. Put $w = u - v$ on $\underline{R}^n \times [0, b/2[$. Then $w(\cdot, 0) = 0$ and

$$\int_0^{b/2} ds \int_{\underline{R}^n} w^+(y, s) \exp(-\|y\|^2/2b) \ dy$$

$$\leq \int_0^{b/2} ds \int_{\underline{R}^n} u(y, s) \exp(-\|y\|^2/2b) \ dy$$

$$\leq (b/2)(4\pi b)^{n/2} u(0, b)$$

by (29). Therefore, by Theorem 2.4, $w \leq 0$ on $\underline{R}^n \times]0, b/2[$. An application of Theorem 2.5 to $-w$ now shows that $w = 0$, and hence $u = v$, on $\underline{R}^n \times]0, b/2[$. It now follows from (31) and Theorem 1.5, applied to u on $\underline{R}^n \times]0, b/2[$, that $d\mu(y) = u(y, 0) \ dy$.

Theorem 2.10, combined with Theorem 1.2, gives a complete characterization of Gauss-Weierstrass integrals as differences of non-negative temperatures.

Theorem 2.11

A function u on $R^n \times]0, a[$ is the Gauss-Weierstrass integral of a signed measure on R^n if and only if $u = v - w$ for some non-negative temperatures v and w on $R^n \times]0, a[$.

Proof. If $u = v - w$, where v and w are non-negative temperatures, then there exist non-negative measures ν and ω such that v and w are the Gauss-Weierstrass integrals of ν and ω, by Theorem 2.10. For all $(x, t) \in \underline{R}^n \times]0, a[$,

$$\int_{\underline{R}^n} W(x - y, t) \, d\nu(y) < \infty$$

and similarly for ω, so that

$$\int_{\underline{R}^n} W(x - y, t) \, d|\nu - \omega|(y) < \infty$$

Therefore the Gauss-Weierstrass integral of $\nu - \omega$ is defined and equal to u on $\underline{R}^n \times]0, a[$.

Conversely, if u is the Gauss-Weierstrass integral of a signed measure μ, then

$$\int_{\underline{R}^n} W(x - y, t) \, d\mu^+(y) + \int_{\underline{R}^n} W(x - y, t) \, d\mu^-(y) < \infty$$

for all $(x, t) \in \underline{R}^n \times]0, a[$. Therefore the Gauss-Weierstrass integrals of μ^+ and μ^- are defined and are temperatures on $\underline{R}^n \times]0, a[$, by Theorem 1.2. They are clearly non-negative, and u is their difference.

7. MAXIMAL RATES OF DECAY OF
    NON-NEGATIVE TEMPERATURES

We show that Theorem 2.10 implies that, if u is a positive temperature on $\underline{R}^n \times]0, a[$, then $u(x, t)$ cannot tend to zero arbitrarily rapidly as $t \to 0$ for a fixed x, or as $\|x\| \to \infty$ for a fixed t. In both cases, the maximal rate is given by W itself.

We consider first the simpler case where $t \to 0$ and x is fixed.

Theorem 2.12

Let u be a non-negative temperature on $\underline{R}^n \times]0, a[$. If there is a point $\xi \in \underline{R}^n$ such that

$$\liminf_{t \to 0^+} (t^{n/2} \exp(\alpha/t) u(\xi, t)) = 0 \tag{32}$$

for every $\alpha > 0$, then $u = 0$ on $\underline{R}^n \times]0, a[$.

Proof. By Theorem 2.10, u is the Gauss-Weierstrass integral of a non-negative measure μ. For each $r > 0$,

$$u(\xi, t) \geq \int_{\|y-\xi\| < r} W(\xi - y, t)\, d\mu(y)$$

$$\geq (4\pi t)^{-n/2} \exp(-r^2/4t) \mu(B(\xi, r))$$

so that

$$t^{n/2} \exp(r^2/4t) u(\xi, t) \geq (4\pi)^{-n/2} \mu(B(\xi, r))$$

Taking $\alpha = r^2/4$ in (32), and making $t \to 0^+$, we deduce that $\mu(B(\xi, r)) = 0$. Since this holds for every r, and μ is non-negative, it follows that μ is the zero measure. Hence $u = 0$.

Remarks. If (32) holds for a particular value α_0, then it holds for all $\alpha < \alpha_0$. It is therefore only arbitrary large values of α that are important, and if (32) fails to hold when $\alpha = \alpha_0$ it fails for all $\alpha \geq \alpha_0$.

The result is sharp, since if (32) fails for one value of α then u may not be zero. For example, if $u = W$ and $\alpha = \|\xi\|^2/4$, we have

$$t^{n/2} \exp(\alpha/t) W(\xi, t) = (4\pi)^{-n/2}$$

for all $t > 0$, so that (32) fails.

Theorem 2.12 shows that $u(\xi, t)$ cannot, for example, be equal to $\exp(-1/t^2)$ for t arbitrarily close to zero, since

$$t^{n/2} \exp(\alpha/t) \exp(-1/t^2) = t^{n/2} \exp((\alpha t - 1)/t^2) \to 0$$

as $t \to 0^+$.

The analogue of Theorem 2.12, in which t is fixed and $\|x\| \to \infty$, will be deduced from our next result, which is of independent interest.

Theorem 2.13

Let u be a positive temperature on $\underline{R}^n \times]0, a[$, let $t_0 \in]0, a[$, and put

$$M(r) = \max\{u(x, t_0) : \|x\| = r\} \qquad (33)$$

for all $r > 0$. Then

$$\liminf_{r \to \infty} \left(\frac{\log M(r)}{r} + \frac{r}{4t_0} \right) > 0$$

unless $u = \kappa W$ for some positive constant κ, in which case

$$\lim_{r \to \infty} \left(\frac{\log M(r)}{r} + \frac{r}{4t_0} \right) = 0$$

Proof. By Theorem 2.10, u is the Gauss-Weierstrass integral of a non-negative measure μ on $\underline{R}^n$. Since $u > 0$, μ is not the zero measure. If μ is concentrated on $\{0\}$, then $u = \kappa W$ for some positive constant κ. Therefore

$$M(r) = \kappa (4\pi t_0)^{-n/2} \exp(-r^2/4t_0)$$

so that

$$\frac{\log M(r)}{r} + \frac{r}{4t_0} = \frac{\log(\kappa (4\pi t_0)^{-n/2})}{r} \to 0$$

as $r \to \infty$.

If μ is not concentrated on $\{0\}$, there is a closed ball $B = \bar{B}(x_0, \rho)$ such that $0 \notin B$ and $\mu(B) > 0$. For all $x \in \underline{R}^n$,

$$u(x, t_0) \geq (4\pi t_0)^{-n/2} \int_B \exp(-\|x - y\|^2/4t_0) \, d\mu(y)$$

so that, since $\|x - y\| \leq \|x - x_0\| + \|x_0 - y\| \leq \|x - x_0\| + \rho$ for all $y \in B$, we have

$$u(x, t_0) \geq (4\pi t_0)^{-n/2} \exp(-(\|x - x_0\| + \rho)^2/4t_0)\mu(B)$$

Furthermore, whenever $\|x\| = r \geq \|x_0\| + \rho$, we have

$$\|x - x_0\| + \rho \geq \|x\| - \|x_0\| + \rho = r - \|x_0\| + \rho$$

with equality when $x = rx_0/\|x_0\|$. Therefore

$$\min\{\|x - x_0\| + \rho : \|x\| = r\} = r - \|x_0\| + \rho$$

and hence

$$M(r) \geq (4\pi t_0)^{-n/2} \exp(-(r - \|x_0\| + \rho)^2/4t_0)\mu(B)$$

It follows that, whenever $r \geq \|x_0\| + \rho$,

$$r^{-1} \log M(r)$$

$$\geq r^{-1} \log((4\pi t_0)^{-n/2}\mu(B)) - (4t_0 r)^{-1}(r^2 + (\|x_0\| - \rho)^2 - 2r(\|x_0\| - \rho))$$

The first term on the right-hand side tends to zero as $r \to \infty$, so that

$$\liminf_{r \to \infty}\left(\frac{\log M(r)}{r} + \frac{r}{4t_0}\right) \geq -\lim_{r \to \infty}\left(\frac{(\|x_0\| - \rho)^2}{4t_0 r}\right) + \frac{\|x_0\| - \rho}{2t_0}$$

$$= \frac{\|x_0\| - \rho}{2t_0}$$

which is positive since $0 \notin \bar{B}(x_0, \rho)$.

Theorem 2.14

Let u be a non-negative temperature on $\underline{R}^n \times {]0, a[}$, and let $t_0 \in {]0, a[}$. If there are constants $\gamma > 1/4t_0$, $\lambda > 0$, and $\rho > 0$, such that

$$u(x, t_0) \leq \lambda \exp(-\gamma\|x\|^2) \tag{34}$$

whenever $\|x\| > \rho$, then $u = 0$ on $\underline{R}^n \times]0,a[$.

   Proof. Suppose that $u(x,t) > 0$ for all $(x,t) \in \underline{R}^n \times]0,a[$, and let $M(r)$ be defined by (33). Then, by (34),

$$M(r) \leq \lambda \exp(-\gamma r^2)$$

whenever $r > \rho$, so that

$$\frac{\log M(r)}{r} + \frac{r}{4t_0} \leq \frac{\log \lambda}{r} - \gamma r + \frac{r}{4t_0} \to -\infty$$

as $r \to \infty$, since $\gamma > 1/4t_0$. By Theorem 2.13, this is impossible. Therefore our supposition that $u(x,t) > 0$ for all (x,t) must be false, so that $u(x_1,t_1) = 0$ for some $(x_1,t_1) \in \underline{R}^n \times]0,a[$. By Theorem 2.7, $u = 0$ throughout $\underline{R}^n \times]0,a[$.

   Remarks. If $u = \lambda W$ for some positive constant λ, then (34) holds with equality when $\gamma = 1/4t_0$, but $u \neq 0$. Therefore the restriction on γ cannot be relaxed.

   Theorems 2.7, 2.12, and 2.14 all show, in different ways, that a positive temperature on $\underline{R}^n \times]0,a[$ cannot be significantly smaller than W.

   Theorem 2.14 is given in a form which can be extended to more general parabolic equations. However, for the case of the heat equation, the following sharper form can be given.

   Let u be a non-negative temperature on $\underline{R}^n \times]0,a[$, and let $t_0 \in]0,a[$.

   (i) If $u(x,t_0) \leq \lambda W(x,t_0)$ for some non-negative constant λ and all $x \in \underline{R}^n$, then $u = \kappa W$ on $\underline{R}^n \times]0,a[$ for some non-negative constant κ.
   (ii) If $u(x,t_0)/W(x,t_0) \to 0$ as $\|x\| \to \infty$, then $u = 0$ on $\underline{R}^n \times]0,a[$.

The proof of (i) is an application of Theorem 2.13, and (ii) follows from (i). The details are left to the reader.

8. SOLUTIONS OF LINEAR PARABOLIC EQUATIONS

Consider the linear parabolic partial differential equation

$$\sum_{i,j=1}^{n} a_{ij}(x,t) D_i D_j u(x,t) + \sum_{j=1}^{n} b_j(x,t) D_j u(x,t) + c(x,t) u(x,t) - D_t u(x,t) = 0 \tag{35}$$

which was discussed in Chapter I, section 6.

Although the maximum principle is often stated only for functions continuous on a closed cylinder, there is no difficulty in extending it to the form given in Theorem 2.1. The domain considered need not be a circular cylinder. For equation (35), if the last coefficient $c(x,t)$ is not identically zero, some care is required. If c is bounded above, then Theorem 2.1 carries over directly, but in applying the result we must remember that constant functions are not solutions of (35) if $c \neq 0$, so that the theorem cannot be applied to the sum of a solution and a constant, as it was in the proof of Theorem 2.2. Despite this, the overall approach of this chapter can be carried over for the equation (35), even when $c \neq 0$, although the proof of the extension of Theorem 2.2 is necessarily more involved than that of the theorem itself.

In extending Lemma 2.3, we have to reduce the range of possible values of the constant c which appears in the statement, following a corresponding change in Lemma 1.4.

The equation adjoint to (35) is

$$\sum_{i,j=1}^{n} D_i D_j (a_{ij}(x,t) u(x,t)) - \sum_{j=1}^{n} D_j (b_j(x,t) u(x,t)) + c(x,t) u(x,t) + D_t u(x,t) = 0 \tag{36}$$

whenever the derivatives $D_i D_j a_{ij}$ and $D_j b_j$ exist and are continuous for all i and j. If $Lu(x,t)$ denotes the left-hand side of (35), and $L^*u(x,t)$ that of (36), we have the identity

$$vLw - wL^*v = \sum_{i=1}^{n} D_i \left(\sum_{j=1}^{n} a_{ij} (v D_j w - w D_j v) + \left(b_i - \sum_{j=1}^{n} D_j a_{ij} \right) vw \right) - D_t (vw)$$

from which Green's formula for the equation (35) can be deduced using Gauss's divergence theorem, analogously to (9). Using this, the proof of Theorem 2.4 can be extended even to equations with unbounded coefficients. The last paragraph of the proof remains valid if given the correct interpretation. For, if the coefficients depend on t, then the function $(x,t) \longmapsto u(x, t + \sigma)$ may not be a solution of the original equation (35) on $\underline{R}^n \times]0, a - \sigma[$. It will, however, satisfy an equation with similar conditions on the coefficients, namely (35) with t replaced by $t + \sigma$ in all coefficients, so that the result established in the earlier parts of the proof is still applicable.

Theorem 2.5 can be extended without difficulty. The lower estimate for the fundamental solution required here is far less sophisticated than the sharp one (given in (37) below).

The extension of Theorem 2.6 is also straightforward, but here the sharp lower estimate is essential to obtain the precise result. This is because Theorem 1.2 is used in the proof. With a cruder estimate, one can show only that (20) holds if $0 < s < t < s + \delta(s)$, where $\delta(s)$ depends on s and $s + \delta(s) < a$.

Theorem 2.7 depends solely on Theorem 2.6, so its extension is trivial.

The results of sections 6 and 7 all carry over to the general case if the sharp lower estimate for Γ holds. With a cruder estimate, an unsatisfactory generalization of Theorem 2.10 can be obtained, in which the representation

$$u(x,t) = \int_{\underline{R}^n} \Gamma(x,t;y,0) \, d\mu(y)$$

holds only on some unspecified substrip $\underline{R}^n \times]0,b[$ with $b < a$. Again, this is caused by the lack of a precise generalization of Theorem 1.2 in this situation.

The exact forms of Theorems 2.13 and 2.14 are slightly changed in extension. If the sharp lower estimate for the fundamental solution Γ is given as

$$\Gamma(x,t;y,s) \geq C_1(t-s)^{-n/2} \exp(-c_1\|x-y\|^2/(t-s)) \tag{37}$$

then the term $r/4t_0$ must be replaced by $c_1 r/t_0$ on both occurrences in Theorem 2.13, and the constant γ of Theorem 2.14 must consequently exceed c_1/t_0.

9. BIBLIOGRAPHICAL NOTES

The weak maximum principle is discussed by Il'in, Kalashnikov, and Oleinik [1]. We shall return to this topic, and give more references, in Chapter VIII.

For temperatures which may take values of either sign, the result of Theorem 2.2 has been shown to be equivalent to the semigroup property, by Watson [8, p. 518].

The general approach to the Gauss-Weierstrass representation of non-negative solutions to the heat equation followed here is due to

Widder [1], when n = 1. A closely related but less general result was
given by Pollard [1] at about the same time. Widder's approach contained
details which could not obviously be extended to more general equations,
or even to a general n. The extension was achieved by Krzyżański [6],
and our approach is modelled on his. Subsequently, his methods were
extended to more general classes of equations by Guenther [1, Theorem 2], Aronson [5, Theorems 11 and 12], Bodanko [2], Chabrowski
[2], and Johnson [1]. A totally new, but less elementary, approach was
recently developed by Mair and Taylor [1].

Theorem 2.2 was proved for the heat equation with n = 1 by Widder
[1], and was extended by Friedman in the proof of [1, Theorem 2].

The convergence of a sequence of non-negative measures in the
sense of (4) is known by several names, such as weak, weak*, vague,
and narrow. Unfortunately, some of these names also have slightly
different meanings.

The proof of Theorem 2.4 is taken from Chabrowski's paper [6],
where it is applied to equations with unbounded coefficients. The result
was obtained earlier for the case of the heat equation with n = 1, by
Gehring [1, Theorem 3]. A different method, for the heat equation with
general n, was given by Watson [1, Theorem 18]. Weaker results for
various classes of equations were obtained prior to Chabrowski's work
in several papers, including those of Krzyżański [1], Besala [1,2], and
Bodanko [1]. Subsequent extensions have been given by Besala [5]; Chen,
Kuroda, and Kusano [1]; and Cosner [1]. We shall return to the extended
maximum principle in Chapter III.

Theorem 2.5 was proved for the heat equation with n = 1 by Widder
[1, Theorem 5], using different techniques from those in section 4.
For more general equations, using a less precise lower estimate for
the fundamental solution than that in (37), the result was obtained by
Friedman [1], Aronson [2], Guenther [2], Aronson [5, p. 646], and
Bodanko [2], under various conditions on the coefficients. Our method
of proof follows Friedman's. An adequate lower estimate for the fundamental solution, under mild conditions on the coefficients, can be found
in the works of Aronson and Besala [3, Lemma 2] and Guenther [1,2].
Theorem 2.5 is closely related to the Cauchy problem, and references
given on that subject, in Chapter III, should also be consulted.

Theorem 2.6 is essential to our approach to the representation of
non-negative solutions, and appears in all papers based on Widder's
method, generally in the weaker form described in section 8. See also
the notes in Chapter I, section 6.

A topological space with a countable dense subset is often called
separable. As usual, the word also has another meaning. Theorem 2.8
is sometimes referred to as a weak, or weak*, compactness theorem,

since it asserts that any sequence, in the set of non-negative measures μ such that $\mu(\underline{R}^n) \leq \kappa$, has a subsequence which converges in the weak (or weak*) sense to a measure in the set. Theorem 2.9 was proved by Krzyżański in [6], where references to related results can be found.

The last part of Theorem 2.10, where u is continuous on $\underline{R}^n \times [0, a[$, was not considered by Widder or Krzyżański. For the heat equation with $n = 1$, it follows from Pollard's result [1]. More generally, it was given by Friedman [2, p. 48], Aronson [5], and Bodanko [2].

Theorem 2.11, for the heat equation, follows from results of Gehring [1, Theorems 7 and 8] when $n = 1$, and from results of Watson [2, Theorems 20 and 21] for general n. For more general equations it is generally known, but does not seem to have been explicitly stated. It often happens that minor results, or results which are easy consequences of important theorems, become generally known without getting published. Others in this category are Theorems 2.7 and 2.12.

Theorem 2.13 is a minor variation of Watson's [7, Theorem 1]. It is related to Krzyżański's [6, Theorem 5]. Theorem 2.14, and the sharper version for the heat equation, also come from Watson [7]. A result similar to Theorem 2.14, but with the non-negativity of u replaced by $|u(x,t)| \leq C \exp(\kappa \|x\|^2)$ on $\underline{R} \times]0, t_0]$, for some constants C and κ, was proved by Gusarov [1, Theorem 6]. In the bibliographical notes in Chapter III, we shall indicate how Gusarov's result and Theorem 2.14 can be included in a more general theorem, at least for the case of the heat equation.

3

The Semigroup Property, Cauchy Problem, and Gauss-Weierstrass Representation

On the strip $\underline{R}^n \times]0, a[$, we introduce a family of integral means M_b indexed by $]0, a]$. Using these, we are able to give weak conditions for a temperature to have the semigroup property, for the Cauchy problem to have at most one solution corresponding to a given function f on $\underline{R}^n \times \{0\}$, and for a temperature to be the Gauss-Weierstrass integral of some measure. Since all this is done from a unified point of view, the relationships between these properties are highlighted. The means M_b are very effective and easy to handle, because they exploit the semigroup property to the full.

1. THE INTEGRAL MEANS M_b AND THE CLASSES Σ_b

If $0 < t < b \leq a$, and v is a function on $\underline{R}^n \times \{t\}$, the integral mean $M_b(v; t)$ is defined by

$$M_b(v; t) = \int_{\underline{R}^n} W(x, b - t) v(x, t) \, dx$$

whenever the integral exists. Observe that, if v is a temperature with the semigroup property on $\underline{R}^n \times]0, a[$, then $M_b(v; t) = v(0, b)$ for all $t \in]0, b[$.

It is convenient to begin by giving a version of Theorem 2.4 in terms of the means M_b.

Theorem 3.1

Suppose that $0 \leq s < b \leq a$, and that u is a continuous real-valued function on $\underline{R}^n \times [s, b[$ and a temperature on $\underline{R}^n \times]s, b[$. If $M_c(u^+; \cdot)$ is locally integrable on $[s, c[$ whenever $s < c < b$, and $u(\cdot, s) \leq 0$, then $u \leq 0$ on $\underline{R}^n \times [s, b[$.

Proof. Whenever $s < r < c$,

$$\int_s^r dt \int_{\underline{R}^n} u^+(x, t) \exp(-\|x\|^2/4(c - r)) \, dx$$

$$\leq \int_s^r dt \int_{\underline{R}^n} u^+(x, t) \exp(-\|x\|^2/4(c - t)) \, dx$$

$$= (4\pi)^{n/2} \int_s^r (c - t)^{n/2} dt \int_{\underline{R}^n} W(x, c - t) u^+(x, t) \, dx$$

$$\leq (4\pi c)^{n/2} \int_s^r M_c(u^+; t) \, dt$$

which is finite since $M_c(u^+; \cdot)$ is locally integrable on $[s, c[$. Therefore, if $v(x, t) = u(x, s + t)$ for all $(x, t) \in \underline{R}^n \times [0, b - s[$, Theorem 2.4 shows that $v \leq 0$. Hence $u \leq 0$ on $\underline{R}^n \times [s, b[$.

One of the nice properties of the means M_b is that they and the strips $\underline{R}^n \times]0, b[$ fit together naturally, as the following lemma indicates.

Lemma 3.1

Suppose that $0 \leq s < b \leq a$, and that μ is a signed measure on $\underline{R}^n$ such that

$$\int_{\underline{R}^n} W(y, b - s) \, d|\mu|(y) < \infty \tag{1}$$

Then the integral u, given by

$$u(x, t) = \int_{\underline{R}^n} W(x - y, t - s) \, d\mu(y)$$

is defined and is a temperature on $\underline{R}^n \times \,]s, b[$. Furthermore, whenever $s < t < b$,

$$M_b(|u| ; t) \le \int_{\underline{R}^n} W(y, b - s) \, d|\mu|(y) \tag{2}$$

with equality if μ is non-negative. Moreover if $d\mu(y) = w(y, s) \, dy$ for some function w, whenever $s < t < b$ we have

$$M_b(|u| ; t) \le M_b(|w| ; s)$$

with equality if $w \ge 0$.

Proof. In view of (1), we have

$$|u(0, b)| \le \int_{\underline{R}^n} W(y, b - s) \, d|\mu|(y) < \infty$$

so that u is defined and a temperature on $\underline{R}^n \times \,]s, b[$, by Theorem 1.2. Next, if $s < t < b$, Fubini's theorem and Theorem 1.4 show that

$$M_b(|u| ; t) \le \int_{\underline{R}^n} W(x, b - t) \, dx \int_{\underline{R}^n} W(x - y, t - s) \, d|\mu|(y)$$

$$= \int_{\underline{R}^n} d|\mu|(y) \int_{\underline{R}^n} W(-x, b - t) W(x - y, t - s) \, dx$$

$$= \int_{\underline{R}^n} W(y, b - s) \, d|\mu|(y)$$

with equality if μ is non-negative. This proves (2). If $d\mu(y) = w(y, s) \, dy$, then $d|\mu|(y) = |w(y, s)| \, dy$, so that (2) implies that

$$M_b(|u| ; t) \le \int_{\underline{R}^n} W(y, b - s) |w(y, s)| \, dy = M_b(|w| ; s)$$

with equality if $w \ge 0$.

Definition. If $b > 0$, u is a temperature on $\underline{R}^n \times]0, b[$, and $M_b(u^+; \cdot)$ is locally integrable on $]0, b[$, we say that $u \in \Sigma_b$.

Note that, since $(u + v)^+ \leq u^+ + v^+$, the sum of two functions in Σ_b is again in Σ_b.

We can immediately show that every Gauss-Weierstrass integral belongs to Σ_b for some b.

Theorem 3.2

If μ is a signed measure on $\underline{R}^n$ whose Gauss-Weierstrass integral u is defined at $(0, b)$, then $u \in \Sigma_b$ and $M_b(|u|; \cdot)$ is bounded on $]0, b[$.

Proof. Since $u(0, b)$ is defined,

$$\int_{\underline{R}^n} W(y, b) \, d|\mu|(y) < \infty$$

so that, by Lemma 3.1, u is a temperature on $\underline{R}^n \times]0, b[$ and

$$M_b(|u|; t) \leq \int_{\underline{R}^n} W(y, b) \, d|\mu|(y)$$

for all $t \in]0, b[$. Hence $M_b(|u|; \cdot)$ is bounded on $]0, b[$, which implies that the same is true of $M_b(u^+; \cdot)$, so that $u \in \Sigma_b$.

Theorem 3.3

Let μ be a signed measure on $\underline{R}^n$ whose Gauss-Weierstrass integral u is defined on $\underline{R}^n \times]0, a[$, and let $b \in]0, a[$. Then $u \in \Sigma_b$, $M_b(|u|; \cdot)$ is bounded on $]0, b[$, and $M_b(u; t) = u(0, b)$ for all $t \in]0, b[$.

Proof. The first two parts follow immediately from Theorem 3.2. Finally, if $0 < t < b$, Theorem 1.4 implies that

$$u(0, b) = \int_{\underline{R}^n} W(y, b - t)u(y, t) \, dy = M_b(u; t)$$

Theorem 3.2, with $b = a$, and Theorem 3.3 together raise the question of the relationship between Σ_a and $\bigcap_{b < a} \Sigma_b$ when $a < \infty$.

Theorem 3.4

The class Σ_a is a proper subclass of $\bigcap_{b<a} \Sigma_b$.

Proof. We first show that every temperature u in Σ_a also belongs to Σ_b whenever $0 < b < a$. If $0 < t < b < a$ and $u \in \Sigma_a$, the easy inequality

$$\exp(-\|x\|^2/4(b-t)) \leq \exp(-\|x\|^2/4(a-t))$$

implies that

$$\int_{\underline{R}^n} W(x, b-t)u^+(x,t)\, dx \leq \left(\frac{a-t}{b-t}\right)^{n/2} \int_{\underline{R}^n} W(x, a-t)u^+(x,t)\, dx$$

Therefore $M_b(u^+; \cdot)$ is locally integrable on $]0, b[$, since $M_a(u^+; \cdot)$ is. Thus $u \in \bigcap_{b<a} \Sigma_b$.

We now give an example of a function in $\bigcap_{b<a} \Sigma_b$ which does not belong to Σ_a. The example following Theorem 1.2, Corollary, together with Theorem 3.3, tells us where to find such a function.

Let $v(x,t) = (a-t)^{-n/2} \exp(\|x\|^2/4(a-t))$ for all $(x,t) \in \underline{R}^n \times]0, a[$. Then v is a positive temperature, so that $M_b(v; t) = v(0, b)$ whenever $0 < t < b < a$, by Theorems 2.10 and 3.3. (The example mentioned in the preceding paragraph can be used instead of Theorem 2.10.) Hence $v \in \Sigma_b$ whenever $0 < b < a$. However,

$$M_a(v; t) = (4\pi)^{-n/2}(a-t)^{-n} \int_{\underline{R}^n} dx = \infty$$

for all $t \in]0, a[$, so that $v \notin \Sigma_a$.

Remark. Comparing the case $b = a$ of Theorem 3.2 with Theorem 3.3, we see that the more important of the two classes in Theorem 3.4 is likely to be the larger one, since it contains all Gauss-Weierstrass integrals defined on $\underline{R}^n \times]0, a[$, whereas the smaller one contains only those which are also defined at $(0, a)$. This is confirmed by most of the results in this chapter. It is therefore desirable to have a briefer notation than $\bigcap_{b<a} \Sigma_b$, and we introduce one at the start of the next section.

2. THE CLASS S_a AND THE SEMIGROUP PROPERTY

If $0 < a \le \infty$, we put

$$S_a = \bigcap_{b<a} \Sigma_b$$

Thus $u \in S_a$ if u is a temperature on $\underline{R}^n \times \,]0, a[$ and, for each $b \in$ $]0, a[$, $M_b(u^+; \cdot)$ is locally integrable on $]0, b[$.

Note that the sum of two functions in S_a is again in S_a.

In Theorem 3.6, we shall show that $u \in S_a$ if and only if u has the semigroup property. Hence the choice of the symbol S. We first give a slightly different characterization of S_a.

Theorem 3.5

Let u be a temperature on $\underline{R}^n \times \,]0, a[$. Then $u \in S_a$ if and only if

$$u^+(x, t) \le \int_{\underline{R}^n} W(x - y, t - s) u^+(y, s) \, dy < \infty \tag{3}$$

whenever $x \in \underline{R}^n$ and $0 < s < t < a$. Furthermore, if $u \in S_a$ and $0 < b < a$, then $M_b(u^+; t) < \infty$ for all $t \in \,]0, b[$.

Proof. Suppose that (3) holds whenever $x \in \underline{R}^n$ and $0 < s < t < a$. Choose $b \in \,]0, a[$. Then, if $0 < s < b$,

$$M_b(u^+; s) = \int_{\underline{R}^n} W(y, b - s) u^+(y, s) \, dy < \infty \tag{4}$$

by hypothesis. It therefore follows from Lemma 3.1, with $d\mu(y) = u^+(y, s) \, dy$, that, if $v_s(x, t)$ denotes the integral in (3),

$$M_b(v_s; t) = M_b(u^+; s)$$

whenever $t \in \,]s, b[$. Hence by (3) and (4),

$$M_b(u^+; t) \le M_b(v_s; t) = M_b(u^+; s) < \infty$$

for all $t \in \,]s, b[$, so that $M_b(u^+; \cdot)$ is bounded on $]s, b[$. Thus $M_b(u^+; \cdot)$ is locally integrable on $]0, b[$, so that $u \in \Sigma_b$. Since b is arbitrary, $u \in S_a$.

Now suppose, conversely, that $u \in S_a$. Let $b \in]0, a[$. We first prove that (3) holds provided that $M_b(u^+; s) < \infty$. It will then follow that $M_b(u^+; r) < \infty$ for all $r \in]0, b[$, so that (3) holds for any $s \in]0, b[$.

Since $M_b(u^+; \cdot)$ is locally integrable on $]0, b[$, it is finite almost everywhere. Choose $s \in]0, b[$ such that

$$\int_{\underline{R}^n} W(y, b - s) u^+(y, s) \, dy = M_b(u^+; s) < \infty$$

If $v_s(x, t)$ denotes the integral in (3), then Lemma 3.1 shows that v_s is a temperature on $\underline{R}^n \times]s, b[$. We must show that $u^+ \leq v_s$ on $\underline{R}^n \times]s, b[$. Since u^+ is continuous, Theorem 1.3, Corollary, shows that

$$\lim_{(x, t) \to (\xi, s^+)} v_s(x, t) = u^+(\xi, s)$$

so that

$$\lim_{(x, t) \to (\xi, s^+)} (u(x, t) - v_s(x, t)) = u(\xi, s) - u^+(\xi, s) \leq 0 \qquad (5)$$

for all $\xi \in \underline{R}^n$. Put $v_s(\cdot, s) = u^+(\cdot, s)$. Then, since $u \in S_a$ and $v_s \geq 0$, $M_c((u - v_s)^+; \cdot)$ is locally integrable on $[s, c[$ whenever $s < c < b$, and $(u - v_s)(\cdot, s) \leq 0$. It therefore follows from Theorem 3.1 that $u \leq v_s$ on $\underline{R}^n \times [s, c[$. Since $v_s \geq 0$, we deduce that $u^+ \leq v_s$ on $\underline{R}^n \times]s, c[$. Hence $u^+ \leq v_s$ on $\underline{R}^n \times]s, b[$, provided that $M_b(u^+; s) < \infty$.

We now prove that $M_b(u^+; r) < \infty$ for all $r \in]0, b[$. Given any such r, since $M_b(u^+; \cdot) < \infty$ almost everywhere, we can find $s \in]0, r[$ such that $M_b(u^+; s) < \infty$. Then $u^+ \leq v_s$ on $\underline{R}^n \times]s, b[$, so that Lemma 3.1, with $d\mu(y) = u^+(y, s) \, dy$, shows that

$$M_b(u^+; r) \leq M_b(v_s; r) = M_b(u^+; s) < \infty$$

It now follows that $u^+ \leq v_s < \infty$ on $\underline{R}^n \times]s, b[$ for every $s \in]0, b[$. Since b is arbitrary, the theorem is proved.

Theorem 3.6

Let u be a temperature on $\underline{R}^n \times]0, a[$. Then $u \in S_a$ if and only if

$$u(x, t) = \int_{\underline{R}^n} W(x - y, t - s) u(y, s) \, dy \qquad (6)$$

whenever $x \in \underline{R}^n$ and $0 < s < t < a$.

Proof. Suppose that u has the semigroup property (6) on $\underline{R}^n \times {]}0, a[$, and let $b \in {]}0, a[$. Whenever $0 < s < b$,

$$u(0, b) = \int_{\underline{R}^n} W(y, b - s)u(y, s) \, dy = M_b(u; s)$$

by (6), so that $M_b(u; s)$ exists and is finite, which means that $M_b(|u|; s) < \infty$. By Lemma 3.1, with $d\mu(y) = u(y, s) \, dy$,

$$M_b(|u|; t) \leq M_b(|u|; s)$$

for all $t \in {]}s, b[$. Since s and t are any numbers such that $0 < s < t < b$, it follows that $M_b(|u|; \cdot)$ is non-increasing on ${]}0, b[$. Therefore, as $u^+ \leq |u|$, $M_b(u^+; \cdot)$ is locally integrable on ${]}0, b[$. Because b is arbitrary, we conclude that $u \in S_a$.

Now suppose, conversely, that $u \in S_a$. By Theorem 3.5,

$$u^+(x, t) \leq \int_{\underline{R}^n} W(x - y, t - s)u^+(y, s) \, dy < \infty \qquad (7)$$

whenever $x \in \underline{R}^n$ and $0 < s < t < a$. Therefore, if $v_s(x, t)$ denotes the integral in (7),

$$v_s - u = v_s - u^+ + u^- \geq u^- \geq 0$$

on $\underline{R}^n \times {]}s, a[$. By Lemma 3.1, $v_s - u$ is a temperature on $\underline{R}^n \times {]}s, a[$. Since u^+ is continuous, Theorem 1.3, Corollary, shows that $v_s - u$ has a continuous extension to $\underline{R}^n \times [s, a[$ if it is defined to be equal to u^- on $\underline{R}^n \times \{s\}$. Hence, by Theorem 2.10,

$$(v_s - u)(x, t) = \int_{\underline{R}^n} W(x - y, t - s)u^-(y, s) \, dy$$

for all $(x, t) \in \underline{R}^n \times {]}s, a[$. Therefore, for such points (x, t),

$$u(x, t) = v_s(x, t) - (v_s - u)(x, t) = \int_{\underline{R}^n} W(x - y, t - s)u(y, s) \, dy$$

Remarks. The hypothesis that u is a temperature can be relaxed, since u needs only to be finite-valued. For, if (6) holds and the integrals are finite, then u is a temperature on $\underline{R}^n \times {]}s, a[$ for all $s \in {]}0, a[$, while if $u \in S_a$ then that u is a temperature is part of the definition of S_a.

By Theorems 3.5 and 3.6, the semigroup property is equivalent to the family of inequalities in (3), given that u is a temperature on $\underline{R}^n \times$]0, a[. It is also equivalent to the collection of inequalities

$$u(x,t) \le \int_{\underline{R}^n} W(x - y, t - s)u(y,s)\, dy < \infty \tag{8}$$

where $x \in \underline{R}^n$ and $0 < s < t < a$. For, the semigroup property obviously implies (8), while if (8) holds the integrals

$$\int_{\underline{R}^n} W(x - y, t - s)u^+(y,s)\, dy$$

are finite and majorize those in (8), which makes them non-negative majorants of $u(x,t)$, and hence no less than $u^+(x,t)$. Thus (8) implies (3), which is equivalent to the semigroup property.

A far more important consequence of Theorem 3.6 is the fact that if $u \in S_a$ then $-u \in S_a$. Since the definition of S_a involves a condition on u^+ alone, this may come as a surprise. But it is a consequence of the fact that non-negative temperatures cannot be too large, since they are all Gauss-Weierstrass integrals, and of our choice of integral means which combine naturally with the semigroup property.

Theorem 3.7

If $u \in S_a$, then $-u \in S_a$.

Proof. If $u \in S_a$, then u has the semigroup property on $\underline{R}^n \times$]0, a[, by Theorem 3.6. Therefore $-u$ also has the semigroup property, so that $-u \in S_a$ by Theorem 3.6.

Remark. The result of Theorem 3.7 would fail if S_a was replaced by Σ_a. For, if $v(x,t) = (a - t)^{-n/2} \exp(\|x\|^2/4(a - t))$ for all $(x,t) \in \underline{R}^n \times$]0, a[, then v is a positive temperature which belongs to $S_a \setminus \Sigma_a$, as was shown in the proof of Theorem 3.4. If $u = -v$, then obviously $u \in \Sigma_a$, but $-u = v \notin \Sigma_a$.

Example. Let u be defined on $\underline{R} \times$ [0, 4[by

$u(x,t)$

$= \dfrac{1}{((4-t)^2+t^2)^{1/4}} \exp\left(\dfrac{x^2(2-t)}{2((4-t)^2+t^2)}\right) \cos\left(\dfrac{x^2}{((4-t)^2+t^2)} - \dfrac{1}{2}\tan^{-1}\left(\dfrac{-t}{4-t}\right)\right)$

Then u is continuous, and it is straightforward, but tedious, to verify
that u is a temperature on $\underline{R} \times]0,4[$. We show that u is the Gauss-
Weierstrass integral of $u(\cdot,0)$ on $\underline{R} \times]0,4[$, that u can be extended to
a temperature on $\underline{R} \times]0,\infty[$ even though the Gauss-Weierstrass integral
of $u(\cdot,0)$ is not defined on $\underline{R} \times]4,\infty[$, and that this extension has the
semigroup property on $\underline{R} \times]0,4[$ and on $\underline{R} \times]2,\infty[$ but not on $\underline{R} \times]0,\infty[$.

We first show that u is the Gauss-Weierstrass integral of $u(\cdot,0)$.
Since

$$u(x,0) = (1/2) \exp(x^2/16) \cos(x^2/16)$$

for all $x \in \underline{R}$, we have $|u(x,0)| \leq (1/2) \exp(x^2/16)$, which implies that
its Gauss-Weierstrass integral v is defined and a temperature on
$\underline{R} \times]0,4[$, by Theorem 1.2, Corollary. Furthermore, if we put $v(\cdot,0)$
$= u(\cdot,0)$, then v is continuous on $\underline{R} \times [0,4[$, by Theorem 1.3, Corollary.
We aim to use Theorem 3.1 to show that $u = v$, so we shall require
estimates for their integral means. If $0 < t < b$ and $2 < b < 4$,

$$\frac{1}{b-t} - \frac{2(2-t)}{((4-t)^2 + t^2)} = \frac{2(b-2)t + 4(4-b)}{((4-t)^2 + t^2)(b-t)}$$

$$\geq \frac{4(4-b)}{((4-t)^2 + t^2)b}$$

$$\geq \frac{4-b}{4b}$$

since $(4-t)^2 + t^2$ has a maximum over $t \in [0,4]$ at $t = 0$. Therefore

$M_b(|u|;t)$

$$\leq (4\pi(b-t))^{-1/2}((4-t)^2 + t^2)^{-1/4} \int_{\underline{R}^n} \exp\left(-\frac{x^2}{4(b-t)} + \frac{x^2(2-t)}{2((4-t)^2 + t^2)}\right) dx$$

$$\leq C(b-t)^{-1/2} \int_{\underline{R}^n} \exp\left(-\frac{x^2(4-b)}{16b}\right) dx$$

$$= C(b-t)^{-1/2}$$

for some positive constants C. Hence, whenever $0 < t < c < b$ and
$2 < b < 4$, we have

$$(c-t)^{1/2} M_c(|u|;t) \leq (b-t)^{1/2} M_b(|u|;t) \leq C$$

from which it follows that $M_c(|u|;\cdot)$ is integrable over $[0,c[$. Since Theorem 3.2 shows that $M_c(|v|;\cdot)$ is bounded on $]0,c[$ whenever $0 < c < 4$, and

$$M_c(|u-v|;\cdot) \le M_c(|u|;\cdot) + M_c(|v|;\cdot)$$

it follows that $M_c(|u-v|;\cdot)$ is integrable over $[0,c[$ whenever $0 < c < 4$. We can now apply Theorem 3.1 to both $u - v$ and $v - u$, and deduce that $u = v$.

By Theorem 1.4, u has the semigroup property on $\underline{R} \times]0,4[$. In particular,

$$u(x,t) = \int_{\underline{R}^n} W(x - y, t - 2)u(y,2)\, dy \tag{9}$$

whenever $(x,t) \in \underline{R} \times]2,4[$. But

$$u(y,2) = 8^{-1/4} \cos((y^2 + \pi)/8)$$

so that $u(\cdot,2)$ is bounded. Therefore the integral in (9) is defined for all $(x,t) \in \underline{R} \times]2,\infty[$, and represents a temperature on $\underline{R} \times]2,\infty[$ which coincides with u on $\underline{R} \times]2,4[$. If we now define $u(x,t)$ by (9) for all $(x,t) \in \underline{R} \times [4,\infty[$, then u is a temperature on $\underline{R} \times]0,\infty[$, and u has the semigroup property not only on $\underline{R} \times]0,4[$, but also on $\underline{R} \times]2,\infty[$, since it is a Gauss-Weierstrass integral there.

But u does not have the semigroup property on $\underline{R} \times]0,\infty[$. For

$$u(y,1) = \frac{1}{10^{1/4}} \exp\left(\frac{y^2}{20}\right) \cos\left(\frac{y^2}{10} - \frac{1}{2}\tan^{-1}\left(-\frac{1}{3}\right)\right)$$

whenever $y \in \underline{R}$, so that

$$\int_{\underline{R}} W(0 - y, 6 - 1)u^+(y,1)\, dy = (20\pi)^{-1/2} \int_{\underline{R}} \exp(-y^2/20)u^+(y,1)\, dy$$

$$= C \int_{\underline{R}} \cos^+((y^2/10) + \gamma)\, dy$$

for some constant C, where $\gamma = (-1/2)\tan^{-1}(-1/3) > 0$. Putting $z = y^2/10$, so that $dy = (5/2z)^{1/2}\, dz$, we have

$$\int_{\underline{R}} \cos^+((y^2/10) + \gamma)\, dy = 2 \int_0^\infty \cos^+(z + \gamma)(5/2z)^{1/2}\, dz$$

Since $\cos(z + \gamma) \geq 2^{-1/2}$ whenever $z + \gamma \in [(7 + 8k)\pi/4, (9 + 8k)\pi/4]$ for any integer k, it follows that

$$\int_{\underline{R}} W(y, 5)u^+(y, 1)\, dy = C \int_0^\infty \cos^+(z + \gamma)\, z^{-1/2}\, dz$$

$$\geq C \sum_{k=\ell}^\infty \int_{-\gamma+(7+8k)\pi/4}^{-\gamma+(9+8k)\pi/4} z^{-1/2}\, dz$$

$$\geq C \sum_{k=\ell}^\infty (-\gamma + (9 + 8k)\pi/4)^{-1/2}$$

$$\geq C \sum_{k=\ell}^\infty (\alpha k)^{-1/2}$$

$$= \infty$$

where ℓ is chosen such that $-\gamma + (7 + 8\ell)\pi/4 \geq 0$, and α such that $\alpha > 2\pi$ and $(\alpha - 2\pi)\ell \geq -\gamma + (9\pi/4)$, which implies that $\alpha k \geq -\gamma + (9 + 8k)\pi/4$ for all $k \geq \ell$. Thus $W(\cdot, 5)u(\cdot, 1)$ is not integrable over $\underline{R}$, so that u does not have the semigroup property on $\underline{R} \times]0, \infty[$.

It follows that u is not a Gauss-Weierstrass integral on $\underline{R} \times]0, \infty[$, in view of Theorem 1.4. In fact, the Gauss-Weierstrass integral of $u(\cdot, 0)$, which coincides with u on $\underline{R} \times]0, 4[$, is not defined on $\underline{R} \times]4, \infty[$. For, if $b > 4$,

$$W(y, b)u(y, 0) = C \exp\left(-\frac{y^2}{4b} + \frac{y^2}{16}\right) \cos\left(\frac{y^2}{16}\right) = C \exp(\beta y^2/16) \cos(y^2/16)$$

where $\beta = (b - 4)/b > 0$. Therefore

$$\int_{\underline{R}} W(y, b)u^+(y, 0)\, dy = C \int_0^\infty \exp(\beta y^2/16) \cos^+(y^2/16)\, dy$$

$$= C \int_0^\infty e^{\beta z} \cos^+(z)\, z^{-1/2}\, dz$$

where we have put $z = y^2/16$. Using elementary calculus, we can show that $e^{\beta z/2} z^{-1/2} \geq (\beta e)^{1/2}$ for all $z > 0$. Therefore

$$\int_{\underline{R}} W(y, b) u^+(y, 0)\, dy \geq C \int_0^\infty e^{\beta z/2} \cos^+ z\, dz$$

Since $\cos z \geq 2^{-1/2}$ whenever $z \in [(7 + 8k)\pi/4, (9 + 8k)\pi/4]$ for any integer k, it follows that

$$\int_{\underline{R}} W(y, b) u^+(y, 0) \geq C \sum_{k=0}^{\infty} \int_{(7+8k)\pi/4}^{(9+8k)\pi/4} e^{\beta z/2}\, dz$$

$$\geq C \sum_{k=0}^{\infty} e^{\beta(7+8k)\pi/8}$$

$$= C \sum_{k=0}^{\infty} e^{\beta k \pi}$$

$$= \infty$$

whenever $b > 4$. Thus the Gauss-Weierstrass integral of $u(\cdot, 0)$ is not defined at $(0, b)$ for any $b > 4$, and is therefore not defined at (x, t) for any $(x, t) \in \underline{R} \times]4, \infty[$, in view of Theorem 1.2.

3. BEHAVIOR OF THE MEANS M_b

The basic properties of the means M_b, given in Lemma 3.1, have already been used in several proofs. We now investigate their behavior in more detail.

Theorem 3.8

If $u \in S_a$ and $0 < b < a$, then $M_b(u^+; \cdot)$ is non-increasing, real-valued, and continuous on $]0, b[$.

Proof. We already know, from Theorem 3.5, that $M_b(u^+; \cdot)$ is real-valued. We now show that it is non-increasing. Let $s \in]0, b[$. By Theorem 3.5,

$$u^+(x,t) \leq \int_{\underline{R}^n} W(x-y, t-s) u^+(y,s) \, dy < \infty \tag{10}$$

whenever $(x,t) \in \underline{R}^n \times]s, a[$. If $v(x,t)$ denotes the integral in (10), then Lemma 3.1, with $d\mu(y) = u^+(y,s) \, dy$, shows that v is a temperature on $\underline{R}^n \times]s, a[$ and $M_b(v; t) = M_b(u^+; s)$ whenever $s < t < b$. By (10), $u^+ \leq v$ on $\underline{R}^n \times]s, a[$, so that $M_b(u^+; t) \leq M_b(v; t) = M_b(u^+; s)$ whenever $s < t < b$. Hence $M_b(u^+; \cdot)$ is non-increasing.

We now prove that $M_b(u^+; \cdot)$ is continuous. Let $r \in]0, b[$, and let $\{r_j\}$ be an arbitrary sequence in $]r/2, b[$ with limit r. Since u^+ is continuous, it follows from Fatou's lemma that

$$\liminf_{j \to \infty} M_b(u^+; r_j) \geq \int_{\underline{R}^n} \lim_{j \to \infty} (W(x, b - r_j) u^+(x, r_j)) \, dx$$

$$= \int_{\underline{R}^n} W(x, b - r) u^+(x, r) \, dx$$

$$= M_b(u^+; r) \tag{11}$$

Next, if w is defined on $\underline{R}^n \times]r/2, a[$ by

$$w(x,t) = \int_{\underline{R}^n} W(x - y, t - (r/2)) u^+(y, r/2) \, dy$$

then w is a temperature and $M_b(w; t) = M_b(u^+; r/2)$ whenever $r/2 < t < b$, by Lemma 3.1. Furthermore, $u^+(\cdot, r_j) \leq w(\cdot, r_j)$ for all j, by (10). It therefore follows, from the continuity of $w - u^+$ and Fatou's lemma, that

$$M_b(u^+; r/2) - M_b(u^+; r) = M_b(w; r) - M_b(u^+; r)$$

$$= M_b(w - u^+; r)$$

$$= \int_{\underline{R}^n} \lim_{j \to \infty} (W(x, b - r_j) \{w(x, r_j) - u^+(x, r_j)\}) \, dx$$

$$\leq \liminf_{j \to \infty} (M_b(w; r_j) - M_b(u^+; r_j))$$

$$= M_b(u^+; r/2) - \limsup_{j \to \infty} M_b(u^+; r_j)$$

Hence

$$\limsup_{j\to\infty} M_b(u^+; r_j) \leq M_b(u^+; r)$$

which combines with (11) to show that

$$M_b(u^+; r_j) \to M_b(u^+; r) \quad \text{as} \quad j \to \infty$$

for any sequence $\{r_j\}$ which converges to r. Thus $M_b(u^+; \cdot)$ is continuous at r. Since r is any point of $]0, b[$, the theorem is proved.

The next theorem shows that the means M_b tell us when a temperature on $R^n \times]0, a[$ becomes non-negative on a substrip. Note that $1 - W$ is a temperature in S_∞ which takes values of either sign while eventually becoming non-negative. The easiest way to see this is to observe that $W(x, t) \leq (4\pi t)^{-n/2}$, with equality when $x = 0$, and that $1 - (4\pi t)^{-n/2}$ is negative for small t (so that $1 - W(0, t) < 0$ for small t) and is positive for large t (so that $1 - W > 0$ for large t).

Of course, if $w \in S_a$ and $w(\cdot, r) \geq 0$ for some $r \in]0, a[$, then the semigroup property shows that $w(x, t) \geq 0$ for all $(x, t) \in R^n \times]r, a[$. Thus a temperature in S_a, non-negative on a hyperplane, remains non-negative.

Theorem 3.9

Suppose that $u \in S_a$ and $0 < r < b < a$. Then u has constant sign on $R^n \times [r, b[$ if and only if $M_b(u^+; \cdot)$ is constant on $[r, b[$.

Proof. If u has constant sign on $R^n \times [r, b[$, then either $u \geq 0$ or $u \leq 0$ there. By Theorem 3.6,

$$u(x, t) = \int_{R^n} W(x - y, t - s)u(y, s)\, dy \tag{12}$$

whenever $x \in R^n$ and $r \leq s < t < a$, so that either $u \geq 0$ or $u \leq 0$ on $R^n \times [r, a[$. Furthermore,

$$u^+(x, t) = \int_{R^n} W(x - y, t - s)u^+(y, s)\, dy$$

whenever $x \in R^n$ and $r \leq s < t < a$, by (12) if $u \geq 0$ on $R^n \times [r, a[$, and obviously if $u \leq 0$ there. Hence

$$u^+(0, b) = \int_{\underline{R}^n} W(y, b - s)u^+(y, s) \, dy = M_b(u^+; s)$$

whenever $s \in [r, b[$, so that $M_b(u^+; \cdot)$ is constant on $[r, b[$.

Conversely, suppose that $M_b(u^+; \cdot)$ is constant on $[r, b[$. By Theorem 3.5,

$$u^+(x, t) \leq \int_{\underline{R}^n} W(x - y, t - r)u^+(y, r) \, dy < \infty \qquad (13)$$

for all $(x, t) \in \underline{R}^n \times]r, a[$. If $v(x, t)$ denotes the integral in (13), then our hypothesis and Lemma 3.1 imply that, whenever $r \leq t < b$,

$$M_b(u^+; r) = M_b(u^+; t) \leq M_b(v; t) = M_b(u^+; r)$$

Hence $M_b(v - u^+; t) = 0$ for all $t \in [r, b[$. Since $v - u^+ \geq 0$ by (13), and $v - u^+$ is continuous on $\underline{R}^n \times]r, b[$ by Lemma 3.1, we deduce that $v = u^+$ on $\underline{R}^n \times]r, b[$. Hence u^+ is a temperature on $\underline{R}^n \times]r, b[$. It follows that $u^- = u^+ - u$ is also a temperature there, and since both u^+ and u^- are non-negative, if either has a zero in $\underline{R}^n \times]r, b[$ it is identically zero there, by Theorem 2.7. But only one of them is non-zero at any point, so one of them is identically zero on $\underline{R}^n \times]r, b[$. Hence u has constant sign on this substrip, and therefore, by continuity, on $\underline{R}^n \times [r, b[$.

Corollary

Suppose that $u \in S_a$ and $0 < r < c < b < a$. Then $M_c(u^+; \cdot)$ is constant on $[r, c[$ if and only if $M_b(u^+; \cdot)$ is constant on $[r, b[$.

Proof. If $M_c(u^+; \cdot)$ is constant on $[r, c[$, then u has constant sign on $\underline{R}^n \times [r, c[$, by Theorem 3.9. Therefore u has constant sign on $\underline{R}^n \times [r, b[$, by the semigroup property, and hence $M_b(u^+; \cdot)$ is constant on $[r, b[$, by Theorem 3.9.

The proof of the converse is similar, but does not require use of the semigroup property.

4. THE CLASS U_a AND THE CAUCHY PROBLEM

Given a continuous function $f : \underline{R}^n \to \underline{R}$, the problem of finding a continuous function u on $\underline{R}^n \times [0, a[$, which is a temperature on $\underline{R}^n \times]0, a[$ and

satisfies $u(\cdot, 0) = f$, is called the <u>Cauchy problem</u> for $\underline{R}^n \times [0, a[$ with
<u>initial function</u> f.

  If f also satisfies the condition

$$\int_{\underline{R}^n} \exp(-\alpha \|x\|^2) |f(x)| \, dx < \infty$$

for some positive constant α, then the Gauss-Weierstrass integral u
of f is defined and a temperature on $\underline{R}^n \times]0, 1/4\alpha[$, by Theorem 1.2,
Corollary, and $u(x, t) \to f(\xi)$ as $(x, t) \to (\xi, 0^+)$ for all $\xi \in \underline{R}^n$, by
Theorem 1.3, Corollary. Thus u is a solution of the Cauchy problem
with initial function f, on $\underline{R}^n \times [0, a[$ for any $a \in]0, 1/4\alpha]$.

  In this section, we introduce a subclass U_a of S_a, in which we shall
prove that the Cauchy problem with a given initial function has at most
one solution. Any class of temperatures with this property is called a
<u>uniqueness class</u> for the Cauchy problem. Hence the choice of the sym-
bol U in U_a. We already know of one uniqueness class, namely the class
of all non-negative temperatures. For if, given f on $\underline{R}^n$, there is a
non-negative solution u of the Cauchy problem on $\underline{R}^n \times [0, a[$ with
$u(\cdot, 0) = f$, then u is the Gauss-Weierstrass integral of f on $\underline{R}^n \times]0, a[$,
by Theorem 2.10. Of course, if f is too large, say $f(x) = \exp(\|x\|^3)$,
then its Gauss-Weierstrass integral is undefined and there is no non-
negative solution of the Cauchy problem.

  Before introducing the class U_a, we give an example to show that
the class of all temperatures is not a uniqueness class for the Cauchy
problem.

  <u>Example</u>. Choose $\beta > 1$, and define a function g on $\underline{R}$ by

$$g(t) = \begin{cases} \exp(-t^{-\beta}) & \text{if } t > 0, \\ 0 & \text{if } t \leq 0 \end{cases}$$

Then g has derivatives of all orders on $\underline{R}$. (This can be proved by an
argument analogous to that outlined for the function F in the proof of
Lemma 2.2.) We show that, if u is defined on $\underline{R}^2$ by

$$u(x, t) = \sum_{k=0}^{\infty} \frac{g^{(k)}(t)}{(2k)!} x^{2k}$$

then u is a temperature on $\underline{R}^2$. Obviously $u(\cdot, t) = 0$ whenever $t \leq 0$,
and $u(0, t) = g(t) \neq 0$ whenever $t > 0$. Hence u is a non-zero solution to

the Cauchy problem on $\underline{R} \times [0, \infty[$ with initial function 0, and so this problem has more than one solution.

To prove that u is a temperature on $\underline{R}^2$, we need good estimates for all the derivatives of g. To get these, we use Cauchy's integral formulae for the derivatives of a complex analytic function. For each complex number z whose real part Re z is positive, let $z^{-\beta}$ denote the principal value of the power. If f is defined by

$$f(z) = \exp(-z^{-\beta})$$

whenever Re $z > 0$, then f is analytic. As $s \to 0$, we have $(1 + s)^\beta \to 1$ and $\cos(\beta \sin^{-1} s) \to 1$. We can therefore choose $\lambda \in]0, 1[$ such that

$$(1 + \lambda)^{-\beta} \cos(\beta \sin^{-1} \lambda) \geq 1/2 \qquad (14)$$

If $t > 0$, and γ denotes the positively described circle in the complex plane with center t and radius λt, Cauchy's integral formulae show that, for every $k \geq 0$,

$$g^{(k)}(t) = f^{(k)}(t) = \frac{k!}{2\pi i} \int_\gamma \frac{\exp(-z^{-\beta})}{(z - t)^{k+1}} \, dz \qquad (15)$$

Each point z on γ can be written in the form $z = t(1 + \lambda e^{i\psi})$ for some $\psi \in \underline{R}$, so that

$$\mathrm{Re}(-z^{-\beta}) = -t^{-\beta} \, \mathrm{Re}((1 + \lambda e^{i\psi})^{-\beta})$$

Since the principal value of the argument of $1 + \lambda e^{i\psi}$ is bounded by $\sin^{-1} \lambda$ for all $\psi \in \underline{R}$, it follows from (14) that

$$\mathrm{Re}((1 + \lambda e^{i\psi})^{-\beta}) = \exp(-\beta \log|1 + \lambda e^{i\psi}|) \cos(-\beta \arg(1 + \lambda e^{i\psi}))$$

$$\geq \exp(-\beta \log(1 + \lambda)) \cos(\beta \sin^{-1} \lambda)$$

$$= (1 + \lambda)^{-\beta} \cos(\beta \sin^{-1} \lambda)$$

$$\geq 1/2$$

where arg denotes the principal value of the argument. Hence

$$\mathrm{Re}(-z^{-\beta}) \leq -t^{-\beta}/2$$

and therefore

$$|\exp(-z^{-\beta})| = \exp(\text{Re}(-z^{-\beta})) \le \exp(-t^{-\beta}/2)$$

for all z on γ. It now follows from (15) that

$$|g^{(k)}(t)| \le \frac{k!}{2\pi} \cdot \frac{\exp(-t^{-\beta}/2)}{(\lambda t)^{k+1}} \cdot 2\pi\lambda t = \frac{k!}{(\lambda t)^k} \exp(-t^{-\beta}/2) \qquad (16)$$

Since $(k!)^2 \le (2k)!$ for all $k \ge 0$, we deduce that, for all $x \in \underline{R}$,

$$\sum_{k=0}^{\infty} \frac{|g_k(t)|}{(2k)!} x^{2k} \le \sum_{k=0}^{\infty} \frac{1}{k!} \left(\frac{x^2}{\lambda t}\right)^k \exp(-t^{-\beta}/2)$$

$$= \exp\left(\frac{x^2}{\lambda t} - \frac{1}{2t^\beta}\right) \qquad (17)$$

Therefore the defining series of u(x, t) is absolutely convergent for all $(x, t) \in \underline{R} \times]0, \infty[$; this is trivially true for other values of (x, t).
Put

$$\phi_k(x, t) = \frac{g^{(k)}(t)}{(2k)!} x^{2k}$$

for all $(x, t) \in \underline{R}^2$. Then, given any $t_0 \in \underline{R}$, $D_1 \phi_k(x, t_0)$ exists for all $x \in \underline{R}$ and all $k \in \underline{N} \cup \{0\}$. Furthermore, by (16), for any $r > 0$ and $t_0 > 0$,

$$|D_1 \phi_k(x, t_0)| = \left| \frac{g^{(k)}(t_0)}{(2k-1)!} x^{2k-1} \right|$$

$$\le \frac{k! \exp(-t_0^{-\beta}/2)|x|^{2k-1}}{(2k-1)!(\lambda t_0)^k}$$

$$\le \frac{\exp(-t_0^{-\beta}/2)r^{2k-1}}{(k-1)!(\lambda t_0)^k}$$

for all $k \in \underline{N}$ and all $x \in \,]-r, r[$. Since

$$\sum_{k=1}^{\infty} \frac{r^{2k-1}}{(k-1)!\,(\lambda t_0)^k} = \frac{r}{\lambda t_0} \sum_{k=1}^{\infty} \frac{(r^2/\lambda t_0)^{k-1}}{(k-1)!} < \infty$$

it follows from Lemma 1.3 (with $\xi = x$, $\eta = t$, $\zeta = k$, $f(\xi, \eta, \zeta) = \phi_k(x, t)$, $]\alpha, \beta[\, = \,]-r, r[$, $A = \underline{R}$, $B = \underline{N} \cup \{0\}$, and μ the counting measure on B) that, for any $x_0 \in \,]-r, r[$, $D_1 u(x_0, t_0)$ exists and is given by

$$D_1 u(x_0, t_0) = \sum_{k=1}^{\infty} \frac{g^{(k)}(t_0)}{(2k-1)!} x_0^{2k-1} \tag{18}$$

the series converging absolutely. Since r is arbitrary, this holds for every $(x_0, t_0) \in \underline{R}^2$.

Next, given any $t_0 \in \underline{R}$, $D_1^2 \phi_k(x, t_0)$ exists for all $x \in \underline{R}$ and all $k \in \underline{N} \cup \{0\}$. Furthermore, by (16), for any $r > 0$ and $t_0 > 0$,

$$|D_1^2 \phi_k(x, t_0)| \leq \frac{k!\, \exp(-t_0^{-\beta}/2) x^{2k-2}}{(2k-2)!\,(\lambda t_0)^k}$$

$$\leq \frac{\exp(-t_0^{-\beta}/2)\, r^{2k-2}}{(k-2)!\,(\lambda t_0)^k}$$

for all $k \in \underline{N} \setminus \{1\}$ and all $x \in \,]-r, r[$. Since

$$\sum_{k=2}^{\infty} \frac{r^{2k-2}}{(k-2)!\,(\lambda t_0)^k} = \left(\frac{r}{\lambda t_0}\right)^2 \sum_{k=2}^{\infty} \frac{(r^2/\lambda t_0)^{k-2}}{(k-2)!} < \infty$$

it follows from Lemma 1.3 (with $f(\xi, \eta, \zeta) = D_1 \phi(x, t)$, but everything else as before) and the absolute convergence of the series in (18), that, for any $x_0 \in \,]-r, r[$, $D_1^2 u(x_0, t_0)$ exists and is given by

$$D_1^2 u(x_0, t_0) = \sum_{k=1}^{\infty} \frac{g^{(k)}(t_0)}{(2k-2)!} x_0^{2k-2} \tag{19}$$

the series converging absolutely. Since r is arbitrary, this holds for every $(x_0, t_0) \in \underline{R}^2$.

The proof that $D_t u$ exists and can be obtained by termwise differentiation of the defining series is much more difficult. Given any $x_0 \in \underline{R}$, $D_t \phi_k(x_0, t)$ exists for all $t \in \underline{R}$ and all $k \in \underline{N} \cup \{0\}$. Furthermore, by (16),

$$|D_t \phi_k(x_0, t)| = \left| \frac{g^{(k+1)}(t)}{(2k)!} x_0^{2k} \right|$$

$$\leq \frac{(k+1)! \exp(-t^{-\beta}/2) x_0^{2k}}{(2k)! (\lambda t)^{k+1}}$$

$$\leq \frac{(k+1)! x_0^{2k}}{(2k)! \lambda^{k+1}} \left(\frac{2(k+1)}{e\beta} \right)^{(k+1)/\beta}$$

for all $t > 0$. Putting $\ell = k + 1$, we obtain

$$|D_t \phi_{\ell-1}(x_0, t)| \leq \frac{\ell! \ell^{\ell/\beta} (2\ell)(2\ell - 1)}{(2\ell)!} \left(\frac{2^{1/\beta}}{\lambda(e\beta)^{1/\beta}} \right)^{\ell} x_0^{2(\ell-1)}$$

Stirling's formula tells us that, as $p \to \infty$,

$$\frac{p!}{p^{p+(1/2)} e^{-p}} \to (2\pi)^{1/2}$$

Therefore, as $\ell \to \infty$,

$$\frac{\ell! \ell^{\ell/\beta} (2\ell)(2\ell - 1)}{(2\ell)!} \sim \frac{4\ell^{\ell+(1/2)+(\ell/\beta)+2} e^{-\ell}}{(2\ell)^{2\ell+(1/2)} e^{-2\ell}}$$

$$= \ell^{-\ell+(\ell/\beta)+2} e^{\ell}/2^{2\ell-(3/2)}$$

$$= 2^{3/2} (e/4)^{\ell} \ell^2 \ell^{-\delta\ell}$$

where $\delta = 1 - (1/\beta) > 0$ since $\beta > 1$, and we write $a_\ell \sim b_\ell$ if $a_\ell/b_\ell \to 1$. Hence there are positive constants C and κ such that

$$|D_t \phi_{\ell-1}(x_0,t)| \leq C \ell^2 \ell^{-\delta \ell} \kappa^\ell x_0^{2(\ell-1)}$$

for all sufficiently large ℓ. Since $\delta > 0$, the series

$$\sum_{\ell=1}^{\infty} \ell^2 \ell^{-\delta \ell} \kappa^\ell x_0^{2\ell}$$

is convergent. We can therefore use Lemma 1.3 (with $\xi = t$, $\eta = x$, $\zeta = k$, $f(\xi, \eta, \zeta) = \phi_k(x,t)$, $]\alpha, \beta[$ any open interval containing t_0, $A = \underline{R}$, $B = \underline{N} \cup \{0\}$, and μ the counting measure on $\underline{N} \cup \{0\}$) to conclude that, for any $t_0 \in \underline{R}$, $D_t u(x_0, t_0)$ exists and is given by

$$D_t u(x_0, t_0) = \sum_{k=0}^{\infty} \frac{g^{(k+1)}(t_0)}{(2k)!} x_0^{2k} \qquad (20)$$

the series converging absolutely.

Since the series in (19) and (20) are the same, u is a temperature on $\underline{R}^2$.

Definition. We say that $u \in U_a$ if u is a temperature on $\underline{R}^n \times]0, a[$ and, for each $b \in]0, a[$, $M_b(u^+; \cdot)$ is locally integrable on $[0, b[$.

Note that the sum of two functions in U_a is again in U_a.

Obviously $U_a \subseteq S_a$. Therefore, by Theorem 3.8, $M_b(u^+; \cdot)$ is non-increasing on $]0, b[$. It follows that a temperature u belongs to U_a if and only if $M_b(u^+; \cdot)$ is integrable over $]0, b[$ for every $b < a$.

Example. We show that U_a is a proper subclass of S_a. Let u be defined on $\underline{R}^n \times]0, \infty[$ by

$$u(x, t) = (\|x\|^2 - 2nt) t^{-(n+4)/2} \exp(-\|x\|^2/4t)$$

Then u is a temperature; in fact, u is a constant multiple of $D_t W$. If $0 < t < b < \infty$, then

$$M_b(u^+; t) = C t^{-(n+4)/2} (b-t)^{-n/2} \int_{\|x\|^2 \geq 2nt} (\|x\|^2 - 2nt) \exp\left(-\frac{b\|x\|^2}{4t(b-t)}\right) dx$$

where $C = (4\pi)^{-n/2}$. We change to polar coordinates, putting $\rho = \|x\|$, so that $dx = \sigma_n \rho^{n-1} d\rho$, where σ_n denotes the surface area of a unit sphere in $\underline{R}^n$. Thus we obtain, for some constant C,

$$M_b(u^+; t) = C t^{-(n+4)/2} (b-t)^{-n/2} \int_{\sqrt{2nt}}^{\infty} (\rho^2 - 2nt) \exp\left(-\frac{b\rho^2}{4t(b-t)}\right) \rho^{n-1} d\rho$$

The substitution $\sigma = b\rho^2/4t(b-t)$ now gives, with $\tau = bn/2(b-t)$,

$$M_b(u^+; t) = C t^{-(n+4)/2} (b-t)^{-n/2} \int_{\tau}^{\infty} \left(\frac{4t(b-t)}{b}\right)^{(n+2)/2} (\sigma - \tau) e^{-\sigma} \sigma^{(n-2)/2} d\sigma$$

$$= C t^{-1}(b-t) \int_{\tau}^{\infty} (\sigma - \tau) e^{-\sigma} \sigma^{(n-2)/2} d\sigma \qquad (21)$$

$$\leq C t^{-1}(b-t) \int_{0}^{\infty} e^{-\sigma} \sigma^{n/2} d\sigma$$

$$= C t^{-1}(b-t)$$

Therefore $M_b(u^+; \cdot)$ is locally integrable on $]0, b[$ whenever $b \in]0, \infty[$, so that $u \in S_a$ for every a. Finally, as t decreases to zero, τ decreases to n/2 and the integral in (21) increases to

$$\int_{n/2}^{\infty} (\sigma - (n/2)) e^{-\sigma} \sigma^{(n-2)/2} d\sigma$$

which is a positive real number. It therefore follows from (21) that $tM_b(u^+; t)$ tends to a positive real number as $t \to 0^+$. Hence $M_b(u^+; \cdot)$ is not integrable over $]0, \epsilon[$ for any positive ϵ, so that $u \notin U_a$ for any a.

The important property that $-u \in S_a$ whenever $u \in S_a$ carries over to the class U_a.

Theorem 3.10

If $u \in U_a$, then $-u \in U_a$.

Proof. If $u \in U_a$ then $u \in S_a$, so that u has the semigroup property on $\underline{R}^h \times]0, a[$. Therefore, whenever $0 < t < b < a$,

$$M_b(u; t) = \int_{\underline{R}^n} W(x, b - t)u(x, t) \, dx = u(0, b)$$

Hence $M_b(|u|; t) < \infty$ and

$$M_b(u^-; t) = M_b(u^+; t) - M_b(u; t) = M_b(u^+; t) - u(0, b) \qquad (22)$$

whenever $0 < t < b < a$. Since $u \in U_a$, it follows that

$$\int_0^b M_b(u^-; t) \, dt = \int_0^b M_b(u^+; t) \, dt - bu(0, b) < \infty$$

for all $b < a$, so that $-u \in U_a$.

In order to prove that U_a is a uniqueness class for the Cauchy problem, we combine Theorem 3.10 with the case $s = 0$, $b = a$ of Theorem 3.1.

Theorem 3.11

Given a continuous function $f : \underline{R}^n \to \underline{R}$, there is at most one solution in U_a to the Cauchy problem on $\underline{R}^n \times [0, a[$ with initial function f.

Proof. Suppose that u and v are continuous on $\underline{R}^n \times [0, a[$, belong to U_a, and satisfy $u(\cdot, 0) = v(\cdot, 0)$. Then $u - v$ and $v - u$ are both solutions of the Cauchy problem on $\underline{R}^n \times [0, a[$ with initial function 0. Furthermore, it follows from Theorem 3.10 and the fact that the sum of two functions in U_a is again in U_a, that $u - v$, $v - u \in U_a$. By the case $s = 0$, $b = a$ of Theorem 3.1, $u - v \leq 0$ and $v - u \leq 0$ on $\underline{R}^n \times [0, a[$. Hence $u = v$.

Corollary

Given a continuous function $f : \underline{R}^n \to \underline{R}$, there is at most one solution u to the Cauchy problem on $\underline{R}^n \times [0, a[$ with initial function f, such that

$$\int_0^a dt \int_{\underline{R}^n} \exp(-\alpha \|x\|^2) u^+(x, t) \, dx < \infty \qquad (23)$$

for some positive constant α.

Proof. It is sufficient to prove uniqueness on some substrip $\underline{R}^n \times]0, d[$ of $\underline{R}^n \times]0, a[$, since the result for the full strip will then follow by repeated application of that for the substrip. For example, if $d < a < 2d$ then, having proved uniqueness on $\underline{R}^n \times]0, d[$, we apply that result to the function v on $\underline{R}^n \times [0, d[$ given by $v(x,t) = u(x, a - d + t)$. Thus we see that v is the unique solution of the Cauchy problem on $\underline{R}^n \times [0, d[$ with initial function $u(\cdot, a - d)$ which satisfies

$$\int_0^d dt \int_{\underline{R}^n} \exp(-\alpha \|x\|^2) v^+(x, t) \, dx < \infty$$

In this way, the uniqueness on $\underline{R}^n \times]0, d[$ is extended to $\underline{R}^n \times]0, a[$.

We may suppose that $\alpha > 1/4a$, since an increase in α does not affect the condition (23). Choose $b > 0$ such that $b < 1/4\alpha < a$. Whenever $0 < t < c < b$ and $x \in \underline{R}^n$,

$$W(x, b - t) \leq (4\pi(b - c))^{-n/2} \exp(-\alpha \|x\|^2)$$

so that, if u is as described in the statement of the theorem,

$$\int_b^c M_b(u^+; t) \, dt \leq (4\pi(b - c))^{-n/2} \int_0^a dt \int_{\underline{R}^n} \exp(-\alpha \|x\|^2) u^+(x, t) \, dx$$

It now follows from (23) that $M_b(u^+; \cdot)$ is integrable over $[0, c[$ whenever $0 < c < b < 1/4\alpha$. Hence $u \in U_{1/4\alpha}$. By Theorem 3.11, u is unique on $\underline{R}^n \times [0, 1/4\alpha[$. The result follows.

Remark. At the beginning of this section, we gave an example of a non-zero solution u of the Cauchy problem on $\underline{R} \times [0, \infty[$ with initial function 0. It follows from (17) that

$$|u(x, t)| \leq \exp\left(\frac{x^2}{\lambda t} - \frac{1}{2t^\beta}\right)$$

for all (x, t), where $\beta > 1$ and $0 < \lambda < 1$. Hence, for each fixed $t > 0$, $u(x, t)$ grows like $\exp(\kappa x^2)$ as $|x| \to \infty$, which is no faster than some members of U_a. It is the fact that $\kappa \to \infty$ as $t \to 0^+$ which makes the difference and ensures that (23) does not hold for any fixed positive α.

5. UNIQUENESS OF FUNDAMENTAL SOLUTIONS

Recall that Γ_0, defined for all (x, t), $(y, s) \in \underline{R}^n \times [0, a[$ by

$$\Gamma_0(x, t; y, s) = \begin{cases} W(x - y, t - s) & \text{if} \quad s < t, \\ 0 & \text{if} \quad t \leq s \end{cases}$$

is a fundamental solution of the heat equation on $\underline{R}^n \times [0, a[$. We are now in a position to prove that Γ_0 is not the only one, but that it has properties possessed by no other. If $0 \leq s < a$, $f \in C_0(\underline{R}^n)$, and u_f is defined on $\underline{R}^n \times [s, a[$ by

$$u_f(x, t) = \int_{\underline{R}^n} \Gamma_0(x, t; y, s) f(y) \, dy \quad \text{if} \quad t > s \tag{24}$$

$$u_f(x, s) = f(x) \tag{25}$$

then u_f is continuous on $\underline{R}^n \times [0, a[$, by definition of a fundamental solution. Furthermore, it follows from Theorem 3.3 that the function $(x, t) \longmapsto u_f(x, s + t)$ belongs to U_{a-s}; we abbreviate this to $u_f \in U_{a, s}$. Thus u_f is the only solution in $U_{a, s}$ of the Cauchy problem on $\underline{R}^n \times [s, a[$ with initial function f, in view of Theorem 3.11. We show that Γ_0 is the only fundamental solution of the heat equation which has this property. Of course, when $t \leq s$, the values $\Gamma(x, t; y, s)$ of a fundamental solution play no part in the definition, so they can either be ignored or defined to be zero.

Theorem 3.12

Let Γ be a fundamental solution of the heat equation on $\underline{R}^n \times [0, a[$ with the following property:
If $0 \leq s < a$, $f \in C_0(\underline{R}^n)$, and v_f is defined on $\underline{R}^n \times]s, a[$ by

$$v_f(x, t) = \int_{\underline{R}^n} \Gamma(x, t; y, s) f(y) \, dy \tag{26}$$

then $v_f \in U_{a, s}$.
Then $\Gamma(x, t; y, s) = \Gamma_0(x, t; y, s)$ whenever $x, y \in \underline{R}^n$ and $0 \leq s < t < a$.

Proof. For any $f \in C_0(\underline{R}^n)$, v_f is a temperature on $\underline{R}^n \times]s, a[$ since $v_f \in U_{a, s}$. Therefore, because Γ is a fundamental solution, v_f

is a solution of the Cauchy problem on $\underline{R}^n \times [s, a[$ with initial function f. By Theorem 3.11, v_f is the only solution in $U_{a,s}$. If u_f is defined on $\underline{R}^n \times [s, a[$ by (24) and (25), then $u_f \in U_{a,s}$ and is a solution of the same Cauchy problem, so that $u_f = v_f$. Let $r > 0$, and let $(x, t) \in \underline{R}^n \times]s, a[$. If $\Lambda_{x,t}$ is defined on $C_c(B(0,r))$ by

$$\Lambda_{x,t}(f) = u_f(x,t) - v_f(x,t)$$

then $\Lambda_{x,t}$ is identically zero and hence is trivially a bounded linear functional on $C_c(B(0,r))$. By the Riesz representation theorem, there is a unique complex measure $\mu_{x,t}$ such that

$$\Lambda_{x,t}(f) = \int_{B(0,r)} f \, d\mu_{x,t}$$

for all $f \in C_c(B(0,r))$. Obviously $\mu_{x,t}$ is the zero measure on $B(0,r)$. But we also have the representation

$$\Lambda_{x,t}(f) = \int_{B(0,r)} f(y)(\Gamma_0(x,t;y,s) - \Gamma(x,t;y,s)) \, dy \qquad (27)$$

for all $f \in C_c(B(0,r))$. If g is any function in $C_c(\underline{R}^n)$ such that $g(x) \geq 1$ for all $x \in B(0,r)$, then the fact that v_g is a temperature on $\underline{R}^n \times]s, a[$ implies that

$$\int_{B(0,r)} |\Gamma(x,t;y,s)| \, dy \leq \int_{\underline{R}^n} |\Gamma(x,t;y,s)g(y)| \, dy < \infty$$

It therefore follows, using Lemma 1.1, that

$$\int_{B(0,r)} |\Gamma_0(x,t;y,s) - \Gamma(x,t;y,s)| \, dy < \infty$$

so that the formula

$$\nu(E) = \int_E (\Gamma_0(x,t;y,s) - \Gamma(x,t;y,s)) \, dy$$

for all Borel sets $E \subseteq B(0,r)$, defines a measure with finite total variation, hence a complex measure. It now follows from (27), and the uniqueness part of the Riesz representation theorem, that $\nu = \mu_{x,t} = 0$. Hence $\Gamma_0(x,t;y,s) = \Gamma(x,t;y,s)$ for all $y \in B(0,r)$, given any $(x,t) \in \underline{R}^n \times]s, a[$ and any $s \in [0, a[$. Since r is arbitrary, the result is proved.

Corollary

Let Γ be a non-negative fundamental solution of the heat equation on $\underline{R}^n \times [0, a[$, with the property that if $0 \le s < a$, $f \in C_c(\underline{R}^n)$, and v_f is defined on $\underline{R}^n \times]s, a[$ by (26), then v_f is a temperature. Then $\Gamma(x, t; y, s) = \Gamma_0(x, t; y, s)$ whenever $x, y \in \underline{R}^n$ and $0 \le s < t < a$.

Proof. For any $s \in [0, a[$ and $f \in C_c(\underline{R}^n)$, v_f can be written in the form

$$v_f(x, t) = \int_{\underline{R}^n} \Gamma(x, t; y, s) f^+(y) \, dy - \int_{\underline{R}^n} \Gamma(x, t; y, s) f^-(y) \, dy$$

for all $(x, t) \in \underline{R}^n \times]s, a[$. Since $\Gamma \ge 0$, v_f is thus expressed as the difference of two non-negative temperatures. Theorems 2.11 and 3.3 together imply that $v_f \in U_{a,s}$, so that the result follows from Theorem 3.12.

Example. Let u_0 be a non-zero solution of the Cauchy problem on $\underline{R}^n \times [0, a[$ with initial function 0. (An example of such a function was given at the beginning of the previous section.) If Γ is defined for all $x, y \in \underline{R}^n$ by

$$\Gamma(x, t; y, s) = \begin{cases} W(x - y, t - s) + u_0(x - y, t - s) & \text{if} \quad 0 \le s < t < a, \\ 0 & \text{if} \quad 0 \le t \le s < a \end{cases}$$

then Γ is a fundamental solution of the heat equation. For each fixed $(y_0, s_0) \in \underline{R}^n \times [0, a[$, it is obvious that $\Gamma(\cdot, \cdot; y_0, s_0)$ is a temperature on $\underline{R}^n \times]s_0, a[$. For each $f \in C_c(\underline{R}^n)$ and each $s \in [0, a[$, Theorem 1.1 shows that

$$\int_{\underline{R}^n} W(x - y, t - s) f(y) \, dy \to f(\xi)$$

as $(x, t) \to (\xi, s^+)$ for all $\xi \in \underline{R}^n$. It is therefore sufficient to prove that

$$\int_{\underline{R}^n} u_0(x - y, t - s) f(y) \, dy \to 0 \tag{28}$$

as $(x, t) \to (\xi, s^+)$ for all $\xi \in \underline{R}^n$.

Given $f \in C_c(\underline{R}^n)$, choose ρ such that the support of f is contained in $B(0, \rho)$. If $\|y\| < \rho$ and $\|x\| < k\rho$ for some $k \in \underline{N}$, then $\|x - y\| <$

$(k + 1)\rho$ so that, since u_0 is continuous on $\bar{B}(0, (k + 1)\rho) \times [0, a - s]$, there is a constant C such that

$$|u_0(x - y, t - s)f(y)| \leq C|f(y)|$$

for all $x \in B(0, k\rho)$, $y \in B(0, \rho)$, and $t \in [s, a]$. Because f is integrable over $B(0, \rho)$, we can now apply the dominated convergence theorem. Hence, as $(x, t) \to (\xi, s^+)$,

$$\int_{B(0, \rho)} u_0(x - y, t - s)f(y)\, dy \to \int_{B(0, \rho)} u_0(\xi - y, 0)f(y)\, dy = 0$$

for all $\xi \in B(0, k\rho)$. Since support of f is contained in $B(0, \rho)$, this shows that (28) holds as $(x, t) \to (\xi, s^+)$ for all $\xi \in B(0, k\rho)$. Because k is arbitrary, there is no restriction on ξ, and our assertion is proved.

6. THE CLASS R_a AND THE GAUSS-WEIERSTRASS REPRESENTATION

We introduce a subclass R_a of U_a which characterizes those temperatures which are Gauss-Weierstrass integrals of signed measures. The R in R_a stands for representation.

Definition. We say that $u \in R_a$ if u is a temperature on $\underline{R}^n \times]0, a[$ and, for each $b \in]0, a[$, $M_b(u^+; \cdot)$ is locally integrable on $]0, b[$ with

$$\liminf_{t \to 0^+} M_b(u^+; t) < \infty$$

Note that the sum of two functions in R_a is also in R_a.

Obviously $R_a \subseteq S_a$, but to show that $R_a \subseteq U_a$ we must use Theorem 3.8. We also show that the inclusion is strict.

Theorem 3.13

If $u \in R_a$ and $b \in]0, a[$, then $M_b(u^+; \cdot)$ is bounded on $]0, b[$. The class R_a is a proper subclass of U_a.

Proof. Let $u \in R_a$ and $b \in]0, a[$. Since $R_a \subseteq S_a$, Theorem 3.8 implies that $M_b(u^+; \cdot)$ is non-increasing and real-valued on $]0, b[$. Therefore, for all $t \in]0, b[$,

$$M_b(u^+; t) \le \lim_{s \to 0^+} M_b(u^+; s) < \infty$$

so that $M_b(u^+; \cdot)$ is bounded on $]0, b[$. Since b is arbitrary, it follows that $u \in U_a$. Thus $R_a \subseteq U_a$.

We now show that the inclusion is strict. Let v be defined on $\underline{R}^n \times]0, \infty[$ by

$$v(x, t) = x_1 t^{-(n+2)/2} \exp\left(-\frac{\|x\|^2}{4t}\right)$$

Then v is a temperature; in fact, v is a constant multiple of $D_1 W$. Whenever $0 < t < b < \infty$,

$$M_b(v^+; t)$$

$$= Ct^{-(n+2)/2}(b-t)^{-n/2} \int_{\underline{R}} \cdots \int_{\underline{R}} \int_0^\infty x_1 \exp\left(-\frac{b\|x\|^2}{4t(b-t)}\right) dx_1\, dx_2 \cdots dx_n$$

for some constant C. Since the integrand is non-negative, we can take the integrals in any order, by Fubini's theorem. Now, for any positive α,

$$\exp(-\alpha\|x\|^2) = \prod_{i=1}^n \exp(-\alpha x_i^2)$$

so that

$$\int_{\underline{R}} \cdots \int_{\underline{R}} \exp(-\alpha\|x\|^2) dx_2 \cdots dx_n = \exp(-\alpha x_1^2) \prod_{i=2}^n \left(\int_{\underline{R}} \exp(-\alpha x_i^2)\, dx_i\right)$$

$$= (\pi/\alpha)^{(n-1)/2} \exp(-\alpha x_1^2)$$

It follows that

$$M_b(v^+; t) = Ct^{-(n+2)/2}(b-t)^{-n/2}\left(\frac{4t(b-t)\pi}{b}\right)^{(n-1)/2} \int_0^\infty x_1 \exp\left(-\frac{bx_1^2}{4t(b-t)}\right) dx_1$$

$$= Ct^{-3/2}(b-t)^{-1/2}\int_0^\infty e^{-s}\left(\frac{4t(b-t)}{b}\right) ds$$

$$= Ct^{-1/2}(b-t)^{1/2}$$

Hence $M_b(v^+; \cdot)$ is integrable over $]0, b[$, but is unbounded. Since b is arbitrary, we deduce that $v \in U_\infty$ but that $v \notin R_\infty$, in view of the first part of this theorem.

As with the other classes defined in terms of the means M_b, the class R_a has the important property that it contains the negatives of all its elements.

Theorem 3.14

If $u \in R_a$, then $-u \in R_a$.

Proof. Since $R_a \subseteq U_a$ by Theorem 3.13, we can use equation (22). Hence, if $u \in R_a$ and $0 < t < b < a$,

$$M_b(u^-; t) = M_b(u^+; t) - u(0, b)$$

By Theorem 3.13, $M_b(u^+; \cdot)$ is bounded on $]0, b[$, so that the same is true of $M_b(u^-; \cdot)$, for every $b \in]0, a[$. Hence $-u \in R_a$.

We already know, from Theorem 3.3, that if w is a Gauss-Weier-strass integral on $\underline{R}^n \times]0, a[$, then $w \in R_a$. In order to prove the converse, we show that any temperature u in R_a can be written as a difference of two non-negative temperatures, and then apply Theorem 2.11. For this we use the idea of a thermic majorant of u^+, which we now define.

Definition. Let u be a temperature on $\underline{R}^n \times]0, a[$. We say that u^+ has a thermic majorant on $\underline{R}^n \times]0, a[$ if there is a temperature v such that $u^+ \leq v$ on $\underline{R}^n \times]0, a[$.

If u^+ has a thermic majorant v, then $v \geq 0$ and $v \geq u$, so that $v - (v - u)$ is an expression for u as a difference of non-negative temperatures.

Theorem 3.15

Let u be a temperature on $\underline{R}^n \times]0, a[$. Then the following statements are equivalent:

(i) $u \in R_a$;

(ii) u^+ has a thermic majorant on $\underline{R}^n \times]0, a[$;

(iii) u can be written as a difference of two non-negative tempera-
        tures on $\underline{R}^n \times]\,0, a[$;

(iv)   u is the Gauss-Weierstrass integral of a signed measure on
        $\underline{R}^n \times]0, a[$.

Proof. We show that (i) $\Rightarrow$ (ii) $\Rightarrow$ (iii) $\Rightarrow$ (iv) $\Rightarrow$ (i).

(i) $\Rightarrow$ (ii). Suppose that $u \in R_a$. For each $r \in]0, a[$, we define v_r
on $\underline{R}^n \times]r, a[$ by

$$v_r(x, t) = \int_{R^n} W(x - y, t - r)u^+(y, r) \, dy$$

Since $R_a \subseteq S_a$, it follows from Theorem 3.5 that v_r is real-valued and
$v_r \geq u^+$ on $\underline{R}^n \times]r, a[$. Theorem 1.2 shows that v_r is a temperature,
and Theorem 1.4 implies that

$$v_r(x, t) = \int_{R^n} W(x - y, t - s)v_r(y, s) \, dy \qquad (29)$$

whenever $x \in \underline{R}^n$ and $r < s < t < a$. Since $v_r \geq u^+$, it follows that

$$v_r(x, t) \geq \int_{R^n} W(x - y, t - s)u^+(y, s) \, dy = v_s(x, t)$$

whenever $(x, t) \in \underline{R}^n \times]s, a[$ and $0 < r < s < a$. Hence, for each $(x, t) \in \underline{R}^n \times]0, a[$, the function $r \longmapsto v_r(x, t)$ is non-increasing on $]0, t[$. Therefore $v_r(x, t)$ tends to a limit $v(x, t)$ as $r \to 0^+$, and the monotone con-
vergence theorem can be used on (29) to deduce that

$$v(x, t) = \int_{R^n} W(x - y, t - s)v(y, s) \, dy \qquad (30)$$

whenever $x \in \underline{R}^n$ and $0 < s < t < a$. However, we don't yet know that
the integrals in (30) are finite. To show that they are, we first note
that, whenever $0 < r < b < a$,

$$v_r(0, b) = M_b(u^+; r)$$

Therefore, by Theorem 3.13,

$$v(0, b) = \lim_{r \to 0^+} v_r(0, b) = \lim_{r \to 0^+} M_b(u^+; r) < \infty$$

Hence, given $s \in]0,a[$, the integral in (30) is finite when $(x,t) = (0,b)$, and is therefore finite for all $(x,t) \in \underline{R}^n \times]s,b[$ and represents a temperature, by Theorem 1.2. Because s and b are arbitrary, v is a temperature on $\underline{R}^n \times]0,a[$. Since $v_r \geq u^+$ on $\underline{R}^n \times]r,a[$ for every r, we see that $v \geq u^+$ on $\underline{R}^n \times]0,a[$, so that v is a thermic majorant of u^+.

(ii) $\Rightarrow$ (iii). Suppose that u^+ has a thermic majorant v on $\underline{R}^n \times]0,a[$. Then $v \geq 0$ and $v \geq u$, so that $u = v - (v - u)$ expresses u as a difference of two non-negative temperatures.

(iii) $\Rightarrow$ (iv). This is part of Theorem 2.11.

(iv) $\Rightarrow$ (i). This follows from Theorem 3.3.

Remarks. It follows from Theorem 3.13 that R_a is a proper subclass of S_a. Hence, in view of Theorems 3.6 and 3.15, a temperature can have the semigroup property on $\underline{R}^n \times]0,a[$ without being the Gauss-Weierstrass integral of a signed measure. Thus the converse of the latter part of Theorem 1.4 is false.

The last part of Theorem 2.10 combines with Theorem 3.15 to show that, if $u \in R_a$ and u is continuous and real-valued on $\underline{R}^n \times [0,a[$, then u is the Gauss-Weierstrass integral of $u(\cdot,0)$.

Corollary

Let u be a temperature on $\underline{R}^n \times]0,a[$. If there is a constant $\alpha \in]0,1/4a]$ such that the function

$$t \longmapsto \int_{\underline{R}^n} \exp(-\alpha\|x\|^2)u^+(x,t) \; dx \tag{31}$$

is locally integrable on $]0,a[$, and

$$\liminf_{t \to 0^+} \int_{\underline{R}^n} \exp(-\alpha\|x\|^2)u^+(x,t) \; dx < \infty \tag{32}$$

then u is the Gauss-Weierstrass integral of a signed measure.

Proof. Since $\exp(-\alpha\|x\|^2)$ does not increase as α increases, we can assume that $\alpha = 1/4a$. If $0 < t < c < b < a$, then

$$W(x, b - t) \leq (4\pi(b - c))^{-n/2} \exp(-\alpha\|x\|^2)$$

for all $x \in \underline{R}^n$, so that

$$M_b(u^+; t) \leq (4\pi(b-c))^{-n/2} \int_{\underline{R}^n} \exp(-\alpha\|x\|^2)u^+(x,t)\, dx$$

It now follows from (31) and (32) that $u \in R_a$, so that u is the Gauss-Weierstrass integral of a signed measure, by Theorem 3.15.

Remark. The condition that $\alpha \leq 1/4a$ in the above corollary cannot be removed. Shortly after Theorem 3.7, we gave an example in which we showed that the Gauss-Weierstrass integral of the function

$$x \longmapsto (1/2)\,\exp\left(\frac{x^2}{16}\right)\cos\left(\frac{x^2}{16}\right)$$

is defined on $\underline{R} \times]0,4[$, and can be extended to a temperature u on $\underline{R} \times]0,\infty[$ but is itself undefined on $\underline{R} \times]4,\infty[$. It follows easily from the explicit formula for u on $\underline{R} \times]0,4[$ that

$$|u(x,t)| \leq C\,\exp\left(\frac{x^2}{16}\right) \tag{33}$$

for all $(x,t) \in \underline{R} \times]0,4[$, and it follows from the formula (9), which is used to extend u to $\underline{R} \times]0,\infty[$, that

$$|u(x,t)| \leq \int_{\underline{R}^n} W(x-y,t-2)|u(y,2)|\, dy \leq C$$

for all $(x,t) \in \underline{R} \times]2,\infty[$, since $u(\cdot,2)$ is bounded. Hence (33) holds throughout $\underline{R}\times]0,\infty[$, and this implies that the conditions of the corollary are satisfied with any $\alpha > 1/16$. But the conclusion of the corollary does not hold on $\underline{R} \times]0,a[$ for any $a > 4$.

7. LINEAR PARABOLIC EQUATIONS

All the main results of this chapter can be proved for a large class of linear parabolic equations by similar methods to those presented above. The means M_b are defined by

$$M_b(v; t) = \int_{\underline{R}^n} \Gamma(0,b;x,t)v(x,t)\, dx$$

where Γ is the fundamental solution. The only items which cannot obviously be extended are the relatively minor ones which we now list.

Theorem 3.4 (which is only to motivate the use of S_a in preference to Σ_a), the remark and example following Theorem 3.7, the first of the two paragraphs which immediately precede Theorem 3.9, the examples in section 4 and the remark at the end of that section, the example in section 5, that part of Theorem 3.13 which tells us that the inclusion of R_a in U_a is strict, and the remark at the end of section 6.

The proof of Theorem 3.8 requires that, for each fixed $(x, b) \in \underline{R}^n \times \,]0, a[$, $\Gamma(0, b; x, \cdot)$ is a continuous function on $]0, b[$. This holds for many fundamental solutions since $\Gamma(0, b; \cdot, \cdot)$ can often be shown to satisfy the equation adjoint to the given one. (See Chapter II, section 8, for the definition of the adjoint equation.)

The methods of this chapter depend essentially on Theorem 1.2, and therefore require the sharp lower estimate of the fundamental solution.

8. BIBLIOGRAPHICAL NOTES

Except for the examples, almost the entire chapter follows Watson's paper [2] on the heat equation. Most of the results were subsequently extended to a wide class of parabolic equations by Watson [3], and the few that weren't can be extended without difficulty using the methods given in the text. The only other exceptions are the part of the proof of Theorem 3.13 which shows that the inclusion of R_a in U_a is strict, and the results on uniqueness of fundamental solutions in section 5. Theorem 3.12 is new, but its proof is essentially the same as one given by Il'in, Kalashnikov and Oleinik [1, p. 79] using a different uniqueness class. The example in section 2 is due to Blackman [1], although he used the techniques of complex function theory to handle the details. The example, in section 4, of a non-zero solution to the Cauchy problem with initial function 0, is due to Tychonoff [1]. The one in section 5 is generally known, but may not have been published before.

In the case of the heat equation, Theorem 3.1 has been improved by Watson [6, Theorem 12], who showed that if $u \in U_a$ and, as $(x, t) \rightarrow (y, 0^+)$, $\limsup u(x, t) < \infty$ for all $y \in \underline{R}^n$ and $\limsup u(x, t) \leq C$ for almost all $y \in \underline{R}^n$, then $u \leq C$. Theorem 3.1 was proved for $n = 1$ by Gehring [1, Theorem 3].

For the heat equation on $\underline{R}^n \times \,]0, a[$, the semigroup property has been characterized in another way by Watson [11, Theorem 10]. It is equivalent to the property that, for each $b \in \,]0, a[$,

$$W(x, b - t)u(x, t) \rightarrow 0 \quad \text{as} \quad \|x\| \rightarrow \infty \tag{34}$$

uniformly for t in any compact subset of $]0, b[$. We can use this result, along with one due to Gusarov [1, Theorem 6], to weaken the hypothesis of non-negativity in Theorem 2.14. Specialized to the case of temperatures on $\underline{R}^n \times]0, a[$, Gusarov's result is as follows.

Suppose that $0 < T < a$ and there are positive constants C and δ such that

$$| u(x,t) | \leq C \exp(\delta \| x \|^2) \tag{35}$$

for all $(x,t) \in \underline{R}^n \times]0,T]$. Then there is a constant β with the property that, if there are positive constants λ and ρ such that

$$| u(x,T) | \leq \lambda \exp(-\beta(\delta + T^{-1}) \| x \|^2)$$

whenever $\| x \| > \rho$, then $u = 0$ on $\underline{R}^n \times]0,T]$.

We now show that (35) can be weakened to include all non-negative solutions, so that the resultant theorem contains Theorem 2.14 (with a slight loss of precision in the constant γ) as well as Gusarov's result.

Let $u \in S_a$ and let $t_0 \in]0, a[$. Then there is a constant γ with the property that, if there are positive constants λ and ρ such that

$$| u(x,t_0) | \leq \lambda \exp(-\gamma \| x \|^2) \tag{36}$$

whenever $\| x \| > \rho$, then $u = 0$ on $\underline{R}^n \times]0, a[$.

Proof. Whenever $0 < c < t_0/2$ and $t_0 < b < a$, it follows from the characterization of temperatures with the semigroup property that (34) holds uniformly for t in $[c, t_0]$. We can therefore find $\rho > 0$ such that, whenever $\| x \| > \rho$ and $t \in [c, t_0]$,

$$| u(x,t) | \leq (4\pi(b - t))^{n/2} \exp(\| x \|^2/4(b - t))$$

$$\leq C \exp(\delta \| x \|^2)$$

with $C = (4\pi(b - c))^{n/2}$ and $\delta = 1/4(b - t_0)$. It now follows from Gusarov's result that, if (36) holds with $\gamma \geq \beta(\delta + (t_0 - c)^{-1})$, then $u = 0$ on $\underline{R}^n \times]c, t_0]$. Since $0 < c < t_0/2$, we have $(t_0 - c)^{-1} \leq (t_0/2)^{-1}$ for all c, so that if (36) holds with $\gamma \geq \beta(\delta + (2/t_0))$, then $u = 0$ on $\underline{R}^n \times]c, t_0]$ for all $c \in]0, t_0/2[$, and hence $u = 0$ on $\underline{R}^n \times]0, t_0]$.

By the semigroup property,

$$u(x,t) = \int_{\underline{R}^n} W(x-y, t-t_0) u(y, t_0) \, dy = 0$$

for all $(x,t) \in \underline{R}^n \times \,]t_0, a[$, and the proof is complete.

This answers a question raised by Watson [7], but only for the case of the heat equation.

The equivalence of (8) and the semigroup property was noted by Watson [8], but only for temperatures.

For the heat equation, Watson [17] has shown that a much weaker condition on the means M_b (than that in the definition of S_a) still suffices to imply the semigroup property.

The fact that the semigroup property implies $M_b(|u|; \cdot)$ is non-increasing was noticed by Rosenbloom and Widder [1, Theorem 9.1], for the heat equation when $n = 1$. They also saw that the local bounded-ness of $M_b(|u|; \cdot)$ on $]0, b[$ for every $b < a$ is equivalent to the semigroup property [1, Theorem 10.2].

Uniqueness classes for the Cauchy problem have been found by many authors under various conditions on the coefficients. The first to get near to the largest uniqueness class, for the heat equation with $n = 1$, was Tychonoff [1], whose class contained those temperatures u which satisfied $|u(x,t)| \le C \exp(c\|x\|^2)$ for some constants C and c. Of the more recent papers on this topic, we mention only those which fall into one of three categories.

The first category consists of papers closely related to the results in the text, with only one-sided conditions on the solution. For those papers concerned only with non-negative solutions, see the bibliographical notes in Chapter II. For equations with smooth, unbounded coefficients, Besala and Krzyżański [1] proved uniqueness under the condition

$$u(x,t) \ge -C \exp(c\|x\|^2) \tag{37}$$

where C and c are arbitrary positive constants. Apparently unaware of this, Foias and Nicolescu [1] proved uniqueness, and the Gauss-Weier-strass representation of the solution, under the same condition, but only for the heat equation with $n = 1$. In a later paper, Foias and Nicolescu [2] improved their results by relaxing (37) to

$$\sup_{0 \le t < a} \int_{\underline{R}^n} u^-(x,t) \exp(-c\|x\|^2) \, dx \le C \tag{38}$$

In this form, the results were extended to arbitrary n by Mustață [1].

At the same time, Foias and Nicolescu [3] proved uniqueness, again
for the heat equation with n = 1, under the condition

$$\int_0^a dt \int_{\underline{R}^n} u^-(x,t) \exp(-c\|x\|^2) \, dx < \infty \tag{39}$$

which is equivalent to (23). This was insufficient to imply that the solution was a Gauss-Weierstrass integral, but they showed that it had the
semigroup property. Uniqueness, under condition (39), was then proved
for equations with smooth, unbounded coefficients by Mustaţă [2], still
with n = 1. For arbitrary n, and with slightly weaker conditions on the
coefficients, Chabrowski [1, 6] proved uniqueness under condition (39)
together with

$$\int_{\underline{R}^n} u^-(x,0) \exp(-c\|x\|^2) \, dx < \infty$$

by a different method. Next came Watson's papers [2,3], in which the
methods presented in the chapter were first used. More recently,
Watson [17] has shown that the condition that the means $M_b(u^+; \cdot)$ are
integrable over $]0,b[$, for all $b \in]0,a[$, can be weakened to

$$\int_0^c t^{-1/2} \log^+ M_b(u^+; t) \, dt < \infty$$

whenever $0 < c < b < a$, but the proof works only for the heat equation.
   The second category consists of papers related to the fundamental
work of Täcklind [1], and this includes some more with one-sided conditions on solutions of the Cauchy problem. Täcklind proved the following result.

   Let h be a positive function on $]1, \infty[$, and let C_h denote the class
of temperatures u on $\underline{R}^n \times]0, a[$ which satisfy

$$|u(x,t)| \leq \exp(\alpha \|x\| h(\|x\|)) \tag{40}$$

whenever $\|x\| > 1$ and $0 < t < a$, for some arbitrary constant α. In
order that C_h is a uniqueness class for the Cauchy problem, it is necessary and sufficient that

$$\int\limits_{1}^{\infty} \bar{h}(r)^{-1} \, dr = \infty \tag{41}$$

where $\bar{h}$ is the greatest non-decreasing minorant of h.

Zolotarev [1] extended the sufficiency part of this theorem to equations with smooth, bounded coefficients. Dressel [1] showed that condition (40) could be replaced by a similar condition on u^+ by proving, via an extended maximum principle (see Theorem 3.1), that such a condition implies that $u \leq 0$. He worked with the heat equation and $n = 1$ only. Stafney [1] showed that condition (40) is necessary for uniqueness for a certain class of equations with some coefficients depending on x, and with $n = 1$. Simultaneously, Hayne [1] and Oleinik and Radkevich [1] showed that smoothness conditions on the coefficients could be discarded when proving sufficiency. Hayne used a one-sided condition like that of Dressel, and substantially modified Täcklind's general approach. Oleinik and Radkevich used a completely different technique. Kamynin and Khimchenko [1, 2, 3] replaced condition (40) by

$$|u(x,t)| \leq C \, \exp(G(\|x\|) h(G(\|x\|))) \tag{42}$$

for all $(x,t) \in \underline{R}^n \times \,]0, a[$, where the function G satisfies one or other of two sets of conditions, and h is non-decreasing and satisfies (41) (with $\bar{h} = h$). The equations they considered could have unbounded coefficients which were not necessarily smooth. They also gave several examples to illustrate the sharpness of their results. The function G in (42) is related to conditions on the coefficients, so that the uniqueness class varies with the equation to some extent. Subsequently, Kamynin and Khimchenko [4] turned the problem around and, given a class of functions on $\underline{R}^n \times \,]0, a[$, asked for which family of equations it was a uniqueness class for the Cauchy problem. Eidel'man and Petrushko [1] showed that, if the initial function grows very rapidly, then there can be a solution of the Cauchy problem on an unbounded subset of the strip, and satisfying Tychonoff's condition there, while at the same time there is no solution on $\underline{R}^n \times \,[0, a[$ which satisfies (40) and (41).

The final category, of which some papers were mentioned in the previous paragraph, consists of uniqueness classes which are related to growth conditions on the coefficients. As an example, we quote an early result of Aronson and Besala [1]. Here we are dealing with a linear parabolic equation in the form

$$\sum_{i,j=1}^{n} D_i D_j(a_{ij}(x,t)u(x,t)) + \sum_{i=1}^{n} D_i(b_i(x,t)u(x,t)) + c(x,t)u(x,t) = D_t u(x,t)$$

$$(43)$$

under appropriate smoothness conditions on the coefficients.

If there are positive constants C and λ such that

$$|a_{ij}(x,t)| \leq C(\|x\|^2 + 1)^{(2-\lambda)/2}$$

$$|b_i(x,t)| \leq C(\|x\|^2 + 1)^{1/2}$$

$$c(x,t) \leq C(\|x\|^2 + 1)^{\lambda/2}$$

for all $(x,t) \in \underline{R}^n \times]0,a[$, then the class of all solutions u of (43) which satisfy

$$\int_0^a dt \int_{\underline{R}^n} \exp(-\alpha(\|x\|^2 + 1)^{\lambda/2}) |u(x,t)| \, dx < \infty$$

for some $\alpha \geq 0$, is a uniqueness class for the Cauchy problem on $\underline{R}^n \times]0,a[$.

Extensions of this result, and variations on the method of proof, were given by Aronson and Besala [1, 2, 3], Besala and Ugowski [1], Chabrowski [3, 4], and Besala [6]. Some of these papers also contain representation theorems for solutions of the Cauchy problem. Other results in this category were obtained by Eidel'man and Porper [1].

Another group of papers is related to both the Cauchy problem and the extended maximum principle (Theorem 3.1). In these, a solution u of the Cauchy problem in some uniqueness class is subjected to an extra condition on $u(\cdot,0)$, but not necessarily the condition $u(\cdot,0) \leq 0$ of the extended maximum principle. Estimates of the solution over the whole strip are then deduced. For example, Krzyżański [4] showed, under certain growth conditions on the coefficients, that if $|u(\cdot,0)| \leq M$ then

$$|u(x,t)| \leq M \exp(-\lambda\|x\|^2 \tanh(\gamma t) + \nu t)$$

for all $(x,t) \in \underline{R}^n \times]0,a[$, where λ, γ, and ν are constants which depend on the coefficients. Results of this kind were subsequently obtained by

Krzyzanski [5], Kusano [1, 2, 3], Chen [1], Chen and Kuroda [1], Eidel'man and Porper [1], Chen, Kuroda and Kusano [1, 2], Chen, Lin and Yeh [1], and Cosner [1].

Representation theorems for solutions of the Cauchy problem imply uniqueness of solution, and references to papers on this topic have been included above. Works dealing with representation theorems for non-negative solutions were mentioned in the bibliographical notes in Chapter II. For the heat equation with $n = 1$, Theorem 3.15 was almost proved by Gehring [1], except that instead of (i) he had the condition that $M_b(|u|; \cdot)$ is bounded on $]0, b[$ for every $b < a$, and in (ii) he had $|u|$ not u^+. His methods were generally similar to those in the text, which come from Watson's papers [2, 3]. Other works on the representation of solutions of equations with variable coefficients include those of Guenther [1], Johnson [1], Chabrowski [8], and Chabrowski and Johnson [1]. These contain results for classes of solutions comparable to R_a. Representation theorems for certain particular classes of solutions will be discussed in Chapter VI. For the heat equation, Watson [17] has shown that the local integrability condition in the definition of R_a can be considerably weakened.

4

Initial Limits of
Gauss-Weierstrass Integrals

We shall prove that the Gauss-Weierstrass integral of a signed measure μ has what is called a 'parabolic limit' at almost every point of $\underline{R}^n \times \{0\}$. To do this, we shall use a notion of differentiability of the initial measure μ. We shall then look at the possibility of improving this result by replacing the parabolic approach to the boundary with a less restricted one. It turns out that no simple improvement is possible, but that a slightly better result can be obtained if approach through a complicated region is allowed, at least, in the case where μ is absolutely continuous with respect to Lebesgue measure. To achieve this, an argument involving what are called 'maximal functions' is used.

1. PARABOLIC LIMITS

We begin by recalling some facts about differentiation of measures. We shall abbreviate 'almost everywhere with respect to μ' to 'μ-a. e.', and 'all x outside a set of μ-measure zero' to 'μ-almost all x'. We denote Lebesgue measure on $\underline{R}^n$ by m.

Let μ be a signed measure on $\underline{R}^n$. For m-almost all $x \in \underline{R}^n$, the limit

$$(D\mu)(x) = \lim_{r \to 0^+} \frac{\mu(B(x, r))}{m(B(x, r))} \tag{1}$$

exists and is finite. This is well-known in the case where μ has finite total variation, and the result in the general case can be deduced from this, as follows. Given $N \in \underline{N}$, define the measure μ_N by putting

$$\mu_N(E) = \mu(E \cap B(0, N))$$

for every Borel set $E \subseteq \underline{R}^n$. Then μ_N has finite total variation, so that $(D\mu_N)(x)$ exists and is finite for m-almost all $x \in B(0, N)$. But if $x \in B(0, N)$, then $\mu_N(B(x, r)) = \mu(B(x, r))$ for all $r < N - \|x\|$, so it follows that $(D\mu)(x)$ exists and is finite for m-almost all $x \in B(0, N)$. Since N is arbitrary, our assertion now follows.

Whenever it exists, $(D\mu)(x)$ is called the <u>symmetric derivative</u> of μ at x. The word 'symmetric' refers to the fact that the sets $B(x, r)$ on the right-hand side of (1) are radially symmetric about the point x. Stronger results, in which the balls $B(x, r)$ in (1) are replaced by more general sets containing x, are not necessary here. The function $D\mu$, defined m-a. e. by (1), is the absolutely continuous part of μ, so that we can write

$$d\mu(y) = (D\mu)(y) \, dy + d\sigma(y)$$

where σ is a signed measure which is singular with respect to m, so that its total variation $|\sigma|$ satisfies

$$(D|\sigma|)(x) = 0 \tag{2}$$

for m-almost all $x \in \underline{R}^n$. The measure σ is called the <u>singular part</u> of μ. Any point x such that

$$m(B(x, r))^{-1} \int_{B(x, r)} |(D\mu)(y) - (D\mu)(x)| \, dy \to 0 \quad \text{as} \quad r \to 0 \tag{3}$$

is called a <u>Lebesgue point</u> of $D\mu$, and the set of all such points is the <u>Lebesgue set</u> of $D\mu$. In fact, (3) holds for m-almost all x.

Suppose that $0 < a \leq \infty$, that u is a function on $\underline{R}^n \times]0, a[$, and that $\ell \in \underline{R}$. We say that u has a <u>parabolic limit</u> ℓ at a point $x \in \underline{R}^n$ if, given any positive numbers ϵ and α, there is $\delta > 0$ such that

$$|u(z,t) - \ell| < \epsilon$$

whenever $\|z - x\|^2 < \alpha t$ and $0 < t < \delta$. Here, δ will generally depend on α as well as ϵ.

We shall prove that, if u is the Gauss-Weierstrass integral of a signed measure μ, then μ has a parabolic limit $(D\mu)(x)$ at x, for every x such that (2) and (3) both hold, and hence for m-almost all $x \in R^n$. However, there is not an exact pointwise relationship between $D\mu$ and the parabolic limits of u, as we shall demonstrate by giving an example where $D\mu$ exists at a point but u has no parabolic limit there, and another where u has a parabolic limit at a point but $D\mu$ does not exist there.

We begin with the positive result.

Theorem 4.1

Let u be the Gauss-Weierstrass integral on $R^n \times]0,a[$ of a signed measure μ, and let σ denote the singular part of μ. At every point x such that

$$(D|\sigma|)(x) = 0 \tag{4}$$

and

$$m(B(x,r))^{-1} \int_{B(x,r)} |(D\mu)(y) - (D\mu)(x)| \, dy \to 0 \quad \text{as} \quad r \to 0 \tag{5}$$

u has a parabolic limit $(D\mu)(x)$. Thus u has a parabolic limit at x for m-almost every $x \in R^n$, and if $d\mu(y) = f(y) \, dy$ then this parabolic limit equals $f(x)$ m-a. e.

Proof. Put $\phi(y) = (D\mu)(y)$ at every point y where the symmetric derivative exists and is finite. Let x be any point where (4) and (5) hold. Then, given a positive ϵ, we can find $r_0 > 0$ such that

$$r^{-n}\left(\int_{B(x,r)} |\phi(y) - \phi(x)| \, dy + |\sigma|(B(x,r)) \right) < \epsilon \tag{6}$$

whenever $0 < r \leq 2r_0$.

The first step is to show that, inside any parabolic region $\{ (z,t) : \|z - x\|^2 < \alpha t \}$, the behavior of u(z,t) as $t \to 0$ is controlled by the behavior as $(x, t) \to (x, 0^+)$ of a related Gauss–Weierstrass integral. A parabolic region is the "largest" for which this is possible, and this is why parabolic limits are studied.

Given $\alpha > 0$, for all $(z, t) \in \underline{R}^n \times]0, a[$ such that $\|z - x\|^2 < \alpha t$, and all $y \in \underline{R}^n$, we have

$$\|y - x\|^2 \le 2(\|z - x\|^2 + \|y - z\|^2) < 2\alpha t + 2\|y - z\|^2$$

Hence

$$\exp(-\|y - x\|^2/8t) \ge \exp(-\|z - y\|^2/4t) e^{-\alpha/4} \tag{7}$$

Furthermore, for any $(z, t) \in \underline{R}^n \times]0, a[$, it follows from the decomposition $d\mu(y) = \phi(y)\, dy + d\sigma(y)$ and Lemma 1.1 that

$$|u(z, t) - \phi(x)| = \left| \int_{\underline{R}^n} W(z - y, t)\phi(y)\, dy - \phi(x) + \int_{\underline{R}^n} W(z - y, t)\, d\sigma(y) \right|$$

$$= \left| \int_{\underline{R}^n} W(z - y, t)(\phi(y) - \phi(x))\, dy + \int_{\underline{R}^n} W(z - y, t)\, d\sigma(y) \right|$$

$$\le \int_{\underline{R}^n} W(z - y, t) |\phi(y) - \phi(x)|\, dy + \int_{\underline{R}^n} W(z - y, t)\, d|\sigma|(y)$$

Therefore, whenever $\|z - x\|^2 < \alpha t$ and $0 < t < a/2$, (7) shows that

$$|u(z, t) - \phi(x)|$$

$$\le 2^{n/2} e^{\alpha/4} \left(\int_{\underline{R}^n} W(x - y, 2t) |\phi(y) - \phi(x)|\, dy + \int_{\underline{R}^n} W(x - y, 2t)\, d|\sigma|(y) \right) \tag{8}$$

Given any $t \in]0, r_0^2[$, let N be the smallest integer which exceeds $r_0 t^{-1/2}$, so that $N\sqrt{t} > r_0$. Then, for any non-negative measure ν whose Gauss–Weierstrass integral is defined on $\underline{R}^n \times]0, a[$, we have

$$\int_{\underline{R}^n} W(x - y, 2t)\, d\nu(y)$$

$$\le \sum_{j=1}^{N} \int_{(j-1)\sqrt{t} \le \|x-y\| \le j\sqrt{t}} W(x - y, 2t)\, d\nu(y) + \int_{\|x-y\| > r_0} W(x - y, 2t)\, d\nu(y) \tag{9}$$

By Lemma 1.5, the last integral tends to zero with t. Furthermore

$$
\sum_{j=1}^{N} \int_{(j-1)\sqrt{t} \le \|x-y\| \le j\sqrt{t}} W(x - y, 2t) \, d\nu(y)
$$

$$
\le (8\pi)^{-n/2} \sum_{j=1}^{N} \exp\left(-\frac{(j-1)^2}{8}\right) t^{-n/2} \int_{(j-1)\sqrt{t} \le \|x-y\| \le j\sqrt{t}} d\nu(y)
$$

$$
\le (8\pi)^{-n/2} \sum_{j=1}^{N} j^n \exp\left(-\frac{(j-1)^2}{8}\right) \nu(B(x, j\sqrt{t}))/(j\sqrt{t})^n \qquad (10)
$$

Since $(N - 1)\sqrt{t} \le r_0 < N\sqrt{t}$ by our choice of N, and $\sqrt{t} < r_0$ by our restriction on t, we see that $N - 1 \ge 1$ and

$$
j\sqrt{t} \le (N - 1)\sqrt{t} + \sqrt{t} \le 2(N - 1)\sqrt{t} \le 2r_0
$$

for all $j \le N$. It therefore follows from (6) that, if $d\nu(y) = |\phi(y) - \phi(x)| \, dy$ or $d\nu(y) = d|\sigma|(y)$,

$$
\nu(B(x, j\sqrt{t}))/(j\sqrt{t})^n < \epsilon
$$

for all $j \le N$. Hence the last expression in (10) does not exceed

$$
(8\pi)^{-n/2} \sum_{j=1}^{\infty} j^n \exp(-(j - 1)^2/8)\epsilon
$$

whenever $t < r_0^2$ and $t < a/2$. Since the series is convergent, we deduce that the sum on the right-hand side of (9) tends to zero with t. It follows that the integrals in (8) both tend to zero also, and hence (8) implies the result of the theorem.

We now give examples to show that there is not an exact pointwise relationship between the existence of a symmetric derivative and that of a parabolic limit.

Example. Let $n = 1$, let $f(y) = 1$ for all $y > 0$ and $f(y) = 0$ for all $y < 0$, and let $d\mu(y) = f(y) \, dy$. Then

$$
\mu(B(0, r)) = \frac{m(B(0, r))}{2}
$$

for all $r > 0$, so that $(D\mu)(0) = 1/2$. However, if u is the Gauss-Weierstrass integral of μ on $\underline{R} \times]0,\infty[$, then u does not have a parabolic limit at 0. For, if u had such a limit ℓ, then we would have

$$u(2\alpha\sqrt{t}, t) \rightarrow \ell \quad \text{as} \quad t \rightarrow 0^+ \tag{11}$$

for every constant α. But, for all $(x,t) \in \underline{R} \times]0,\infty[$,

$$u(x,t) = (4\pi t)^{-1/2} \int_0^\infty \exp\left(-\frac{(x-y)^2}{4t}\right) dy$$

$$= (4\pi t)^{-1/2} \int_{-x/2\sqrt{t}}^\infty \exp(-z^2)2\sqrt{t} \, dz$$

(where we have made the substitution $z = (y-x)/2\sqrt{t}$), so that

$$u(2\alpha\sqrt{t}, t) = \pi^{-1/2} \int_{-\alpha}^\infty \exp(-z^2) \, dz$$

which varies with α but not t. Thus (11) cannot hold for any fixed ℓ, and so u does not have a parabolic limit at 0.

Example. Let $n = 1$, and choose a sequence $\{b_k\}$ such that $0 < b_k < 2^{-k}$ for all $k \in \underline{N}$ and the series $\Sigma \, 2^k b_k$ converges. Put $\alpha(k) = 2^{-k}$ and $\beta(k) = 2^{-k} + b_k$ for all k, and define a measure μ on $\underline{R}$ by

$$\mu = \sum_{k=1}^\infty 2^{-k}(\delta_{\alpha(k)} - \delta_{\beta(k)})$$

where δ_x denotes the Dirac δ-measure concentrated at x. Then

$$|\mu|(\underline{R}) = 2 \sum_{k=1}^\infty 2^{-k} = 2$$

The symmetric derivative of μ does not exist at 0, since on the one hand we have

$$\frac{\mu(]-\alpha(k), \alpha(k)[)}{m(]-\alpha(k), \alpha(k)[)} = \frac{1}{2\alpha(k)} \sum_{j=k+1}^\infty (2^{-j} - 2^{-j}) = 0$$

while on the other hand we have

$$\frac{\mu(\,]-\beta(k)\,,\,\beta(k)\,[\,)}{m(\,]-\beta(k)\,,\,\beta(k)\,[\,)} \;=\; \frac{1}{2\beta(k)}\left(2^{-k} + \sum_{j=k+1}^{\infty} (2^{-j} - 2^{-j})\right)$$

$$=\; \frac{2^{-k-1}}{2^{-k} + b_k}$$

$$>\; \frac{2^{-k-1}}{2^{-k+1}}$$

$$=\; 1/4$$

for all k, and both $\{\alpha(k)\}$ and $\{\beta(k)\}$ are null sequences.

Despite this, the Gauss-Weierstrass integral u of μ has a parabolic limit zero at 0. To see this, we first use the mean value theorem to show that, given any $(x,t) \in \underline{R} \times \,]0,\infty[$ there are $\gamma(k) \in \,]\alpha(k)\,,\,\beta(k)[$ such that

$$u(x,t) \;=\; \sum_{k=1}^{\infty} 2^{-k}(W(x - \alpha(k)\,,\,t) - W(x - \beta(k)\,,\,t))$$

$$=\; \sum_{k=1}^{\infty} 2^{-k}(\beta(k) - \alpha(k))D_1 W(x - \gamma(k)\,,\,t)$$

$$=\; \sum_{k=1}^{\infty} 2^{-k} b_k D_1 W(x - \gamma(k)\,,\,t) \tag{12}$$

We now require the fact that, given any positive κ,

$$\sup\{\,|\,D_1 W(x - 2y,t)\,|\,:\,x^2 \le \kappa t\} \;=\; 4^{-1}\sup\{|D_1 W(x - y,t)\,|\,:x^2 \le \kappa t\} \tag{13}$$

for an arbitrary $y \in \underline{R} \setminus \{0\}$. To prove (13) we first note that, if $\eta \ne 0$ and we put $\zeta = (x - \eta)/2\sqrt{t}$ in the formula

$$D_1 W(x - \eta,t) \;=\; -(4\pi)^{-1/2} 2^{-1}(x - \eta)t^{-3/2}\exp(-(x - \eta)^2/4t)$$

we get

$$|\,D_1 W(x - \eta,t)\,| \;=\; (4\pi)^{-1/2}t^{-1}|\,\zeta\,|\,\exp(-\zeta^2) \tag{14}$$

For any fixed ζ, the right-hand side of (14) is a decreasing function of t. Given any point (x_0, t_0) such that $x_0^2 < \kappa t_0$, there is a value ζ_0 such that $x_0 - \eta = 2\zeta_0\sqrt{t_0}$ and a point (x_1, t_1) such that $x_1 - \eta = 2\zeta_0\sqrt{t_1}$, $x_1^2 = \kappa t_1$, and $t_1 < t_0$, so that

$$|D_1 W(x_1 - \eta, t_1)| \geq |D_1 W(x_0 - \eta, t_0)|$$

Hence, for any $\eta \in \underline{R} \setminus \{0\}$,

$$\sup\{|D_1 W(x - \eta, t)| : x^2 \leq \kappa t\} = \sup\{|D_1 W(x - \eta, t)| : x^2 = \kappa t\}$$

It follows that, for any $y \in \underline{R} \setminus \{0\}$,

$$\sup\{|D_1 W(x - y, t)| : x^2 \leq \kappa t\}$$

$$= \sup\{(4\pi)^{-1/2} 2^{-1} |x - y| t^{-3/2} \exp(-(x - y)^2/4t) : x^2 = \kappa t\}$$

$$= \sup\{(4\pi)^{-1/2} 2^{-1} |x - y| (x^2/\kappa)^{-3/2} \exp(-(x - y)^2 \kappa/4x^2) : x \in \underline{R}\}$$

and similarly

$$\sup\{|D_1 W(x - 2y, t)| : x^2 \leq \kappa t\}$$

$$= \sup\{(4\pi)^{-1/2} 2^{-1} |x - 2y| (x^2/\kappa)^{-3/2} \exp(-(x - 2y)^2 \kappa/4x^2) : x \in \underline{R}\}$$

$$= \sup\{(4\pi)^{-1/2} 2^{-1} |2z - 2y| (4z^2/\kappa)^{-3/2} \exp(-(2z - 2y)^2 \kappa/16z^2) : z \in \underline{R}\}$$

$$= 4^{-1} \sup\{(4\pi)^{-1/2} 2^{-1} |z - y| (z^2/\kappa)^{-3/2} \exp(-(z - y)^2 \kappa/4z^2) : z \in \underline{R}\}$$

which proves (13). Using (13), and the fact that

$$2^{-k} = \alpha(k) < \gamma(k) < \beta(k) < 2^{1-k}$$

which implies that $2^k \gamma(k) \in]1, 2[$ for all $k \in \underline{N}$, we have

$$\sup\{|D_1 W(x - \gamma(k), t)| : x^2 \leq \kappa t\}$$

$$= 4^k \sup\{|D_1 W(x - 2^k \gamma(k), t)| : x^2 \leq \kappa t\}$$

$$\leq 4^k \sup\{|D_1 W(x - z, t)| : x^2 \leq \kappa t, \ 1 < z < 2\}$$

$$= C 4^k$$

where C depends only on κ. Since $\{b_k\}$ was chosen such that $\Sigma\, 2^k b_k$ is convergent, it follows from Weierstrass's M-test that the series in (12) is uniformly convergent on $\{(x,t):x^2 \le \kappa t\}$. Therefore, as $(x,t) \to (0,0^+)$ through this parabolic region, we have

$$\lim u(x,t) = \sum_{k=1}^{\infty} 2^{-k} b_k \, \lim D_1 W(x - \gamma(k),t) = 0$$

since $\gamma(k) \ne 0$ for all k. Since κ is arbitrary, we have proved that u has a parabolic limit zero at 0.

Remark. In Chapter V we shall establish a pointwise relationship between the upper and lower limits as $r \to 0$ of $\mu(B(x,r))/m(B(x,r))$, and those as $t \to 0^+$ (with x fixed) of $u(x,t)$, where u is the Gauss-Weierstrass integral of μ. Most of the chapter will be devoted to exploring the consequences of this relationship.

2. NEGATIVE RESULTS ON IMPROVING THE PARABOLIC LIMIT THEOREM

We show that Theorem 4.1 is almost the best result possible, in the sense that the limits from within (parabolic) regions of the form $\{(z,t): \|z-x\|^2 < \alpha t\}$ cannot in general be replaced by limits from within sets like $\{(z,t):\omega(\|z-x\|) < t\}$ if $\omega(r)/r^2 \to 0$ as $r \to 0$. The parabolic regions can be expressed in the form $\{(z,t):\omega(\|z-x\|) < t\}$ with $\omega(r)/r^2 = 1/\alpha$, and between this condition on ω and $\omega(r)/r^2 \to 0$ as $r \to 0$, there is a small gap within which some improvement on Theorem 4.1 is possible, as we shall show in section 4.

The first result of this section shows that, if the parabolic regions are widened as described, then we can no longer expect limits to exist m-a.e. The second result shows that, in fact, such limits can exist m-almost nowhere. For simplicity, we take n = 1 throughout.

Theorem 4.2

Let ω be a non-decreasing, positive function on $]0,1[$ such that $\omega(r)/r^2 \to 0$ as $r \to 0$, and for each $\xi \in \underline{R}$ let L_ξ denote the graph $\{(x,t):t = \omega(x-\xi),\ 0 < x-\xi < 1\}$. Given any $\lambda < 1$, we can find a compact set $A \subseteq [0,1]$ with $m(A) > \lambda$ such that, if

$$u(x,t) = \int_{\underline{R}\backslash A} W(x-y,t)\, dy$$

for all $(x,t) \in \underline{R} \times]0,\infty[$, then for m-almost every $\xi \in A$ we have

$$\lim \sup u(x,t) \geq (\pi^2 e)^{-1/4}$$

as $(x,t) \to (\xi,0)$ along L_ξ, and

$$u(\xi,0^+) = 0 \tag{15}$$

Proof. We first note that, whenever $0 < t \leq \eta$,

$$\int_{-\sqrt{\eta}}^{\sqrt{\eta}} W(y,t)\, dy \geq (4\pi t)^{-1/2} \int_{-\sqrt{t}}^{\sqrt{t}} \exp\left(-\frac{y^2}{4t}\right) dy > \pi^{-1/2} e^{-1/4} = \kappa \tag{16}$$

say. Thus, on the line segment $\{0\} \times]0,\eta]$, the Gauss-Weierstrass integral of the characteristic function of $]-\sqrt{\eta},\sqrt{\eta}[$ exceeds a positive quantity which does not depend on η.

Let A be a closed subset of $\underline{R}$. Suppose that, for some $\xi \in A$, there is a sequence of intervals $\{J_k\}$ such that

$$J_k =]\xi + \alpha_k, \xi + \beta_k[\subseteq \underline{R} \setminus A$$

where $0 < \alpha_k < \beta_k$ for all $k \in \underline{N}$, $\beta_k \to 0$ as $k \to \infty$, and

$$\omega((\alpha_k + \beta_k)/2) \leq (\beta_k - \alpha_k)^2/4 \tag{17}$$

for all $k \in \underline{N}$. We put

$$\gamma_k = (\alpha_k + \beta_k)/2$$

so that $\xi + \gamma_k$ is the mid-point of J_k. Then, whenever $0 < t \leq (\beta_k - \alpha_k)^2/4$, (16) implies that

$$\int_{\underline{R}\setminus A} W(\xi + \gamma_k - y,t)\, dy > \int_{J_k} W(\xi + \gamma_k - y,t)\, dy$$

$$= \int_{-(\beta_k-\alpha_k)/2}^{(\beta_k-\alpha_k)/2} W(z,t)\, dz$$

$$> \kappa$$

In particular, in view of (17), this holds when $t = \omega(\gamma_k)$. Thus, at the point $(\xi + \gamma_k, \omega(\gamma_k))$ where the perpendicular bisector of J_k meets L_ξ, the Gauss-Weierstrass integral of the characteristic function of $\underline{R} \setminus A$ exceeds κ, provided that (17) holds. Since this is true for all $k \in \underline{N}$, and $\gamma_k \to 0$ as $k \to \infty$, so that $(\xi + \gamma_k, \omega(\gamma_k)) \to (\xi, 0)$ along L_ξ, we have

$$\limsup_{\underline{R} \setminus A} \int W(x - y, t) \, dy \geq \kappa \qquad (18)$$

as $(x, t) \to (\xi, 0)$ along L_ξ.

Given $\lambda < 1$, we now construct a compact set $A \subseteq [0, 1]$ with $m(A) > \lambda$, such that every $\xi \in A$ has the property described in the previous paragraph. This will complete the proof of the theorem, since (18) will then hold as $(x, t) \to (\xi, 0)$ along L_ξ for every $\xi \in A$, and it follows from Theorem 4.1 that (15) holds for m-almost $\xi \in A$ because the characteristic function of $\underline{R} \setminus A$ is zero on A.

Since $\omega(r)/r^2 \to 0$ as $r \to 0$, we can find $N \in \underline{N}$ such that

$$n^2 \omega(1/n) < 1/4 \qquad (19)$$

for all $n \geq N$. Given any $n \geq N$, for each $i \in \underline{N}$ such that $i \leq n$, we denote by J_n^i the open interval with center $(2i - 1)/2n$ and length $2\omega(1/n)^{1/2}$. Since the distance between consecutive centers is $1/n$, and $2\omega(1/n)^{1/2} < 1/n$ by (19), the intervals $J_n^1, \ldots, J_n^j$ are pairwise disjoint subsets of $[0, 1]$. The set

$$C_n = \bigcup_{i=1}^{n} J_n^i$$

is open, and

$$m(C_n) = 2n\omega(1/n)^{1/2} = 2(\omega(1/n)/(1/n)^2)^{1/2}$$

This holds for every $n \geq N$. Since $\omega(r)/r^2 \to 0$ as $r \to 0$, we see that $m(C_n) \to 0$ as $n \to \infty$. We can therefore choose a sequence $\{n(j)\}$ in $[N, \infty[$, such that

$$m(C_{n(j)}) < (1 - \lambda)2^{-j}$$

for all $j \in \underline{N}$. Then the open set

$$C = \bigcup_{j=1}^{\infty} C_{n(j)}$$

has measure

$$m(C) \le \sum_{j=1}^{\infty} m(C_{n(j)}) < (1 - \lambda) \sum_{j=1}^{\infty} 2^{-j} = 1 - \lambda$$

Hence the set $A = [0, 1] \setminus C$ is compact, and $m(A) > \lambda$. We now show that every $\xi \in A$ has the required property. Let $\xi \in A$, let k be a fixed natural number, and let $]\xi + \alpha_k, \xi + \beta_k[$ denote that one of the intervals $J^1_{n(k)}, \ldots, J^{n(k)}_{n(k)}$ which is nearest to ξ and lies in $]\xi, \infty[$. Its length $\beta_k - \alpha_k$ is $2\omega(1/n(k))^{1/2}$, and the distance γ_k from its mid-point to ξ does not exceed the distance $1/n(k)$ from its mid-point to that of the next of the intervals $J^i_{n(k)}$. Therefore, since ω is non-decreasing

$$2\omega(\gamma_k)^{1/2} \le 2\omega(1/n(k))^{1/2} = \beta_k - \alpha_k$$

Hence (17) holds for each $k \in \underline{N}$. Hence the arbitrary $\xi \in A$ has the required property, and the result follows.

Theorem 4.3

Let ω be a non-decreasing positive function on $]0, 1[$ such that $\omega(r)/r^2 \to 0$ as $r \to 0$, and for each $\xi \in \underline{R}$ let L_ξ denote the graph $\{(x, t) : t = \omega(x - \xi), 0 < x - \xi < 1\}$. There is a bounded function f on $\underline{R}$ whose Gauss-Weierstrass integral u satisfies both

$$\lim \sup u(x, t) > f(\xi) \tag{20}$$

as $(x, t) \to (\xi, 0)$ along L_ξ, and

$$u(\xi, 0^+) = f(\xi) \tag{21}$$

for m-almost all $\xi \in \underline{R}$.

Proof. We first show that (20) and (21) can both hold for m-almost all $\xi \in [0, 1]$. By Theorem 4.2, for each $k \in \underline{N}$ we can find a compact set $A_k \subseteq [0, 1]$ with $m(A_k) > 1 - k^{-1}$, such that if

$$w_k(x, t) = 4^{-k} \int_{\underline{R} \backslash A_k} W(x - y, t) \, dy$$

for all $(x, t) \in \underline{R} \times]0, \infty[$, then for m-almost every $\xi \in A_k$

$$\limsup w_k(x, t) \geq 4^{-k} (\pi^2 e)^{-1/4} \tag{22}$$

as $(x, t) \to (\xi, 0)$ along L_ξ, and

$$w_k(\xi, 0^+) = 0 \tag{23}$$

Let χ_k denote the characteristic function of $\underline{R} \backslash A_k$ for each k, and put

$$g = \sum_{k=1}^{\infty} 4^{-k} \chi_k$$

on $\underline{R}$. Since $4^{-k} \chi_k \leq 4^{-k}$ for all k, and

$$\sum_{k=1}^{\infty} 4^{-k} = 1/3 \tag{24}$$

the series defining g is uniformly convergent on $\underline{R}$, by Weierstrass's M-test, and $0 \leq g(x) \leq 1/3$ for all x. Therefore the Gauss-Weierstrass integral w of g is defined on $\underline{R} \times]0, \infty[$ and

$$w = \sum_{k=1}^{\infty} w_k$$

For each $k \in \underline{N}$, let

$$E_k = \{\xi \in A_k : (22) \text{ or } (23) \text{ does not hold}\}$$

and put

$$E = \bigcup_{k=1}^{\infty} E_k, \qquad S = \left(\bigcup_{k=1}^{\infty} A_k\right) \backslash E$$

Then $m(E_k) = 0$ for all k, so that $m(E) = 0$. Therefore

$$m(S) = m\left(\bigcup_{k=1}^{\infty} A_k\right) \geq m(A_\ell) > 1 - \ell^{-1}$$

for all $\ell \in \underline{N}$, so that $m(S) = 1$. Given any $\eta \in S$, let

$$j = \min\{k : \eta \in A_k\}$$

Then η belongs to the open set

$$\underline{R} \setminus \left(\bigcup_{k=1}^{j-1} A_k\right)$$

on which each of the functions $\chi_1, \cdots, \chi_{j-1}$ is identically equal to 1. Therefore, for all $k \leq j - 1$, $w_k(x, t) \to 4^{-k}$ as $(x, t) \to (\eta, 0)$ along L_η, in view of Theorem 1.3, Corollary. Furthermore, since $\eta \in A_j \setminus E$, we have

$$\limsup_{} w_j(x, t) \geq 4^{-j}(\pi^2 e)^{-1/4}$$

as $(x, t) \to (\eta, 0)$ along L_η, and $w_j(\eta, 0^+) = 0$. Using the non-negativity of w_k, it therefore follows that, as $(x, t) \to (\eta, 0)$ along L_η,

$$\limsup w(x, t) = \limsup\left(\sum_{k=1}^{j-1} w_k(x, t) + w_j(x, t) + \sum_{k=j+1}^{\infty} w_k(x, t)\right)$$

$$\geq \lim \sum_{k=1}^{j-1} w_k(x, t) + \limsup w_j(x, t)$$

$$\geq \sum_{k=1}^{j-1} 4^{-k} + 4^{-j}(\pi^2 e)^{-1/4} \qquad (25)$$

Also, using the fact that $w_k \leq 4^{-k}$ for all k (by Lemma 1.1), we have

$$\limsup_{t \to 0^+} w(\eta, t) \le \sum_{k=1}^{j-1} 4^{-k} + w_j(\eta, 0^+) + \sum_{k=j+1}^{\infty} 4^{-k}$$

$$= \sum_{k=1}^{j-1} 4^{-k} + 4^{-j}3^{-1} \tag{26}$$

Since $\pi^2 e < 3^4$, the right-hand side of (26) is less than that of (25). Hence, as $(x, t) \to (\eta, 0)$ along L_η,

$$\limsup w(x, t) > \limsup_{s \to 0^+} w(\eta, s)$$

By Theorem 4.1, $w(\xi, 0^+) = g(\xi)$ for m-almost all $\xi \in \underline{R}$. Since η is an arbitrary member of S, it follows that

$$\limsup w(x, t) > g(\eta) \tag{27}$$

as $(x, t) \to (\eta, 0)$ along L_η, for m-almost all $\eta \in S$, and hence for m-almost all $\eta \in [0, 1]$, since $m(S) = 1$.

To obtain an example where (20) and (21) hold for m-almost all $\xi \in \underline{R}$, we construct a function with period 1 which coincides with g on $]0, 1[$, as follows. Since $A_k \subseteq [0, 1]$ for each k, $\chi_k(x) = 1$ for all $x \in \underline{R} \setminus [0, 1]$, so it follows from (24) that $g(x) = 1/3$ for all $x \in \underline{R} \setminus [0, 1]$. Therefore, if we define f on the non-integer real numbers by

$$f(x) = 3^{-1} - \sum_{i=-\infty}^{\infty} (3^{-1} - g(x + i))$$

then for each x only one term of the series is non-zero. Since $0 \le g \le 1/3$ on $\underline{R}$, it follows that $0 \le f \le 1/3$ m-a. e. on $\underline{R}$. Given any integer ℓ, for all $y \in]\ell, \ell + 1[$ we have

$$f(y) = 3^{-1} - (3^{-1} - g(y - \ell)) = g(y - \ell) \tag{28}$$

so that

$$\int_{\underline{R}} W(x - y, t)(f(y) - g(y - \ell)) \, dy = \int_{\underline{R} \setminus]\ell, \ell+1[} W(x - y, t)(f(y) - g(y - \ell)) \, dy$$

If $\xi \in]\ell, \ell + 1[$, then since $0 \le f, g \le 3^{-1}$ on $\underline{R}$,

$$\left| \int_{\underline{R} \backslash]\ell,\, \ell+1[} W(x - y, t)(f(y) - g(y - \ell))\, dy \right|$$

$$\leq 3^{-1} \int_{|\xi - y| > \min\{\xi - \ell,\, \ell + 1 - \xi\}} W(x - y, t)\, dy$$

$$\to 0$$

as $(x, t) \to (\xi, 0^+)$, by Lemma 1.2. Hence, as $(x, t) \to (\xi, 0)$ along L_ξ, we have

$$\limsup_{\underline{R}} \int W(x - y, t) f(y)\, dy = \limsup_{\underline{R}} \int W(x - y, t) g(y - \ell)\, dy$$

$$= \limsup_{\underline{R}} \int W(x - \ell - z, t) g(z)\, dz$$

$$= \limsup\, w(x - \ell, t)$$

$$> g(\xi - \ell)$$

$$= f(\xi)$$

for m-almost all $\xi \in\,]\ell, \ell + 1[$, by (27) and (28). Since ℓ is arbitrary, and the set of integers has Lebesgue measure zero, we have proved that (20) holds for m-almost all $\xi \in \underline{R}$. The fact that (21) holds m-a. e. on $\underline{R}$ follows from Theorem 4.1.

3. MAXIMAL FUNCTIONS

Section 4 will contain an improvement on the parabolic limit theorem for Gauss-Weierstrass integrals of functions. The proof will require results about certain maximal functions, and these are the subject of the present section. Theorem 4.1 can be proved using maximal functions, but the proof we have given is shorter and more direct.

For the present, it is sufficient to consider the half-space $\underline{R}^n \times\,]0, \infty[$.

To each set $\Omega \subseteq \underline{R}^n \times\,]0, \infty[$, we associate the following maximal functions M_Ω and M_Ω^W.

(i) For each locally integrable function f on $\underline{R}^n$, and each $x_0 \in \underline{R}^n$, we put

$$M_\Omega f(x_0) = \sup \left\{ (4\pi t)^{-n/2} \int_{B(x, \sqrt{t})} |f(x_0 + y)|\, dy : (x, t) \in \Omega \right\}$$

(ii) For each function f on $\underline{R}^n$ whose Gauss-Weierstrass integral is defined on $\underline{R}^n \times]0, \infty[$, and each $x_0 \in \underline{R}^n$, we put

$$M_\Omega^W f(x_0) = \sup \left\{ \int_{\underline{R}^n} W(x_0 + x - y, t) \, | f(y) | \, dy : (x, t) \in \Omega \right\}$$

Notation. For each $t > 0$, we put

$$\Omega(t) = \{ x \in \underline{R}^n : (x, t) \in \Omega \}$$

where Ω is any subset of $\underline{R}^n \times]0, \infty[$.

Lemma 4.1

If $\Omega \subseteq \underline{R}^n \times]0, \infty[$, and $\Omega(t) \subseteq \Omega(\tau)$ whenever $0 < t < \tau$, then there is a constant C, which depends only on n, such that

$$M_\Omega^W f(x_0) \leq C M_\Omega f(x_0) \tag{29}$$

for all $x_0 \in \underline{R}^n$.

Proof. Let $(x, t) \in \Omega$. Then, for any $x_0 \in \underline{R}^n$,

$$\int_{\underline{R}^n} W(x_0 + x - z, t) \, | f(z) | \, dz$$

$$= \int_{\underline{R}^n} W(y, t) \, | f(x_0 + x - y) | \, dy$$

$$= \sum_{k=0}^{\infty} \int_{k\sqrt{t} \leq \|y\| < (k+1)\sqrt{t}} W(y, t) \, | f(x_0 + x - y) | \, dy \tag{30}$$

Now, for each $k \geq 0$,

$$\int_{k\sqrt{t} \leq \|y\| < (k+1)\sqrt{t}} W(y, t) \, | f(x_0 + x - y) | \, dy$$

$$\leq (4\pi t)^{-n/2} \exp(-(k\sqrt{t})^2/4t) \int_{\|y\| < (k+1)\sqrt{t}} | f(x_0 + x - y) | \, dy$$

$$= (k + 1)^n \exp(-k^2/4)(4\pi(k + 1)^2 t)^{-n/2} \int\limits_{\|x-z\|< (k+1)\sqrt{t}} |f(x_0 + z)| \, dz$$

$$\leq (k + 1)^n \exp(-k^2/4) M_\Omega f(x_0)$$

where $(x, (k + 1)^2 t) \in \Omega$ since $(x, t) \in \Omega$ and we have assumed that $\Omega(t) \subseteq \Omega(\tau)$ whenever $t < \tau$. It now follows from (30) that

$$\int\limits_{\underline{R}^n} W(x_0 + x - z, t)|f(z)| \, dz \leq M_\Omega f(x_0) \sum_{k=0}^\infty (k + 1)^n \exp(-k^2/4)$$

$$= C M_\Omega f(x_0)$$

where C denotes the sum of the series, which depends only on n. Taking the supremum over Ω of the left-hand side, we obtain (29).

Definitions. Let Ω be an open subset of $\underline{R}^n \times \,]0, \infty[$.

(i) We put $t_0 = \inf\{t > 0 : \Omega(t) \neq \emptyset\}$.

(ii) For each $t > 0$, we put

$$Q(t) = \bigcup_{x \in \Omega(t)} B(x, \sqrt{t}).$$

(iii) We say that Ω satisfies a <u>para–cone condition with aperture</u> α if, whenever $(x, t) \in \Omega$, $\{(\xi, \tau) : \|\xi - x\| < \alpha(\sqrt{\tau} - \sqrt{t}), \tau > t\} \subseteq \Omega$.

The sets $Q(t)$ will be used in the definition of a third maximal function, Nf, below. When considering Nf in relationship to M_Ω^W, we shall need the following properties of the sets $Q(t)$ and $\Omega(t)$.

Lemma 4.2

Let Ω be an open subset of $\underline{R}^n \times \,]0, \infty[$. Suppose that there is a positive constant β such that

$$m(\Omega(t)) \leq \beta t^{n/2} \quad \text{for all} \quad t > 0 \tag{31}$$

and that Ω satisfies a para–cone condition with aperture α. Put $\gamma = (\alpha + 1)/\alpha$ and $\omega_n = m(B(0, 1))$.

(i) If $0 < t < \tau$, then $\Omega(t) \subseteq \Omega(\tau)$ and $Q(t) \subseteq Q(\tau)$.

(ii) If $t > 0$, then $Q(t) \subseteq \Omega(\gamma^2 t)$.

(iii) If $t > 0$, then $m(Q(t)) \leq \beta(\gamma^2 t)^{n/2}$.

(iv) For all $t > t_0$, $\Omega(t) \neq \emptyset$ and $m(Q(t)) \geq \omega_n t^{n/2}$.

(v) If M denotes the greatest integer not exceeding $(\beta/\omega_n)\gamma^n$, then for each $t > 0$ there are points $x_1, \ldots, x_M \in \Omega(t)$ such that

$$Q(t) \subseteq \bigcup_{j=1}^{M} B(x_j, 3\sqrt{t})$$

(vi) There exists ξ_0 such that $\xi_0 \in Q(t)$ for all $t > t_0$.

Proof.

(i) Since Ω satisfies a para-cone condition, if $(x,t) \in \Omega$ and $\tau > t$ then $(x, \tau) \in \Omega$. Thus $\Omega(t) \subseteq \Omega(\tau)$ if $t < \tau$, and hence $Q(t) \subseteq Q(\tau)$.

(ii) If $y \in Q(t)$, then there exists $x \in \Omega(t)$ such that $y \in B(x, \sqrt{t})$. Then

$$\|y - x\| < \sqrt{t} = \alpha(\sqrt{\gamma^2 t} - \sqrt{t})$$

Therefore, because $(x,t) \in \Omega$ and Ω satisfies a para-cone condition with aperture α, $(y, \gamma^2 t) \in \Omega$. Since y is arbitrary in $Q(t)$, we have proved that $Q(t) \subseteq \Omega(\gamma^2 t)$.

(iii) It follows from (ii) and the condition (31) that

$$m(Q(t)) \leq m(\Omega(\gamma^2 t)) \leq \beta(\gamma^2 t)^{n/2}$$

(iv) By definition of t_0, there is a decreasing sequence $\{t_j\}$ such that $\Omega(t_j) \neq \emptyset$ and $t_j \to t_0$. If $t > t_0$, then $t > t_j$ for some $j \in \underline{N}$, so that, since $\Omega(t_j) \neq \emptyset$, it follows from (i) that $\Omega(t) \neq \emptyset$. Therefore, whenever $t > t_0$, $Q(t)$ contains a ball of radius $\sqrt{t}$, so that $m(Q(t)) \geq \omega_n t^{n/2}$.

(v) If $x_1, \ldots, x_p$ are points of $\Omega(t)$ such that $\|x_i - x_j\| \geq 2\sqrt{t}$ whenever $i \neq j$, then the balls $B(x_1, \sqrt{t}), \ldots, B(x_p, \sqrt{t})$ are disjoint subsets of $Q(t)$. Therefore, by (iii),

$$p\omega_n t^{n/2} = m\left(\bigcup_{j=1}^{p} B(x_j, \sqrt{t})\right) \leq m(Q(t)) \leq \beta(\gamma^2 t)^{n/2}$$

so that

$$p \le (\beta/\omega_n)\gamma^n \tag{32}$$

In particular, there can be only finitely many such points, so that we can find a collection $\{x_1, \ldots, x_k\}$ of points of $\Omega(t)$ such that $\|x_i - x_j\| \ge 2\sqrt{t}$ whenever $i \ne j$, and no collection of $k + 1$ points in $\Omega(t)$ has this property. It follows from (32) that $k \le (\beta/\omega_n)\gamma^n$. Given any $y \in Q(t)$, we can find $x \in \Omega(t)$ such that $y \in B(x, \sqrt{t})$. Then there is x_ℓ such that $\|x - x_\ell\| < 2\sqrt{t}$, for otherwise $\{x_1, \ldots, x_k, x\}$ would be a collection of $k + 1$ points in $\Omega(t)$ each of which is distant at least $2\sqrt{t}$ from all the others, which is impossible. Hence

$$\|y - x_\ell\| \le \|y - x\| + \|x - x_\ell\| < 3\sqrt{t}$$

that is, $y \in B(x_\ell, 3\sqrt{t})$. Since y is arbitrary in $Q(t)$, it follows that

$$Q(t) \subseteq \bigcup_{i=1}^{k} B(x_i, 3\sqrt{t})$$

and since $k \le M$ the result is proved.

(vi) It follows from the definition of t_0 that there is a sequence $\{(x_i, t_i)\}$ in Ω such that $\{t_i\}$ decreases to t_0. We first show that the sequence $\{x_i\}$ is bounded. Suppose, on the contrary, that it is unbounded. Then we can find a subsequence $\{(x_{i(j)}, t_{i(j)})\}$ of $\{(x_i, t_i)\}$ such that

$$\|x_{i(j+1)}\| \ge 2\|x_{i(j)}\| > 4\alpha \tag{33}$$

and

$$t_0 < t_{i(j)} < t_0 + j^{-1} \tag{34}$$

for all $j \in \underline{N}$. For each such j, the set

$$P_j = \{(\xi, \tau) : \|\xi - x_{i(j)}\| < \alpha(\sqrt{\tau} - \sqrt{t_{i(j)}}), \tau > t_{i(j)}\}$$

is contained in Ω, because Ω satisfies a para-cone condition with aperture α, and meets $\underline{R}^n \times \{t_0 + 1\}$ in the ball $B_j = B(x_{i(j)}, \alpha(\sqrt{t_0 + 1} - \sqrt{t_{i(j)}}))$. In view of (34),

$$\alpha(\sqrt{t_0+1}-\sqrt{t_{i(j)}}) \le \alpha\sqrt{t_0+1-t_{i(j)}} < \alpha$$

and in view of (33) the points of the sequence $\{x_{i(j)}\}$ are more than 2α apart. Therefore the balls B_j are disjoint, so that the sets $P_j \cap (R^n \times]t_0,t_0+1[)$ are also disjoint. This implies that, for each $t \in]t_0,t_0+1[$, if k denotes the smallest integer such that $t_{i(k)} < t$,

$$m(\Omega(t)) \ge m\left(\bigcup_{j=k}^{\infty} B(x_{i(j)}, \alpha(\sqrt{t}-\sqrt{t_{i(j)}}))\right)$$

$$= \sum_{j=k}^{\infty} m(B(x_{i(j)}, \alpha(\sqrt{t}-\sqrt{t_{i(j)}})))$$

Therefore, because $\{t_{i(j)}\}$ is decreasing,

$$m(\Omega(t)) \ge \omega_n \alpha^n \sum_{j=k}^{\infty} (\sqrt{t}-\sqrt{t_{i(j)}})^n \ge \omega_n \alpha^n \sum_{j=k}^{\infty} (\sqrt{t}-\sqrt{t_{i(k)}})^n = \infty$$

This contradicts (31), and so the sequence $\{x_i\}$ must be bounded. Hence $\{x_i\}$ has a subsequence $\{x_{i(\ell)}\}$ which converges to a point $\xi_0 \in \underline{R}^n$. By the para-cone condition, each set

$$\{(\xi,\tau): \|\xi - x_{i(\ell)}\| < \alpha(\sqrt{\tau}-\sqrt{t_{i(\ell)}}), \tau > t_{i(\ell)}\}$$

is contained in Ω, and it contains $\{\xi_0\} \times](\sqrt{t_{i(\ell)}} + \alpha^{-1}\|\xi_0 - x_{i(\ell)}\|)^2, \infty[$. Therefore, since $t_{i(\ell)} \to t_0$ and $x_{i(\ell)} \to \xi_0$, Ω contains $\{\xi_0\} \times]t_0,\infty[$.

Definition. Let Ω be an open subset of $R^n \times]0,\infty[$. For each locally integrable function f on $\underline{R}^n$, and each $x_0 \in \underline{R}^n$, we put

$$Nf(x_0) = \sup\left\{m(Q(t))^{-1} \int_{Q(t)} |f(x_0+y)| \, dy : t > t_0\right\}$$

Here, of course, t_0 and $Q(t)$ are as defined just before Lemma 4.2.

We now use some parts of Lemma 4.2 to show that Nf dominates the other maximal function $M_\Omega^W f$.

Lemma 4.3

Let Ω be an open subset of $\underline{R}^n \times \ ]0, \infty[$. Suppose that there is a positive constant β such that

$$m(\Omega(t)) \leq \beta t^{n/2} \quad \text{for all} \quad t > 0$$

and that Ω satisfies a para-cone condition with aperture α. Then there is a constant κ, which depends only on α, β, and n, such that

$$M_\Omega^W f(x_0) \leq \kappa N f(x_0) \tag{35}$$

for all $x_0 \in \underline{R}^n$.

Proof. If $0 < t < \tau$, then $\Omega(t) \subseteq \Omega(\tau)$ by Lemma 4.2(i), so that Lemma 4.1 is applicable. Thus (35) will follow if we prove that

$$M_\Omega f(x_0) \leq C N f(x_0) \tag{36}$$

for all $x_0 \in \underline{R}^n$, where C depends only on α, β and n. If $(x,t) \in \Omega$, then $B(x, \sqrt{t}) \subseteq Q(t)$, so that, for all $x_0 \in \underline{R}^n$,

$$(4\pi t)^{-n/2} \int_{B(x,\sqrt{t})} |f(x_0 + y)| \ dy \leq (4\pi t)^{-n/2} \int_{Q(t)} |f(x_0 + y)| \ dy$$

By Lemma 4.2(iii), $m(Q(t)) \leq \beta(\gamma^2 t)^{n/2}$, where $\gamma = (\alpha + 1)/\alpha$. Furthermore, since Ω is open and $(x,t) \in \Omega$, $t > t_0$. Hence

$$(4\pi t)^{-n/2} \int_{B(x,\sqrt{t})} |f(x_0 + y)| \ dy$$

$$\leq (4\pi t)^{-n/2} \beta(\gamma^2 t)^{n/2} \sup \left\{ m(Q(\tau))^{-1} \int_{Q(\tau)} |f(x_0 + y)| \ dy : \tau > t_0 \right\}$$

$$= (\gamma^2/4\pi)^{n/2} \beta N f(x_0)$$

for all $x_0 \in \underline{R}^n$. Taking the supremum over Ω of the left-hand side, we obtain (36), which implies (35).

One final lemma about the sets $Q(t)$, and then we can prove the estimate for $M_\Omega^W f$ which is essential for improving the parabolic limit theorem.

Lemma 4.4

Let Ω be an open subset of $R^n \times]0,\infty[$, such that $\Omega(s) \subseteq \Omega(t)$ whenever $0 < s < t$, and let $x,y \in R^n$. Suppose that $0 < s \leq t$, that $\{x\} + Q(t)$ meets $\{y\} + Q(s)$, and that there are points $z_1, \dots, z_p$ in $\Omega(t)$ such that

$$Q(t) \subseteq \bigcup_{\ell=1}^{p} B(z_\ell, 3\sqrt{t})$$

Then, if we put

$$\tilde{Q}(t) = \bigcup_{i=1}^{p} \bigcup_{j=1}^{p} \bigcup_{k=1}^{p} B(z_i - z_j + z_k, 9\sqrt{t}) \tag{37}$$

we have the inclusion

$$\{y\} + Q(s) \subseteq \{x\} + \tilde{Q}(t)$$

Proof. We have to show that, to each $\xi \in Q(s)$, there corresponds $\eta \in \tilde{Q}(t)$ such that $y + \xi = x + \eta$. Since $\{x\} + Q(t)$ meets $\{y\} + Q(s)$, there are points $a \in Q(t)$ and $b \in Q(s)$ such that $x + a = y + b$, or $y - x = a - b$. Therefore the result will follow if we show that, given any $\xi \in Q(s)$,

$$\xi + a - b \in \tilde{Q}(t) \tag{38}$$

Since $\Omega(s) \subseteq \Omega(t)$ our hypothesis concerning the points $z_1, \dots, z_p$ implies that

$$Q(s) \subseteq Q(t) \subseteq \bigcup_{\ell=1}^{p} B(z_\ell, 3\sqrt{t})$$

since $0 < s \leq t$. Therefore, since $a \in Q(t)$ and $\xi, b \in Q(s)$, we can find indices $\ell(0)$, $\ell(1)$ and $\ell(2)$ such that

$$\| \xi - z_{\ell(0)} \| < 3\sqrt{t}, \quad \| a - z_{\ell(1)} \| < 3\sqrt{t}, \quad \| b - z_{\ell(2)} \| < 3\sqrt{t}$$

The point

$$\xi + a - b = (\xi - z_{\ell(0)}) + z_{\ell(0)} + (a - z_{\ell(1)}) + z_{\ell(1)} - (b - z_{\ell(2)}) - z_{\ell(2)}$$

$$= z_{\ell(1)} - z_{\ell(2)} + z_{\ell(0)} + (\xi - z_{\ell(0)}) + (a - z_{\ell(1)}) - (b - z_{\ell(2)})$$

can now be seen to belong to $\tilde{Q}(t)$, because its distance from $z_{\ell(1)} - z_{\ell(2)} + z_{\ell(0)}$ is

$$\| (\xi - z_{\ell(0)}) + (a - z_{\ell(1)}) - (b - z_{\ell(2)}) \| < 9\sqrt{t}$$

This proves (38), and the result now follows.

Theorem 4.4

Let Ω be an open subset of $\underline{R}^n \times]0, \infty[$. Suppose that there is a positive constant β such that

$$m(\Omega(t)) \le \beta t^{n/2} \quad \text{for all} \quad t > 0$$

and that Ω satisfies a para-cone condition with aperture α. Then there is a constant C, which depends only on α, β and n, such that

$$m(\{x \in \underline{R}^n : M_\Omega^W f(x) > \ell\}) \le C \ell^{-1} \int_{\underline{R}^n} |f(y)| \, dy \tag{39}$$

for every $\ell > 0$ and every function f which is integrable over $\underline{R}^n$.

Proof. It is sufficient to prove the result with $M_\Omega^W f$ replaced by Nf, for if this is done, then Lemma 4.3 gives

$$m(\{x : M_\Omega^W f(x) > \ell\}) \le m(\{x : Nf(x) > \ell \kappa^{-1}\}) \le C \kappa \ell^{-1} \int_{\underline{R}^n} |f(y)| \, dy$$

for every $\ell > 0$ and integrable f. (The sole purpose of proving Lemma 4.3 is to show that we can work with Nf rather than $M_\Omega^W f$ here, and the only reason for proving Lemma 4.1 was to use it to obtain Lemma 4.3.)

By Lemma 4.2(vi), there exists ξ_0 such that $\xi_0 \in Q(t)$ for all $t > t_0$. The result is unaffected if we replace Ω by $\Omega - \{(\xi_0, 0)\}$ throughout, so there is no loss of generality in assuming that $0 \in Q(t)$ for all $t > t_0$.

Let f be integrable over $\underline{R}^n$, and for each $\ell > 0$ put

$$E_\ell = \{x \in \underline{R}^n : Nf(x) > \ell\}$$

We must show that

$$m(E_\ell) \leq C\ell^{-1} \int_{\underline{R}^n} |f(y)| \, dy \tag{40}$$

for some constant C which depends only on α, β and n. If $x \in E_\ell$, then the definition of Nf shows that there is $t_x > t_0$ such that

$$\int_{Q(t_x)} |f(x + y)| \, dy > m(Q(t_x))\ell$$

Therefore, if we put

$$\Phi(x) = \{x\} + Q(t_x)$$

we have

$$\int_{\Phi(x)} |f(y)| \, dy > m(Q(t_x))\ell \tag{41}$$

Since $t_x > t_0$, we have $0 \in Q(t_x)$, and hence $x \in \Phi(x)$. Therefore E_ℓ is covered by the family of open sets $\{\Phi(x) : x \in E_\ell\}$. Let F be an arbitrary compact subset of E_ℓ. Then we can find a finite subfamily

$$\mathscr{F} = \{\Phi(x(j)) : j = 1, \ldots, k\}$$

which covers F.

We select a disjoint subfamily of $\mathscr{F}$ in the following way. Since $\mathscr{F}$ contains finitely many sets, there is at least one set $\Phi(x)$ in $\mathscr{F}$ such that $t_x \geq t_{x(j)}$ for $j = 1, \ldots, k$. Choose one such set, and label it $\Phi(x(j(1)))$. From those members of $\mathscr{F}$ which do not meet $\Phi(x(j(1)))$, we choose $\Phi(x(j(2)))$ with the largest value of $t_{x(j)}$. Continuing this process, we obtain a collection of disjoint sets

$$\mathscr{G} = \{\Phi(x(j(r))) : r = 1, \ldots, q\}$$

such that each set $\Phi(x(j))$ in $\mathscr{F}$ meets some set $\Phi(x(j(r)))$ with $t_{x(j)} \leq t_{x(j(r))}$.

Let $i \in \{1, \ldots, k\}$, and denote by $\Phi(x(j(s)))$ a set, chosen from $\mathscr{G}$, such that $\Phi(x(i))$ meets $\Phi(x(j(s)))$ and $t_{x(i)} \leq t_{x(j(s))}$. Applying Lemma 4.2(v) with $t = t_{x(j(s))}$, we see that, if p denotes the greatest integer not exceeding $(\beta/\omega_n)(\alpha + 1)^n \alpha^{-n}$, there are points $z_1, \ldots, z_p \in \Omega(t_{x(j(s))})$ such that

$$Q(t_{x(j(s))}) \subseteq \bigcup_{\lambda=1}^{p} B(z_\lambda, 3\sqrt{t_{x(j(s))}})$$

Let $\widetilde{Q}(t_{x(j(s))})$ be defined by (37) with $t = t_{x(j(s))}$. We can now apply Lemma 4.4 (with $s = t_{x(i)}$ and $t = t_{x(j(s))}$), and deduce that

$$\Phi(x(i)) = \{x(i)\} + Q(t_{x(i)}) \subseteq \{x(j(s))\} + \widetilde{Q}(t_{x(j(s))})$$

Since this holds for every i and the sets in $\mathscr{F}$ cover F, it follows that

$$F \subseteq \bigcup_{r=1}^{q} (\{x(j(r))\} + \widetilde{Q}(t_{x(j(r))}))$$

Furthermore, each set $\widetilde{Q}(t_{x(j(r))})$ is, by (37), the union of p^3 balls of radius $9\sqrt{t_{x(j(r))}}$, so that

$$m(F) \leq \sum_{r=1}^{q} m(\widetilde{Q}(t_{x(j(r))})) \leq \sum_{r=1}^{q} p^3 \omega_n (9\sqrt{t_{x(j(r))}})^n$$

It now follows from Lemma 4.2(iv) and (41) that

$$m(F) \leq p^3 9^n \sum_{r=1}^{q} m(Q(t_{x(j(r))}))$$

$$\leq p^3 9^n \sum_{r=1}^{q} \left(\ell^{-1} \int_{\Phi(x(j(r)))} |f(y)| \, dy\right)$$

Since the sets in $\mathscr{G}$ are disjoint, we deduce that

$$m(F) \leq p^3 9^n \ell^{-1} \int_{\underline{R}^n} |f(y)| \, dy$$

Now, F is an arbitrary compact subset of E_ℓ, so it follows that (40) holds. The proof is now complete.

4. AN IMPROVEMENT OF THE PARABOLIC LIMIT THEOREM

We use Theorem 4.4 to obtain an improvement of Theorem 4.1 for the special case of Gauss-Weierstrass integrals of functions. The general case of Gauss-Weierstrass integrals of measures is dealt with in Chapter 8.

Definition. Let S be a set, (ξ, τ) a limit point of S, and u a function on S. If $u(x, t) \to \ell$ as $(x, t) \to (\xi, \tau)$ with $(x, t) \in S$, we say that $u(x, t) \to \ell$ as $(x, t) \to (\xi, \tau)$ through S.

As in section 3, if $\Omega \subseteq \underline{R}^n \times]0, \infty[$ and $t > 0$, we put $\Omega(t) = \{x \in \underline{R}^n : (x, t) \in \Omega\}$.

Theorem 4.5

Let Ω be an open subset of $\underline{R}^n \times]0, \infty[$. Suppose that there is a positive number β such that

$$m(\Omega(t)) \leq \beta t^{n/2} \qquad \text{for all} \quad t > 0$$

that Ω satisfies a para-cone condition with aperture α, and that $(0, 0)$ is a limit point of Ω. Suppose also that $a > 0$, and that u is the Gauss-Weierstrass integral on $\underline{R}^n \times]0, a[$ of a function f on $\underline{R}^n$. Then

$$u(x, t) \to f(\xi)$$

as $(x, t) \to (\xi, 0)$ through $(\Omega + \{(\xi, 0)\}) \cap (\underline{R}^n \times]0, a[)$, for m-almost all $\xi \in \underline{R}^n$.

Proof. We first show that it is sufficient to prove the result in the special case where f is integrable over $\underline{R}^n$. Suppose that this has been done, and that ϕ is a function on $\underline{R}^n$ whose Gauss-Weierstrass integral is defined at $(0, 1/4b)$ for some $b > 0$. Then

$$\int_{\underline{R}^n} \exp(-b\|y\|^2) |\phi(y)| \, dy < \infty$$

so that ϕ is locally integrable on $\underline{R}^n$. For each $N \in \underline{N}$, let ϕ_N denote the restriction of ϕ to $B(0, N)$, and put $\psi_N = \phi - \phi_N$. Then

$$\int_{\underline{R}^n} W(x - y, t) \phi(y) \, dy = \int_{\underline{R}^n} W(x - y, t) \phi_N(y) \, dy + \int_{\underline{R}^n} W(x - y, t) \psi_N(y) \, dy$$

for all $(x, t) \in \underline{R}^n \times \,]0, 1/4b[$, in view of Theorem 1.2. Since ϕ is locally integrable on $\underline{R}^n$, ϕ_N is integrable over $\underline{R}^n$, so that

$$\int_{\underline{R}^n} W(x - y, t) \phi_N(y) \, dy \rightarrow \phi_N(\xi) = \phi(\xi)$$

as $(x, t) \rightarrow (\xi, 0)$ through $\Omega + \{(\xi, 0)\}$, for m-almost all $\xi \in B(0, N)$. Furthermore, since $\psi_N(\xi) = 0$ for all $\xi \in B(0, N)$, given any such ξ we have

$$\int_{\underline{R}^n} W(x - y, t) \psi_N(y) \, dy = \int_{\|y - \xi\| > N - \|\xi\|} W(x - y, t) \psi_N(y) \, dy$$

and the latter integral tends to 0 as $(x, t) \rightarrow (\xi, 0^+)$, by Lemma 1.5. Hence

$$\int_{\underline{R}^n} W(x - y, t) \phi(y) \, dy \rightarrow \phi(\xi) \tag{42}$$

as $(x, t) \rightarrow (\xi, 0)$ through $\Omega + \{(\xi, 0)\}$, for m-almost all $\xi \in B(0, N)$. Since the integral in (42) does not depend on N, N is arbitrary, and a countable union of sets of measure zero has measure zero, we deduce that (42) holds as $(x, t) \rightarrow (\xi, 0)$ through $\Omega + \{(\xi, 0)\}$ for m-almost all $\xi \in \underline{R}^n$. We may therefore suppose that f is integrable over $\underline{R}^n$.

Let g be any function which is integrable over $\underline{R}^n$, and let v be its Gauss-Weierstrass integral, which is defined on $\underline{R}^n \times \,]0, \infty[$ in view of Theorem 1.2, Corollary. For each $\xi \in \underline{R}^n$, we denote by $\omega g(\xi)$ the oscillation of $v(x, t)$ as $(x, t) \rightarrow (\xi, 0)$ through $\Omega + \{(\xi, 0)\}$, that is

$$\omega g(\xi) = \lim \sup v(x, t) - \lim \inf v(x, t)$$

Given any $\delta > 0$, we can find functions h and h_c on $\underline{R}^n$ such that

$$f = h + h_c$$

$h_c \in C_c(\underline{R}^n)$, and

$$\int_{\underline{R}^n} |h(y)| \ dy < \delta \qquad (43)$$

(Rudin [1], p. 68). Let w and w_c denote the Gauss–Weierstrass integrals on $\underline{R}^n \times]0,\infty[$ of h and h_c, respectively. By Theorem 1.1, $w_c(x,t) \to h_c(\xi)$ as $(x,t) \to (\xi,0^+)$ for every $\xi \in \underline{R}^n$, so that

$$\omega h_c(\xi) = 0 \quad \text{for all} \quad \xi \in \underline{R}^n \qquad (44)$$

In view of (43), we can apply Theorem 4.4 to h and deduce that

$$m(\{\xi \in \underline{R}^n : M_\Omega^W h(\xi) > \ell\}) \leq C \ell^{-1}\delta \qquad (45)$$

for every $\ell > 0$, where C depends only on α, β and n. Next, for all $\xi \in \underline{R}^n$,

$$\omega h(\xi) \leq 2 \sup\{|w(x,t)| : (x,t) \in \Omega + \{(\xi,0)\}\}$$
$$\leq 2 \sup\left\{\int_{\underline{R}^n} W(x-y,t)|h(y)| \ dy : (x,t) \in \Omega + \{(\xi,0)\}\right\}$$
$$= 2 M_\Omega^W h(\xi)$$

Therefore (45) implies that, for all $\ell > 0$

$$m(\{\xi \in \underline{R}^n : \omega h(\xi) > \ell\}) \leq m(\{\xi : M_\Omega^W h(\xi) > \ell/2\}) \leq C \ell^{-1}\delta$$

Since (44) shows that, for all $\xi \in \underline{R}^n$,

$$\omega f(\xi) \leq \omega h(\xi) + \omega h_c(\xi) = \omega h(\xi)$$

it follows that

$$m(\{\xi \in \underline{R}^n : \omega f(\xi) > \ell\}) \leq m(\{\xi : \omega h(\xi) > \ell\}) \leq C \ell^{-1}\delta$$

for every $\ell > 0$. Since ωf does not depend on δ, and δ can be made arbitrarily small, we deduce that

$$m(\{\xi \in \underline{R}^n : \omega f(\xi) > \ell\}) = 0$$

for every $\ell > 0$. Hence

$$m(\{\xi \in \underline{R}^n : \omega f(\xi) > 0\}) \leq \sum_{k=1}^{\infty} m(\{\xi \in \underline{R}^n : \omega f(\xi) > 1/k\}) = 0$$

so that $\omega f(\xi) = 0$ for m-almost all ξ. Thus $u(x,t)$ tends to a limit as $(x,t) \to (\xi,0)$ through $\Omega + \{(\xi,0)\}$ for m-almost every $\xi \in \underline{R}^n$.

We conclude the proof by showing that these limits equal $f(\xi)$ m-a.e. on $\underline{R}^n$. Since we have assumed that $(0,0)$ is a limit point of Ω, there is a sequence $\{(x_j, t_j)\}$ in Ω which converges to $(0,0)$. For each j,

$$\{0\} \times](\sqrt{t_j} + \alpha^{-1}\|x_j\|)^2, \infty[\subseteq \{(z,s) : \|z - x_j\| < \alpha(\sqrt{s} - \sqrt{t_j})\} \subseteq \Omega$$

because Ω satisfies a para-cone condition with aperture α. Therefore, since $x_j \to 0$ and $t_j \to 0$, $\{0\} \times]0,\infty[\subseteq \Omega$. Hence $\{\xi\} \times]0,\infty[\subseteq \Omega + \{(\xi,0)\}$ for all $\xi \in \underline{R}^n$, and so it follows from Theorem 4.1 that, as $(x,t) \to (\xi,0)$ through $\Omega + \{(\xi,0)\}$,

$$\lim u(x,t) = u(\xi,0^+) = f(\xi)$$

for m-almost all $\xi \in \underline{R}^n$.

Remark. In the last paragraph, we showed that $\{0\} \times]0,\infty[\subseteq \Omega$ whenever Ω satisfies the conditions in Theorem 4.5. In fact, Ω contains a paraboloid with vertex at $(0,0)$, because

$$\{(z,s) : \|z\| < \alpha\sqrt{s}\} = \bigcup_{i=1}^{\infty} \{(z,s) : \|z\| < \alpha(\sqrt{s} - \sqrt{i^{-1}}), \ s > i^{-1}\}$$

and, since $(0, i^{-1}) \in \Omega$ for all $i \in \underline{N}$ and Ω satisfies a para-cone condition with aperture α, the union on the right-hand side is a subset of Ω. Because the choice of α is unrestricted, Theorem 4.5 therefore implies that u has a parabolic limit $f(\xi)$ for m-almost all $\xi \in \underline{R}^n$. Thus Theorem 4.5 implies the result of Theorem 4.1 for the case of Gauss-Weierstrass integrals of functions.

We shall conclude this section with examples to show that Ω need not be contained in any paraboloid. First, we give a lemma which contains the essential points.

Lemma 4.5

Let $\{(x_k, t_k)\}$ be a sequence in $\underline{R}^n \times]0, \infty[$ such that $\{t_k\}$ decreases to 0 and $\|x_{k+1}\| = \sqrt{t_k}$ for all $k \in \underline{N}$. If

$$\Omega = \bigcup_{k=1}^{\infty} \{(y, s) : \|y - x_k\| < \sqrt{s} - \sqrt{t_k}, \ s > t_k\} \tag{46}$$

then there is a positive constant β such that

$$m(\Omega(t)) \leq \beta t^{n/2} \quad \text{for all} \quad t > 0 \tag{47}$$

and Ω satisfies a para-cone condition with aperture 1.

Proof. Given any positive t, let $N = \min\{k : t_k < t\}$. If

$$(x, t) \in \{(y, s) : \|y - x_k\| < \sqrt{s} - \sqrt{t_k}, \ s > t_k\}$$

for some $k \in \underline{N}$, then $t > t_k$ and so $k \geq N$. Therefore

$$\Omega(t) \subseteq \bigcup_{k=N}^{\infty} \{y : \|y - x_k\| < \sqrt{t} - \sqrt{t_k}\} \tag{48}$$

But, whenever $k > N$, the monotonicity of $\{t_j\}$ implies that

$$\|x_k\| = \sqrt{t_{k-1}} \leq \sqrt{t_N} < \sqrt{t}$$

Therefore, whenever $\|y - x_k\| < \sqrt{t} - \sqrt{t_k}$ and $k > N$,

$$\|y\| \leq \|y - x_k\| + \|x_k\| < (\sqrt{t} - \sqrt{t_k}) + \sqrt{t} < 2\sqrt{t}$$

It now follows from (48) that

$$\Omega(t) \subseteq B(x_N, \sqrt{t} - \sqrt{t_N}) + \bigcup_{k=N+1}^{\infty} B(x_k, \sqrt{t} - \sqrt{t_k})$$

$$\subseteq B(x_N, \sqrt{t}) + B(0, 2\sqrt{t})$$

Hence

$$m(\Omega(t)) \le m(B(x_N, \sqrt{t})) + m(B(0, 2\sqrt{t})) = \omega_n(1 + 2^n)t^{n/2}$$

and (47) is established.

Finally, if $(y, s) \in \Omega$, there is $k \in \underline{N}$ such that

$$\|y - x_k\| < \sqrt{s} - \sqrt{t_k}, \qquad s > t_k$$

Therefore, whenever $\xi \in \underline{R}^n$, $\tau > s$, and $\|\xi - y\| < \sqrt{\tau} - \sqrt{s}$, we have

$$\|\xi - x_k\| \le \|\xi - y\| + \|y - x_k\| < (\sqrt{\tau} - \sqrt{s}) + (\sqrt{s} - \sqrt{t_k}) = \sqrt{\tau} - \sqrt{t_k}$$

and $\tau > s > t_k$. Thus

$$\{(\xi, \tau) : \|\xi - y\| < \sqrt{\tau} - \sqrt{s}, \ \tau > s\} \subseteq \{(\xi, \tau) : \|\xi - x_k\| < \sqrt{\tau} - \sqrt{t_k}, \ \tau > t_k\} \subseteq \Omega$$

so that Ω satisfies a para-cone condition with aperture 1.

Examples. In Lemma 4.5, the sequence $\{(x_k, t_k)\}$ can be chosen to approach $(0, 0)$ with any prescribed degree of tangency, despite Theorem 4.3. Let ω be a positive function on $]0, 1[$ such that $\omega(r)/r^2 \to 0$ as $r \to 0$. Then there is $r_0 > 0$ such that $\omega(r) \le r^2/4$ for all $r \in]0, r_0]$. Choose x_1 such that $\|x_1\| = r_0$, and put $t_1 = \omega(\|x_1\|)$. We now choose (x_{k+1}, t_{k+1}) inductively, for all $k \in \underline{N}$. Given (x_k, t_k), we choose x_{k+1} such that $\|x_{k+1}\| = \sqrt{t_k}$, and then put $t_{k+1} = \omega(\|x_{k+1}\|)$. For $k = 1$, we then have

$$\|x_2\|^2 = t_1 = \omega(\|x_1\|) = \omega(r_0) \le r_0^2/4 = \|x_1\|^2/4$$

so that $\|x_2\| \le \|x_1\|/2 \le r_0$. This is extended inductively, as follows. If $\|x_k\| \le r_0$, then

$$\|x_{k+1}\|^2 = t_k = \omega(\|x_k\|) \le \|x_k\|^2/4$$

so that $\|x_{k+1}\| \le \|x_k\|/2 \le r_0$. It follows that $\{\|x_k\|\}$ decreases to 0, and therefore the same is true of $\{t_k\} = \{\|x_{k+1}\|^2\}$. Hence the conditions in Lemma 4.5 are satisfied, and $(x_k, t_k) \to (0, 0)$. Therefore, if Ω is defined by (46), Ω satisfies the conditions in Theorem 4.5. But $t_k = \omega(\|x_k\|)$ for all k, and ω can be chosen such that $\omega(r) \to 0$ arbitrarily rapidly as $r \to 0$.

As a concrete example let $\|x_k\| = \exp(-2^{k-2})$ and $t_k = \exp(-2^k)$ for all $k \in \underline{N}$. Then $\{t_k\}$ decreases to 0, and $\|x_{k+1}\| = \exp(-2^{k-1}) = \sqrt{\exp(-2^k)} = \sqrt{t_k}$ for all k. Hence $\{(x_k, t_k)\}$ satisfies the conditions in Lemma 4.5, but $t_k = \|x_k\|^4$.

5. LINEAR PARABOLIC EQUATIONS

The proof of Theorem 4.1 can be extended to more general equations, and requires only the standard upper estimate for the fundamental solution Γ, namely

$$0 \leq \Gamma(x, t; y, s) \leq C(t - s)^{-n/2} \exp(-c\|x - y\|^2/(t - s)) \tag{49}$$

Negative results are usually proved only for the heat equation, as their extension to other equations seems superfluous. Therefore the examples in section 1, and the whole of section 2, have not been generalized.

In the extension of the results of sections 3 and 4, only the standard upper estimate for the fundamental solution (49) is required. The maximal function $M_\Omega^W f(x_0)$ is replaced by

$$\sup \left\{ \int_{\underline{R}^n} C t^{-n/2} \exp(-c \|x_0 + x - y\|^2/t) \,|\, f(y)\,| \; dy : (x, t) \in \Omega \right\}$$

where the constants C and c are the same as in (49). The methods in the text can then be carried over almost unchanged.

6. BIBLIOGRAPHICAL NOTES

Results which assert the existence of limits at the boundary, such as Theorems 4.1 and 4.5, are often called Fatou theorems, after the mathematician who proved such a result for complex analytic functions.

The existence m-a. e. of parabolic limits of temperatures was first established by Doob [1, 2] and Gehring [2]. When n = 1, Doob [1] considered the quotient of two Gauss–Weierstrass integrals u and v, of a signed measure μ and a positive measure ν respectively, and proved that u/v has a parabolic limit ℓ at any point ξ such that $\mu(I_\xi)/\nu(I_\xi) \to \ell$ as $m(I_\xi) \to 0$, where I_ξ denotes an arbitrary interval containing ξ. Simultaneously, Gehring [2] proved the special case where $\nu = m$ of this result, again with n = 1, and also showed that if the symmetric

derivative $(D\mu)(\xi) = \ell$, then u has a non-tangential limit ℓ at ξ; that is,
given any positive numbers ϵ and α, there is $\delta > 0$ such that $|u(x,t) - \ell|$
$< \epsilon$ whenever $\|x - \xi\| < \alpha t$ and $0 < t < \delta$. Using a still stronger definition of the derivative of one measure with respect to another, Doob [2]
considered the case of a general n, and proved that, if u and v are the
Gauss-Weierstrass integrals of signed measures μ and ν, then u/v has
parabolic limits equal to the derivative of μ with respect to ν, $|\nu|$-a. e.
on $\underline{R}^n$. These papers were apparently overlooked for many years, and
Kato [1] subsequently proved that any positive temperature on $\underline{R}^n \times$
$]0, a[$ has a normal limit (i.e., a limit of $u(\xi, t)$ as $t \rightarrow 0^+$ with ξ fixed)
m-a. e. on $\underline{R}^n$, using a method given earlier by Carleson [1] for proving
a similar result on harmonic functions. This was, however, a good
thing for the study of general parabolic equations, because Kato's paper
was noticed by Chabrowski [2, 7] who observed that this method could
be applied much more generally, using the sharp lower estimate for
the fundamental solution. This was also observed by B. F. Jones, Jr.,
according to Aronson [6]. The first to prove the existence m-a. e. of
parabolic limits for solutions of a large class of parabolic equations,
was Johnson [1], who used a maximal function argument. The method
in the text was given by Chabrowski [8], and incorporates that of Kato.
For the heat equation, Korányi and Taylor [1] proved the parabolic
limit theorem using the potential theoretic concept of a fine limit. This
and related methods were extended to more general parabolic equations
by Mair [1], using the sharp lower estimate for the fundamental solution.
Details of potential theoretic concepts for the heat equation can be found
in Doob's book [3].

Watson [6, Theorem 11] proved that, if Z is any subset of $\underline{R}^n$ with
$m(Z) = 0$, then there exists a positive temperature u on $\underline{R}^n \times]0, \infty[$
such that $u(x,t) \rightarrow \infty$ as $(x,t) \rightarrow (\xi, 0^+)$ for all $\xi \in Z$. Thus finite parabolic limits can only exist m-a. e. on $\underline{R}^n$, and Theorem 4.1 is therefore
sharp in this respect.

The results in section 2 have not previously been published. They
are analogous to theorems obtained for harmonic functions by Littlewood [1] and Zygmund [1], and their proofs use adaptations of the methods of these two authors. The function ω need not be decreasing (compare
Zygmund [1]), but this extra hypothesis simplifies the proof.

The results in section 3 are special cases of those proved by Nagel
and Stein [1] in their study of the corresponding problem for harmonic
functions, except only that we have square-rooted the exceptional variable to adapt them to the parabolic case. Those in section 4 are based
on the same paper, but more substantial changes have been made. In
particular, Nagel and Stein considered only the Poisson integrals
(harmonic analogs of Gauss-Weierstrass integrals) of functions in L^p

with $1 \le p \le \infty$, whereas we have dealt with all Gauss-Weierstrass integrals. The work in these sections was done jointly with J. Chabrowski, and the details have not previously been published. The methods can similarly be applied to the solutions of systems of equations which are uniformly parabolic in the sense of Petrowskii, a concept defined in Chabrowski's survey [9].

For special initial functions, the parabolic limit theorem can be substantially improved. The obvious example is that of continuous functions. Chabrowski [10] has found other examples.

5

Normal Limits and Representation Theorems

We begin by proving a pointwise relationship between the upper and lower limits as $r \to 0$ of $\mu(B(x,r))/m(B(x,r))$, and those as $t \to 0^+$ of $u(x,t)$, where μ is a non-negative measure on $\underline{R}^n$, u is its Gauss-Weierstrass integral, and x is fixed. This motivates a study of the former limits, in which we establish that, if μ is singular with respect to Lebesgue measure, then $\mu(B(x,r))/m(B(x,r)) \to \infty$ as $r \to 0$, for μ-almost all $x \in \underline{R}^n$. This involves a sophisticated covering theorem, which we shall use again in Chapter VII. Our property of singular measures is then used to prove some special representation theorems for non-negative temperatures. These results then interact with the notion of a thermic majorant of the positive part of a temperature, to produce other representation theorems and a variant of the generalized maximum principle, both for temperatures which are not necessarily non-negative.

1. A FUNDAMENTAL RELATIONSHIP

The word <u>normal</u> is used in the sense of orthogonal. Thus, if u is a function on $\underline{R}^n \times]0, a[$ and $u(x, 0^+)$ exists for some x, then u is said to have a <u>normal limit</u> at x.

Given a non-negative measure μ on $\underline{R}^n$, and a point $x \in \underline{R}^n$, the <u>upper symmetric derivative</u> $(\bar{D}\mu)(x)$ of μ at x is defined by

$$(\bar{D}\mu)(x) = \lim_{r \to 0} \sup \frac{\mu(B(x, r))}{m(B(x, r))} \qquad (1)$$

and the <u>lower symmetric derivative</u> $(\underline{D}\mu)(x)$ by

$$(\underline{D}\mu)(x) = \lim_{r \to 0} \inf \frac{\mu(B(x, r))}{m(B(x, r))} \qquad (2)$$

Here, as usual, m denotes Lebesgue measure on $\underline{R}^n$. If the upper and lower symmetric derivatives of μ at x are equal, their common value is called the <u>symmetric derivative</u> of μ at x, and is denoted by $(D\mu)(x)$, as in Chapter IV.

Before proving the main result of this section, we show that the open balls in (1) and (2) can be replaced by closed balls, without affecting $\bar{D}\mu$ and $\underline{D}\mu$. We do this because the proof of the results in section 2 require that the balls are closed, whereas we used open balls in Chapter IV. Of course, we could have used closed balls throughout, but some authors use closed balls while others use open ones, so it is necessary to realize that this makes no difference to the results.

Lemma 5.1

Given a non-negative measure μ on $\underline{R}^n$, and a point $x \in \underline{R}^n$,

$$(\bar{D}\mu)(x) = \lim_{r \to 0} \sup \frac{\mu(\bar{B}(x, r))}{m(\bar{B}(x, r))}$$

and

$$(\underline{D}\mu)(x) = \lim_{r \to 0} \inf \frac{\mu(\bar{B}(x, r))}{m(\bar{B}(x, r))}$$

Proof. For brevity, we put $\mu_m(E) = \mu(E)/m(E)$ whenever E is a Borel subset of $\underline{R}^n$. The result will follow if we prove the following statement, in which $\{r_k\}$ and $\{s_k\}$ are positive null sequences, and ℓ, λ can be finite or infinite: If $\{r_k\}$ is such that $\mu_m(\bar{B}(x, r_k)) \to \ell$, then there is $\{s_k\}$ such that $\mu_m(B(x, s_k)) \to \ell$, and conversely, if $\{s_k\}$ is such that $\mu_m(B(x, s_k)) \to \lambda$, then there is $\{r_k\}$ such that $\mu_m(\bar{B}(x, r_k)) \to \lambda$.

Suppose that $\mu_m(\bar{B}(x, r_k)) \to \ell$. For each $k \in \underline{N}$ and $\nu \in \{m, \mu\}$,

$$\nu(\bar{B}(x, r_k)) = \nu\left(\bigcap_{i=1}^{\infty} B(x, r_k + i^{-1})\right) = \lim_{i \to \infty} \nu(B(x, r_k + i^{-1}))$$

Therefore we can choose s_k such that $r_k < s_k < r_k + k^{-1}$ and

$$|\mu_m(B(x, s_k)) - \mu_m(\bar{B}(x, r_k))| < k^{-1}$$

If $\ell < \infty$, given any positive ϵ we can find $N \in \underline{N}$ such that

$$|\mu_m(\bar{B}(x, r_k)) - \ell| < \epsilon$$

for all $k > N$. Hence, for all $k > \max\{N, \epsilon^{-1}\}$,

$$|\mu_m(B(x, s_k)) - \ell| < \epsilon + k^{-1} < 2\epsilon$$

so that $\mu_m(B(x, s_k)) \to \ell$. On the other hand, if $\ell = \infty$, given any positive ϵ we can find $K \in \underline{N}$ such that

$$\mu_m(\bar{B}(x, r_k)) > \epsilon^{-1}$$

whenever $k > K$, so that

$$\mu_m(B(x, s_k)) > \epsilon^{-1} - k^{-1} \geq \epsilon^{-1} - 1$$

for all such k. Hence $\mu_m(B(x, s_k)) \to \infty$.

Conversely, suppose that $\mu_m(B(x, s_k)) \to \lambda$. For each $k \in \underline{N}$ and $\nu \in \{m, \mu\}$,

$$\nu(B(x, s_k)) = \nu\left(\bigcup_{i=1}^{\infty} \bar{B}(x, s_k - i^{-1})\right) = \lim_{i \to \infty} \nu(\bar{B}(x, s_k - i^{-1}))$$

where $I > s_k^{-1}$. Techniques similar to those used above can now be employed to show that there is a positive null sequence $\{r_k\}$ such that $\mu_m(\bar{B}(x, r_k)) \to \lambda$.

Theorem 5.1

Let u be the Gauss-Weierstrass integral on $\underline{R}^n \times]0, a[$ of a non-negative measure μ on $\underline{R}^n$. There is a positive constant κ, which depends only on n, such that

$$\kappa (\underline{D}\mu)(x) \leq \lim_{t \to 0^+} \inf u(x, t) \leq \lim_{t \to 0^+} \sup u(x, t) \leq (\bar{D}\mu)(x) \tag{3}$$

for all $x \in \underline{R}^n$.

<u>Proof</u>. We begin with the left-hand inequality. Since μ is non-negative, for all $(x, t) \in \underline{R}^n$ we have

$$u(x, t) \geq \int_{B(x, \sqrt{t})} W(x - y, t) \, d\mu(y)$$

$$\geq (4\pi t)^{-n/2} e^{-1/4} \mu(B(x, \sqrt{t}))$$

$$= (4\pi)^{-n/2} e^{-1/4} \omega_n \mu(B(x, \sqrt{t}))/m(B(x, \sqrt{t}))$$

where $\omega_n = m(B(0, 1))$. Putting $\kappa = (4\pi)^{-n/2} e^{-1/4} \omega_n$, and taking the lower limit as $t \to 0^+$, we obtain

$$\lim_{t \to 0^+} \inf u(x, t) \geq \kappa \lim_{t \to 0^+} \inf \mu(B(x, \sqrt{t}))/m(B(x, \sqrt{t}))$$

as required.

We now prove the last inequality in (3). Let $x \in \underline{R}^n$. If $(\bar{D}\mu)(x) = \infty$ the inequality is obvious, so we suppose that $(\bar{D}\mu)(x) < \infty$. Given any $A > (\bar{D}\mu)(x)$, Lemma 5.1 implies that we can find $\delta > 0$ such that

$$\mu(\bar{B}(x, \gamma)) \leq A\omega_n \gamma^n$$

whenever $0 \leq \gamma \leq \delta$. We now write

$$u(x, t) = I(t) + J(t)$$

where

$$I(t) = \int_{\|x-y\| > \delta} W(x - y, t) \, d\mu(y), \quad J(t) = \int_{\|x-y\| \le \delta} W(x - y, t) \, d\mu(y)$$

Lemma 1.5 implies that $I(t) \to 0$ as $t \to 0^+$, so that it is sufficient to prove that

$$\limsup_{t \to 0^+} J(t) \le (\bar{D}\mu)(x) \tag{4}$$

If we put

$$F(r) = \mu(\bar{B}(x, r)) \quad \text{for all} \quad r \in [0, \delta]$$

then for all $\gamma \in [0, \delta]$ we have

$$\int_{[0,\gamma]} dF(r) = \mu(\bar{B}(x, \gamma)) = F(\gamma) \le A\omega_n \gamma^n \tag{5}$$

Furthermore, putting $V_t(\gamma) = (4\pi t)^{-n/2} \exp(-\gamma^2/4t)$ for all $\gamma \in [0, \delta]$, we obtain

$$J(t) = \int_{[0,\delta]} V_t(\gamma) \, dF(\gamma) \tag{6}$$

In this form, we can use integration by parts to estimate $J(t)$. This is justified by Fubini's theorem, as we now show. Let

$$\Delta = \{(r, \gamma) : 0 \le r \le \delta, \ r \le \gamma \le \delta\}$$

and consider the integral

$$N = \iint_\Delta V_t'(\gamma) \, d\gamma \, dF(r)$$

Integrating with respect to r, we obtain

$$N = \int_0^\delta V_t'(\gamma) \, d\gamma \int_{[0,\gamma]} dF(r) = \int_0^\delta V_t'(\gamma) F(\gamma) \, d\gamma$$

On the other hand, integration with respect to γ gives

$$N = \int_{[0,\delta]} dF(r) \int_r^\delta V_t'(\gamma) \, d\gamma$$

$$= \int_{[0,\delta]} (V_t(\delta) - V_t(r)) \, dF(r)$$

$$= V_t(\delta) F(\delta) - J(t)$$

by (6). Since $V_t' < 0$ and

$$|V_t(\delta) F(\delta) - J(t)| \le V_t(\delta) |\mu| (\bar{B}(x,\delta)) + \int_{\|x-y\| \le \delta} W(x-y,t) \, d|\mu|(y)$$

$$< \infty$$

Fubini's theorem shows that the different expressions for N are equal and finite. Hence

$$J(t) = V_t(\delta) F(\delta) - \int_0^\delta V_t'(\gamma) F(\gamma) \, d\gamma$$

It now follows from (5), and the inequalities $V_t(\delta) > 0$ and $V_t'(\gamma) < 0$, that

$$J(t) \le V_t(\delta)(A\omega_n \delta^n) - \int_0^\delta V_t'(\gamma)(A\omega_n \gamma^n) \, d\gamma$$

$$= A\omega_n n \left(\left[V_t(\gamma)\gamma^n n^{-1} \right]_0^\delta - \int_0^\delta V_t'(\gamma)\gamma^n n^{-1} \, d\gamma \right)$$

$$= A\omega_n n \int_0^\delta V_t(\gamma)\gamma^{n-1} \, d\gamma$$

Making the substitution $\gamma = (4t\rho)^{1/2}$, we obtain

$$J(t) \le A\omega_n n \int_0^{\delta^2/4t} (4\pi t)^{-n/2} e^{-\rho} (4t\rho)^{(n-1)/2} (t/\rho)^{1/2} \, d\rho$$

$$= A\omega_n \pi^{-n/2} (n/2) \int_0^{\delta^2/4t} e^{-\rho} \rho^{(n/2)-1} \, d\rho$$

$$\le A$$

since $\omega_n = \pi^{n/2}/(n/2)\Gamma(n/2)$, where Γ denotes the gamma function. Hence

$$\limsup_{t \to 0^+} J(t) \le A$$

where A is any number exceeding $(\bar{D}\mu)(x)$, so that (4) holds and the theorem is proved.

2. A COVERING THEOREM

Let μ be a non-negative measure on $\underline{R}^n$, and let E be a Borel subset of $\underline{R}^n$. The main result of this section, Theorem 5.2, tells us that, if E is covered in a particular way by a family of closed balls, then E can be approximated by the union G of a disjoint sequence of those balls, in the sense that $\mu(E\backslash G) = 0$. Some consequences of this result appear in later sections, and also in Chapter VII. The applications to temperatures in the present chapter are made through Theorem 5.1. The other results in this section are all needed to prove the main theorem.

Lemma 5.2

For each $\eta \ge 0$, let $\mathscr{F}_\eta$ be a family of closed balls in $\underline{R}^n$ with the following properties:

(i) whenever $\bar{B}(x,r), \bar{B}(y,\rho) \in \mathscr{F}_\eta$,

$$\|x - y\| \ge \max\{r,\rho\} - \eta \min\{r,\rho\}; \tag{7}$$

(ii) $\mathscr{F}_\eta$ contains a ball $\bar{B}(x_0,r_0)$ such that $r_0 \le r$ whenever $\bar{B}(x,r) \in \mathscr{F}_\eta$.

Then there exist $\delta > 0$ and $N \in \underline{N}$, such that no more than N of the balls in $\mathscr{F}_\eta$ meet $\bar{B}(x_0,r_0)$, for all $\eta \in [0,\delta]$. Here N is independent of η.

Proof. Since a change of scale does not affect the result in any way, we can assume that $r_0 = 1$. Let $\eta \ge 0$, and let $\bar{B}(x_1,r_1)$ and $\bar{B}(x_2,r_2)$ be arbitrary balls in $\mathscr{F}_\eta$ which meet $\bar{B}(x_0,1)$. By (ii), $r_1 \ge 1$ and $r_2 \ge 1$. We shall consider the triangle with vertices x_0,x_1,x_2, with a view to finding a positive lower bound for the angle θ at x_0.

Put $a = \|x_0 - x_1\|$, $b = \|x_0 - x_2\|$, $c = \|x_1 - x_2\|$. It follows from (7) that

$$a \ge \max\{1,r_1\} - \eta \min\{1,r_1\} = r_1 - \eta$$
$$b \ge \max\{1,r_2\} - \eta \min\{1,r_2\} = r_2 - \eta$$

and hence that

$$
\begin{aligned}
c &\geq \max\{r_1, r_2\} - \eta \min\{r_1, r_2\} \\
&= \max\{r_1 - \eta r_2, r_2 - \eta r_1\} \\
&\geq \max\{r_1 - \eta(b + \eta), r_2 - \eta(a + \eta)\}
\end{aligned}
$$

Since $\bar{B}(x_1, r_1)$ and $\bar{B}(x_2, r_2)$ both meet $\bar{B}(x_0, 1)$, we have $a \leq 1 + r_1$ and $b \leq 1 + r_2$, so that $r_1 \geq \max\{1, a - 1\}$ and $r_2 \geq \max\{1, b - 1\}$. Hence

$$
a \geq r_1 - \eta \geq \max\{1, a - 1\} - \eta \tag{8}
$$

$$
b \geq r_2 - \eta \geq \max\{1, b - 1\} - \eta \tag{9}
$$

and

$$
\begin{aligned}
c &\geq \max\{r_1 - \eta(b + \eta), \; r_2 - \eta(a + \eta)\} \\
&\geq \max\{1 - \eta(b+\eta), \; a - 1 - \eta(b+\eta), \; 1 - \eta(a+\eta), \; b - 1 - \eta(a+\eta)\} \\
&= \max\{1 - \eta b, \; a - 1 - \eta b, \; 1 - \eta a, \; b - 1 - \eta a\} - \eta^2 \tag{10}
\end{aligned}
$$

It is trivial that $a \geq a - 1 - \eta$, so that we can reduce (8) to $a \geq 1 - \eta$. Similarly, (9) becomes $b \geq 1 - \eta$. Furthermore, since the problem is symmetric in a and b, we can suppose that $a \geq b$, so that $a - 1 - \eta b \geq b - 1 - \eta a$ and $1 - \eta b \geq 1 - \eta a$. Thus (10) can be reduced to

$$
c \geq \max\{1 - \eta b, \; a - 1 - \eta b\} - \eta^2 \tag{11}
$$

and (8), (9) become simply

$$
a \geq b \geq 1 - \eta \tag{12}
$$

Now consider the angle θ at x_0. Obviously $\theta \in [0, \pi]$, so that a lower bound for θ is equivalent to an upper bound for $\cos \theta$, which is given in terms of a, b and c by the elementary formula

$$
\cos \theta = \frac{a^2 + b^2 - c^2}{2ab}
$$

However, if we put $\eta = 0$, $b = 1$, and $a \geq 2$, so that $a - 1 \geq 1$ and $c \geq a - 1$ by (11), all we get is the trivial result

$$
\cos \theta \leq \frac{a^2 + 1 - (a - 1)^2}{2a} = 1
$$

(Geometrically, it is obvious that $\theta = 0$ is a possibility in this case, with $r_1 = r_2 = 1$ and $a = 2$.) We can still progress with the idea of maximizing $\cos \theta$ if we look separately at those balls $\bar{B}(x, r)$ which meet $\bar{B}(x_0, 1)$ and have $\|x - x_0\| \leq 3/2$, and those which meet $\bar{B}(x_0, 1)$ and have $\|x - x_0\| \geq 3/2$. We are thus led, in view of (11) and (12), to consider the following cases:

(α) $a \geq b \geq 3/2$ and $a \geq 2$, so that $c \geq a - 1 - \eta b - \eta^2$;

(β) $2 \geq a \geq b \geq 3/2$, $c \geq 1 - \eta b - \eta^2$;

(γ) $3/2 \geq a \geq b \geq 1 - \eta$, $c \geq 1 - \eta b - \eta^2$.

$\underline{\text{Case}}$ (α). Using the inequality for c, for all sufficiently small η we have

$$\cos \theta \leq (a^2 + b^2 - (a - \eta b - (1 + \eta^2))^2)/2ab = \phi(\eta)$$

say, where ϕ is a continuous function so that, in particular, $\phi(0) = \phi(0^+)$. Now

$$\phi(0) = (b^2 + 2a - 1)/2ab$$

and elementary calculus shows that, in the region where $a \geq b \geq 3/2$ and $a \geq 2$, $\phi(0) \leq 7/8$. Hence $\phi(0^+) \leq 7/8$, so that there exists $\delta_1 > 0$ such that $\cos \theta \leq \phi(\eta) \leq 8/9$ for all $\eta \in [0, \delta_1]$.

$\underline{\text{Case}}$ (β). Here $c \geq 1 - \eta b - \eta^2$, so that, for all sufficiently small η,

$$\cos \theta \leq (a^2 + b^2 - ((1 - \eta^2) - \eta b)^2)/2ab = \phi(\eta) \qquad (13)$$

say, where $\phi(0) = \phi(0^+)$. Now

$$\phi(0) = (a^2 + b^2 - 1)/2ab \leq 7/8$$

because of the constraints on a and b. As in case (α), there exists $\delta_2 > 0$ such that $\cos \theta \leq \phi(\eta) \leq 8/9$ for all $\eta \in [0, \delta_2]$.

$\underline{\text{Case}}$ (γ). Here (13) again holds, but this time the conditions on a and b vary with η. Since they do so in a continuous manner, it is still sufficient to consider the case $\eta = 0$. This time we have

$$\phi(0) = (a^2 + b^2 - 1)/2ab \leq 7/9$$

whenever $3/2 \geq a \geq b \geq 1$. Therefore there exists $\delta_3 > 0$ such that $\cos \theta \leq \phi(\eta) \leq 8/9$ whenever $3/2 \geq a \geq b \geq 1 - \eta$.

We have thus established that there exists $\delta = \min\{\delta_1, \delta_2, \delta_3\} > 0$ such that, for all $\eta \in [0,\delta]$, whenever $\bar{B}(x_1,r_1)$, $\bar{B}(x_2,r_2) \in \mathcal{F}_\eta$ and meet $\bar{B}(x_0, 1)$, the angle $x_1 x_0 x_2$ is at least $\cos^{-1}(8/9)$, provided that either $\|x_0 - x_1\| \geq 3/2$ and $\|x_0 - x_2\| \geq 3/2$ (cases (α) and (β)) or $\|x_0 - x_1\| \leq 3/2$ and $\|x_0 - x_2\| \leq 3/2$ (case (γ)). Hence there can be only a finite number of such balls in each case, the number depending only on $\cos^{-1}(8/9)$ and therefore independent of η. The result follows.

Lemma 5.3

Let δ and N be the numbers whose existence is assured by Lemma 5.2. Choose $\eta \in [0,\delta]$, and let $\mathcal{F}_\eta$ be a family of closed balls in $\underline{R}^n$ such that property (i) of Lemma 5.2 holds, and such that the union of all the balls in $\mathcal{F}_\eta$ is a bounded set. Then $\mathcal{F}_\eta$ can be divided into $N + 1$ subfamilies $\mathcal{G}_1, \ldots, \mathcal{G}_{N+1}$, such that the members of each $\mathcal{G}_i$ are disjoint.

Proof. We first show that the balls in $\mathcal{F}_\eta$ can be arranged as a sequence in order of non-increasing radii. We can assume that $\eta < 1/4$. Since the union of the balls in $\mathcal{F}_\eta$ is a bounded set, there is a number s such that $r \leq s$ whenever $\bar{B}(x, r) \in \mathcal{F}_\eta$. For each $k \in \underline{N}$, put

$$I_k = [(k + 1)\eta^k s, \ k\eta^{k-1} s]$$

noting that $(k + 1)\eta < (k + 1)/4 < k$. If $\bar{B}(x, r)$, $\bar{B}(y,\rho) \in \mathcal{F}_\eta$ and $r,\rho \in I_k$, then, by property (i) of Lemma 5.2,

$$\|x - y\| \geq \max\{r,\rho\} - \eta \min\{r,\rho\} \geq (k + 1)\eta^k s - \eta(k\eta^{k-1} s) = \eta^k s > 0$$

Therefore for each k, there are only finitely many balls in $\mathcal{F}_\eta$ with radii in I_k, since their union is bounded. Furthermore, since $\eta < 1/4$, $(k + 1)\eta^k \to 0$ as $k \to \infty$. It follows that, given any positive λ, there are only finitely many balls in $\mathcal{F}_\eta$ whose radii exceed λ, namely those with radii in

$$\bigcup_{k=1}^{\ell} I_k = [(\ell + 1)\eta^\ell s, \ s]$$

where ℓ is the smallest integer such that $(\ell + 1)\eta^\ell s < \lambda$. Therefore the members of $\mathscr{F}_\eta$ can be arranged as a sequence $\{B_i\} = \{\bar{B}(x_i, r_i)\}$ such that $r_{i+1} \le r_i$ for all $i \in \underline{N}$.

We now select subfamilies $\mathscr{S}_1, \ldots, \mathscr{S}_{N+1}$ in the following way. For each $i \le N + 1$, we put B_i into $\mathscr{S}_i$. Consider B_{N+2}. Since $r_{N+2} \le r_i$ for all $i \le N + 1$, we can apply Lemma 5.2 to the family $\{B_1, \ldots, B_{N+2}\}$, and deduce that B_{N+2} cannot meet more than N of $B_1, \ldots, B_{N+1}$. Therefore we can choose $j \le N + 1$ such that $B_{N+2} \cap B_j = \emptyset$, and having done so we put B_{N+2} into $\mathscr{S}_j$. Next, since $r_{N+3} \le r_i$ for all $i \le N + 2$, we can apply Lemma 5.2 to the family $\{B_1, \ldots, B_{N+3}\}$ and deduce that B_{N+3} cannot meet more than N of $B_1, \ldots, B_{N+2}$. Therefore we can choose $k \le N + 2$ such that no ball yet placed in $\mathscr{S}_k$ meets B_{N+3}, and we put B_{N+3} into $\mathscr{S}_k$. We continue this process indefinitely. At the Mth stage, each of $B_1, \ldots, B_{N+M}$ has been assigned to one of $\mathscr{S}_1, \ldots, \mathscr{S}_{N+1}$ in such a way that no two balls that meet are in the same one. Since $r_{N+M+1} \le r_i$ for all $i \le N + M$, we can apply Lemma 5.2 to the family $\{B_1, \ldots, B_{N+M+1}\}$ and deduce that B_{N+M+1} cannot meet more than N of $B_1, \ldots, B_{N+M}$. As there are $N + 1$ of the subfamilies $\mathscr{S}_i$, we can choose $p \le N + 1$ such that B_{N+M+1} does not meet any ball yet assigned to $\mathscr{S}_p$, and then put B_{N+M+1} into $\mathscr{S}_p$. This proves the lemma.

We can now prove the covering theorem, which employs the following terminology. Given a subset E of $\underline{R}^n$, and a family $\mathscr{F}$ of closed balls such that, given any $x \in E$ and $\epsilon > 0$, there is a ball $\bar{B}(x, r) \in \mathscr{F}$ with $r < \epsilon$, we say that $\mathscr{F}$ covers E in the <u>narrow Vitali sense.</u>

Theorem 5.2

Let μ be a non-negative measure on $\underline{R}^n$, and let E be a bounded Borel subset of $\underline{R}^n$. If a family $\mathscr{F}$ of closed balls covers E in the narrow Vitali sense, then there is a countable subfamily $\mathscr{S}$ of $\mathscr{F}$, which consists of disjoint balls whose union G satisfies $\mu(E \setminus G) = 0$.

<u>Proof.</u> Since $\mathscr{F}$ covers E in the narrow Vitali sense, we can suppose that the set of radii of balls in $\mathscr{F}$ is bounded above, and hence has a supremum ρ_1, because otherwise we could discard some members of $\mathscr{F}$. We can also suppose that every ball in $\mathscr{F}$ has its center in E. This implies that the union of all members of $\mathscr{F}$ is a bounded set, because E is itself bounded. Let δ be the number whose existence is assured by Lemma 5.2, and choose $\eta \in \,]0, \delta[$. Let $\bar{B}(x_1, r_1)$ be a ball in $\mathscr{F}$ with $r_1 > \rho_1/(1 + \eta)$. Let ρ_2 be the supremum of the radii of those balls in $\mathscr{F}$ whose centers lie outside $\bar{B}(x_1, r_1)$, and let $\bar{B}(x_2, r_2)$ be one such ball with $r_2 > \rho_2/(1 + \eta)$. We repeat this selection process indefinitely. At

the Mth stage, let ρ_{M+1} be the supremum of the radii of those balls in $\mathscr{F}$ whose centers lie outside the set

$$U_M = \bigcup_{j=1}^{M} \bar{B}(x_j, r_j)$$

and let $\bar{B}(x_{M+1}, r_{M+1})$ be one such ball with $r_{M+1} > \rho_{M+1}/(1 + \eta)$. Put

$$U = \bigcup_{j=1}^{\infty} \bar{B}(x_j, r_j)$$

We show that $E \subseteq U$. Suppose, on the contrary, that there is a point $y \in E \setminus U$. If $y \in \bar{U}$, then since each set U_M is closed, $y \notin U_M$ for any M. Therefore there is a sequence of balls $\{\bar{B}(x_{j(i)}, r_{j(i)})\}$ such that $x_{j(i)} \to y$ and $r_{j(i)} \to 0$. Since $\mathscr{F}$ covers E in the narrow Vitali sense, there is a ball $\bar{B}(y, s)$ in $\mathscr{F}$. By choice of the balls whose union is U, $r_{j(i)} > \rho_{j(i)}/(1 + \eta)$ for all i, so that $\rho_{j(i)} \to 0$, and hence there is i with $\rho_{j(i)} < s$. Since the center of $\bar{B}(y, s)$ lies outside $U_{j(i)-1}$, this contradicts the definition of $\rho_{j(i)}$. It follows that y cannot belong to $\bar{U}$.

Therefore, since $\mathscr{F}$ covers E in the narrow Vitali sense, we can find $\sigma > 0$ such that $\bar{B}(y, \sigma) \in \mathscr{F}$ and $\bar{B}(y, \sigma) \cap U = \emptyset$. By definition of the ρ_M, we have $\rho_M \geq \sigma$ for all M. It therefore follows from the fact that $x_{M+k} \notin \bar{B}(x_M, r_M)$, for each M and all $k \in \underline{N}$, that

$$\|x_{M+k} - x_M\| > r_M > \rho_M/(1 + \eta) \geq \sigma/(1 + \eta)$$

Since $x_{M+k} \in E$ for all k, and E is bounded, this is impossible. It follows that there is no point $y \in E \setminus U$, so that $E \subseteq U$.

Let $\mathscr{H}$ denote the family of balls $\bar{B}(x_j, r_j)$ chosen above, so that $\mathscr{H}$ covers E. We now show that $\mathscr{H}$ satisfies the conditions on the class $\mathscr{F}_\eta$ in Lemma 5.3. Since we know that U is bounded, we have only to prove that

$$\|x_k - x_\ell\| \geq \max\{r_k, r_\ell\} - \eta \min\{r_k, r_\ell\} \qquad (14)$$

for all distinct k, $\ell \in \underline{N}$. We can suppose that $k > \ell$. By our choice of the balls $\bar{B}(x_j, r_j)$, we have

$$\rho_{M+1}/(1 + \eta) < r_{M+1} \leq \rho_{M+1} \tag{15}$$

and $x_{M+1} \notin U_M$, for all $M \in \underline{N}$. Hence, in particular, $x_k \notin \bar{B}(x_\ell, r_\ell)$, so that $\|x_k - x_\ell\| > r_\ell$. There are three cases to consider. Firstly, if $r_\ell \geq r_k$ we have

$$\max\{r_k, r_\ell\} - \eta \min\{r_k, r_\ell\} = r_\ell - \eta r_k < r_\ell < \|x_k - x_\ell\|$$

so that (14) holds in this case. Secondly, if $r_\ell < r_k \leq \|x_k - x_\ell\|$ we have

$$\max\{r_k, r_\ell\} - \eta \min\{r_k, r_\ell\} = r_k - \eta r_\ell < r_k \leq \|x_k - x_\ell\|$$

so that (14) again holds. Thirdly, if $r_\ell < \|x_k - x_\ell\| < r_k$ it follows from (15) that

$$\begin{aligned}
\max\{r_k, r_\ell\} - \eta \min\{r_k, r_\ell\} &= r_k - \eta r_\ell \\
&= r_k + r_\ell - (1 + \eta)r_\ell \\
&< r_k + r_\ell - \rho_\ell \\
&\leq r_\ell + r_k - \rho_k \\
&\leq r_\ell \\
&< \|x_k - x_\ell\|
\end{aligned}$$

so that (14) holds in this case also. Hence $\mathcal{H}$ satisfies the conditions on the class $\mathcal{F}_\eta$ in Lemma 5.3.

It now follows from Lemma 5.3 that $\mathcal{H}$ can be divided into $N + 1$ subfamilies $\mathcal{H}_1, \ldots, \mathcal{H}_{N+1}$, such that the members of each $\mathcal{H}_i$ are disjoint. For each i, let V_i denote the union of the balls in $\mathcal{H}_i$. Since $\mathcal{H}$ covers E, there exists i such that

$$\mu(V_i \cap E) \geq \mu(E)/(N + 1) \tag{16}$$

for otherwise we would have the contradiction

$$\mu(E) \leq \sum_{i=1}^{N+1} \mu(V_i \cap E) < \sum_{i=1}^{N+1} \mu(E)/(N + 1) = \mu(E)$$

By relabelling if necessary, we can assume that (16) holds when $i = 1$. The subfamily $\mathcal{H}_1$ may be infinite, in which case the sum of the μ-measures of the intersections with E of the (disjoint) balls in $\mathcal{H}_1$ equals $\mu(V_1 \cap E)$, which is finite since E is bounded. Hence $\mathcal{H}_1$ has a finite subfamily $\mathcal{I}_1$, the union of whose elements, denoted by I_1, satisfies

$$\mu(I_1 \cap E) \geq \mu(E)/(N + 2) \tag{17}$$

On the other hand, if $\mathcal{H}_1$ is finite we can take $\mathcal{I}_1 = \mathcal{H}_1$, and then (17) follows immediately from (16). Hence, in either case, we have

$$\mu(E \setminus I_1) \leq \left(\frac{N + 1}{N + 2}\right) \mu(E) \tag{18}$$

Now denote by $\mathcal{F}_1$ the family of all balls in $\mathcal{F}$ which do not meet I_1. Since $\mathcal{I}_1$ consists of finitely many closed balls, I_1 is closed. Therefore the fact that $\mathcal{F}$ covers E in the narrow Vitali sense, implies that $\mathcal{F}_1$ covers $E \setminus I_1$ in the same sense. Hence the proof as far as (18) can equally well be applied to $\mathcal{F}_1$ and $E \setminus I_1$ as to $\mathcal{F}$ and E. It follows that there is a finite subfamily $\mathcal{I}_2$ of $\mathcal{F}_1$, whose elements are disjoint balls with union I_2, which does not meet I_1 and satisfies

$$\mu(I_2 \cap (E \setminus I_1)) \geq \mu(E \setminus I_1)/(N + 2) \tag{19}$$

corresponding to (17). This, together with (18), gives

$$\mu(E \setminus (I_1 \cup I_2)) = \mu(E \setminus I_1) - \mu(I_2 \cap (E \setminus I_1))$$

$$\leq \left(\frac{N + 1}{N + 2}\right) \mu(E \setminus I_1)$$

$$\leq \left(\frac{N + 1}{N + 2}\right)^2 \mu(E) \tag{20}$$

We apply this process repeatedly. At the pth stage, where $p \geq 2$, we denote by $\mathcal{F}_p$ the family of those balls in $\mathcal{F}_{p-1}$ which do not meet $J_p = I_1 \cup \cdots \cup I_p$. For each $q \leq p$, $\mathcal{I}_q$ consists of finitely many closed balls, so that each I_q (the union of the elements of $\mathcal{I}_q$) is closed, and hence the same is true of J_p. Hence $\mathcal{F}_p$ covers $E \setminus J_p$ in the narrow Vitali sense. Therefore, as above, there is a finite subfamily $\mathcal{I}_{p+1}$ of $\mathcal{F}_p$, whose elements are disjoint balls whose union I_{p+1} satisfies

$$\mu(I_{p+1} \cap (E \setminus J_p)) \geq \frac{\mu(E \setminus J_p)}{N + 2}$$

corresponding to (17) and (19). Hence

$$\mu(E \setminus J_{p+1}) = \mu(E \setminus J_p) - \mu(I_{p+1} \cap (E \setminus J_p))$$

$$\leq \left(\frac{N+1}{N+2}\right) \mu(E \setminus J_p)$$

$$\leq \left(\frac{N+1}{N+2}\right)^{p+1} \mu(E)$$

which corresponds to (20).

We now put

$$\mathscr{G} = \bigcup_{q=1}^{\infty} \mathscr{G}_q$$

and let G be the union of the elements of $\mathscr{G}$. For each $q > 1$, $\mathscr{G}_q$ is a subfamily of $\mathscr{G}_{q-1}$, and so consists of balls that do not meet I_{q-1}. Therefore the sets I_q, $q \geq 1$, are disjoint. Since each $\mathscr{G}_q$, $q \geq 1$, is a finite family of disjoint balls, it follows that $\mathscr{G}$ is a countable family of disjoint balls. Finally,

$$\mu(E \setminus G) = \mu\left(E \setminus \bigcup_{q=1}^{\infty} I_q\right) = \lim_{p \to \infty} \mu(E \setminus J_p) \leq \lim_{p \to \infty} \left(\frac{N+1}{N+2}\right)^p \mu(E) = 0$$

3. NON-NEGATIVE SINGULAR MEASURES AND THE REPRESENTATION OF NON-NEGATIVE TEMPERATURES

We shall use the covering theorem of the previous section to prove that, if μ is a non-negative measure which is singular with respect to m, then $D\mu = \infty$ μ-a. e. This result will then be combined with Theorem 5.1, to prove representation theorems for non-negative temperatures on $\underline{R}^n \times$]0, a[in terms of their behavior near $\underline{R}^n \times \{0\}$.

Theorem 5.3

Let μ be a non-negative measure on $\underline{R}^n$ which is singular with respect to Lebesgue measure. Then

$$D\mu = \infty \quad \mu\text{-a. e. on } \underline{R}^n \tag{21}$$

Proof. Since μ is singular, there is a Borel set A such that m(A) = 0 and $\mu(E) = \mu(A \cap E)$ for every Borel set E. For each $k \in \underline{N}$, put

$$A_k = \left\{ x \in A : \liminf_{r \to 0} \frac{\mu(\bar{B}(x,r))}{m(\bar{B}(x,r))} < k \right\}$$

Then, if $S = \{x \in \underline{R}^n : (\underline{D}\mu)(x) < \infty\}$, Lemma 5.1 implies that

$$A \cap S = \bigcup_{k=1}^{\infty} A_k$$

so that

$$\mu(S) = \mu(A \cap S) \leq \sum_{k=1}^{\infty} \mu(A_k)$$

To prove the theorem, we must show that $\mu(S) = 0$, and we now see that this will follow if we prove that $\mu(A_k) = 0$ for all k.

Given any $k \in \underline{N}$, let K be an arbitrary compact subset of A_k. Then $K \subseteq A$, so that m(K) = 0. Therefore, given $\epsilon > 0$, there is an open set V such that $K \subseteq V$ and m(V) < ϵ. Since $K \subseteq A_k$, given any $x \in K$ we can find a decreasing null sequence $\{r_i\}$ such that

$$\bar{B}(x, r_i) \subseteq V, \quad \mu(\bar{B}(x, r_i)) < km(\bar{B}(x, r_i)) \tag{22}$$

for all i. Consider the family $\mathscr{F}$ of all these balls $\bar{B}(x, r_i)$ for every x in K. Because each sequence $\{r_i\}$ is null, $\mathscr{F}$ covers K in the narrow Vitali sense. By Theorem 5.2, there is a disjoint sequence $\{B_j\}$ of members of $\mathscr{F}$, whose union U satisfies $\mu(K \setminus U) = 0$. It follows that

$$\mu(K) = \mu(K \setminus U) + \mu(K \cap U) \leq \mu(U) = \sum_{j=1}^{\infty} \mu(B_j)$$

Furthermore, each B_j is one of the balls $\bar{B}(x, r_i)$ which satisfy (22), so that

$$\mu(K) \leq \sum_{j=1}^{\infty} \mu(B_j) \leq k \sum_{j=1}^{\infty} m(B_j) = km(U) \leq km(V) < k\epsilon$$

Since ϵ is arbitrary, $\mu(K) = 0$. Since K is an arbitrary compact subset of A_k, $\mu(A_k) = 0$. Since k is arbitrary, the result follows.

Remark. It is well-known that, in the situation described in Theorem 5.3, $D\mu = 0$ m-a. e. The theorem therefore completes the picture by providing information μ-a. e.

Recall from Theorem 2.10 that, if u is a non-negative temperature on $\underline{R}^n \times]0, a[$, then u is the Gauss-Weierstrass integral of a non-negative measure μ on $\underline{R}^n$. Using Theorems 5.1 and 5.3, we can now give a description of μ in terms of the limits $u(x, 0^+)$ whenever they exist.

Theorem 5.4

Let u be a non-negative temperature on $\underline{R}^n \times]0, a[$, and let $Z = \{x \in \underline{R}^n : u(x, 0^+) = \infty\}$. Then $m(Z) = 0$, and u has the representation

$$u(x, t) = \int_{\underline{R}^n} W(x - y, t) u(y, 0^+) \, dy + \int_{\underline{R}^n} W(x - y, t) \, d\sigma(y) \qquad (23)$$

for all $(x, t) \in \underline{R}^n \times]0, a[$, where σ is singular with respect to m and $\sigma(\underline{R}^n \setminus Z) = 0$.

Proof. By Theorem 2.10, u is the Gauss-Weierstrass integral of a non-negative measure μ. Furthermore, $D\mu$ is the absolutely continuous part of μ, so that there is a non-negative measure σ, singular with respect to m, such that

$$d\mu(y) = (D\mu)(y) \, dy + d\sigma(y)$$

By Theorem 4.1, $u(y, 0^+) = (D\mu)(y)$ and is finite for m-almost all $y \in \underline{R}^n$, so that u has the representation (23) and $m(Z) = 0$. If $x \notin Z$, then

$$\liminf_{t \to 0^+} u(x, t) < \infty$$

so that Theorem 5.1 implies that $(\underline{D}\mu)(x) < \infty$. Therefore, since $D\mu \geq 0$, we have $(\underline{D}\sigma)(x) \leq (\underline{D}\mu)(x) < \infty$ for all $x \in \underline{R}^n \setminus Z$. By Theorem 5.3, $\sigma(\underline{R}^n \setminus Z) = 0$.

Theorem 5.4 has some important consequences, the first of which is a test for the absolute continuity of the initial measure.

Theorem 5.5

Let u be a non-negative temperature on $\underline{R}^n \times]0, a[$, and suppose that

$$\liminf_{t \to 0^+} u(x, t) < \infty$$

for all $x \in \underline{R}^n$. Then u is the Gauss-Weierstrass integral of $u(\cdot, 0^+)$.

Proof. If $Z = \{x \in \underline{R}^n : u(x, 0^+) = \infty\}$, then our hypothesis implies that $\overline{Z} = \emptyset$. Therefore, by Theorem 5.4, u has the representation (23) for all $(x, t) \in \underline{R}^n \times]0, a[$, where $\sigma(\underline{R}^n) = 0$. Since σ is non-negative, the result is proved.

Remarks. Theorem 5.5 is a significant improvement on the latter part of Theorem 2.10, which gives a similar conclusion under the hypothesis that u is continuous and real-valued on $\underline{R}^n \times [0, a[$. It also contains the following generalization of Theorem 2.5, where u was again assumed to be continuous on $\underline{R}^n \times [0, a[$.

Corollary

If u is a non-negative temperature on $\underline{R}^n \times]0, a[$ such that

$$\liminf_{t \to 0^+} u(x, t) < \infty$$

for all $x \in \underline{R}^n$, and

$$\liminf_{t \to 0^+} u(x, t) = 0$$

for m-almost all $x \in \underline{R}^n$, then $u = 0$ on $\underline{R}^n \times]0, a[$.

Proof. By Theorem 5.5, u is the Gauss-Weierstrass integral of 0.

Another consequence of Theorem 5.4 is the following characterization of W.

Theorem 5.6

Let u be a non-negative temperature on $\underline{R}^n \times]0, a[$, and let $\xi \in \underline{R}^n$.
If

$$\liminf_{t \to 0^+} u(x, t) < \infty \qquad \text{whenever } x \neq \xi \tag{24}$$

and

$$\liminf_{t \to 0^+} u(x, t) = 0 \tag{25}$$

for m-almost all $x \in \underline{R}^n$, then there is a non-negative constant κ such that

$$u(x, t) = \kappa W(x - \xi, t)$$

for all $(x, t) \in \underline{R}^n \times]0, a[$.

Proof. If $Z = \{x \in \underline{R}^n : u(x, 0^+) = \infty\}$, then our hypothesis (24) implies that $Z \subseteq \{\xi\}$. Since (25) implies that $u(\cdot, 0^+) = 0$ m-a. e. on $\underline{R}^n$, it follows from Theorem 5.4 that u is the Gauss-Weierstrass integral of a measure σ such that $\sigma(\underline{R}^n \setminus \{\xi\}) \leq \sigma(\underline{R}^n \setminus Z) = 0$. Hence $\sigma = \kappa \delta_\xi$ for some non-negative constant κ, where δ_ξ is the Dirac δ-measure concentrated at ξ, and

$$u(x, t) = \kappa \int_{\underline{R}^n} W(x - y, t) \, d\delta_\xi(y) = \kappa W(x - \xi, t)$$

for all $(x, t) \in \underline{R}^n \times]0, a[$.

Remark. Theorem 5.6 effectively asserts the minimality of W, since it shows that if a non-negative temperature is not much bigger than W near $\underline{R}^n \times \{0\}$, in the sense that (24) and (25) are satisfied, then it must be a multiple of W. Theorem 2.13 implies another assertion of this type, as follows. Let u be a positive temperature on $\underline{R}^n \times]0, a[$, let $t_0 \in]0, a[$, and put $M(r) = \max\{u(x, t_0) : \|x\| = r\}$ for all $r > 0$. If

$$\liminf_{r \to \infty} \left(\frac{\log M(r)}{r} + \frac{r}{4t_0} \right) = 0 \tag{26}$$

then Theorem 2.13 implies that $u = \kappa W$ for some positive constant κ.

Since (26) holds when u is replaced by W in M(r), a temperature u which satisfies (26) cannot be much bigger than W on $\underline{R}^n \times \{t_0\}$.

4. TEMPERATURES IN THE CLASS R_a

The class R_a was studied in section 6 of Chapter III. In particular, it was shown in Theorem 3.15 that the following statements are equivalent, for a temperature u on $\underline{R}^n \times]0, a[$: (i) $u \in R_a$; (ii) u^+ has a thermic majorant on $\underline{R}^n \times]0, a[$; (iii) u can be written as a difference of two non-negative temperatures on $\underline{R}^n \times]0, a[$; (iv) is the Gauss-Weierstrass integral of a signed measure on $\underline{R}^n \times]0, a[$.

In this section, we consider to what extent the results of Theorems 5.4, 5.5, and 5.6 can be generalized to other temperatures in R_a. The ideas embodied in statements (ii) and (iii) above play a fundamental role, and we begin by considering the thermic majorization of u^+, for an arbitrary $u \in R_a$. Note that, if u is the Gauss-Weierstrass integral of a measure μ on $\underline{R}^n \times]0, a[$, then the Gauss-Weierstrass integral of μ^+ is a positive thermic majorant of u on that strip, and is therefore a thermic majorant of u^+ there. It is, in fact, the best such majorant in the sense that any other is larger, as Theorem 5.7 shows.

We use the following terminology. If $u \in R_a$, and there is a thermic majorant w of u^+ on $\underline{R}^n \times]0, a[$, such that $w \leq v$ for every thermic majorant v of u^+ on $\underline{R}^n \times]0, a[$, we call w the <u>least thermic majorant</u> of u^+ on $\underline{R}^n \times]0, a[$.

Theorem 5.7

Let u be the Gauss-Weierstrass integral on $\underline{R}^n \times]0, a[$ of a signed measure μ. Then u^+ has a least thermic majorant on $\underline{R}^n \times]0, a[$, which is the Gauss-Weierstrass integral of μ^+.

<u>Proof.</u> By Theorem 3.15, u^+ has a thermic majorant on $\underline{R}^n \times]0, a[$. Let v be any such majorant. Since v is a non-negative temperature, it is the Gauss-Weierstrass integral on $\underline{R}^n \times]0, a[$ of some non-negative measure ν on $\underline{R}^n$, by Theorem 2.10. Because μ is a signed measure in the sense of Chapter I, section 2, there is a positive constant γ such that the formula

$$d\lambda(y) = \exp(-\gamma \|y\|^2) \, d\mu(y)$$

defines a signed measure λ of finite total variation. Since an increase in γ does not alter the fact that $|\lambda|\,(\underline{R}^n) < \infty$, we can suppose that $\gamma > 1/4a$. Then, if the non-negative measure η is defined by

$$d\eta(y) = \exp\left(-\gamma\|y\|^2\right)\,d\nu(y)$$

we have

$$\eta(\underline{R}^n) = \int_{\underline{R}^n} \exp(-\gamma\|y\|^2)\,d\nu(y) = \left(\frac{\pi}{\gamma}\right)^{n/2} \int_{\underline{R}^n} W\left(y, \frac{1}{4\gamma}\right)\,d\nu(y) = \left(\frac{\pi}{\gamma}\right)^{n/2} v\left(0, \frac{1}{4\gamma}\right)$$

Thus λ and η both have finite total variation.

We can write $\lambda = \lambda^+ - \lambda^-$, and can also write $\lambda = \eta - (\eta - \lambda)$. The measure η is non-negative, and we shall prove that $\eta - \lambda$ is too. It will then follow from the minimum property of the Jordan decomposition that $\lambda^+ \le \eta$ and $\lambda^- \le \eta - \lambda$, and the result of the theorem will be deduced from these inequalities.

Since $v - u \ge 0$, we know from Theorem 2.10 that $v - u$ is the Gauss-Weierstrass integral on $\underline{R}^n \times \,]0,a[$ of a non-negative measure ω. But we also have

$v(x,t) - u(x,t)$

$$= \int_{\underline{R}^n} W(x - y, t)\,d\nu(y) - \int_{\underline{R}^n} W(x - y, t)\,d\mu(y)$$

$$= \int_{\underline{R}^n} W(x - y, t)\exp\left(\gamma\|y\|^2\right)\,d\eta(y) - \int_{\underline{R}^n} W(x - y, t)\exp\left(\gamma\|y\|^2\right)\,d\lambda(y)$$

$$= \int_{\underline{R}^n} W(x - y, t)\exp\left(\gamma\|y\|^2\right)\,d(\eta - \lambda)\,(y)$$

for all $(x,t) \in \underline{R}^n \times \,]0,a[$, where $\eta - \lambda$ is well-defined since both η and λ have finite total variation. It now follows from the uniqueness of the Gauss-Weierstrass representation (Theorem 1.5) that

$$d\omega(y) = \exp\left(\gamma\|y\|^2\right)\,d(\eta - \lambda)\,(y)$$

Since ω is non-negative, the same is true of $\eta - \lambda$.

It now follows that $\lambda^+ \le \eta$ and $\lambda^- \le \eta - \lambda$. Therefore, whenever $(x,t) \in \underline{R}^n \times \,]0,a[$,

$$v(x,t) = \int_{\underline{R}^n} W(x - y, t)\exp\left(\gamma\|y\|^2\right)\,d\eta(y)$$

$$\geq \int_{\underline{R}^n} W(x - y, t) \exp(\gamma \|y\|^2) \, d\lambda^+(y)$$

$$= \int_{\underline{R}^n} W(x - y, t) \, d\mu^+(y)$$

$$\geq u^+(x, t)$$

Thus, if w denotes the Gauss-Weierstrass integral of μ^+ on $\underline{R}^n \times]0, a[$, we have proved that $v \geq w \geq u^+$ for any thermic majorant v of u^+ on $\underline{R}^n \times]0, a[$, as required.

Remarks. If v is a thermic majorant of u^+, then $u = v - (v - u)$ is an expression for u as the difference of two non-negative temperatures. If w is the least such majorant, then $w \leq v$ and $w - u \leq v - u$, so that $u = w - (w - u)$ is a minimal expression for u as the difference of two non-negative temperatures.

If u is the Gauss-Weierstrass integral of μ, then one simple way of expressing u as a difference of non-negative temperatures, is to write it as the Gauss-Weierstrass integral of μ^+ minus that of μ^-. Theorem 5.7 tells us that this is the same as writing $u = w - (w - u)$, as in the preceding paragraph.

We cannot give a complete extension of Theorem 5.4 to arbitrary temperatures in R_a, since we do not have any inequalities like those of Theorem 5.1 for signed measures. However, we can still give a useful generalization.

Theorem 5.8

Let $u \in R_a$, let w be the least thermic majorant of u^+ on $\underline{R}^n \times]0, a[$, and let $Z = \{x \in \underline{R}^n : u(x, 0^+) = -\infty\}$. If

$$\limsup_{t \to 0^+} w(x, t) < \infty \tag{27}$$

for all $x \in \underline{R}^n$, then u has the representation

$$u(x, t) = \int_{\underline{R}^n} W(x - y, t) u(y, 0^+) \, dy - \int_{\underline{R}^n} W(x - y, t) \, d\sigma(y)$$

for all $(x, t) \in \underline{R}^n \times]0, a[$, where σ is a non-negative measure, singular with respect to m, such that $\sigma(\underline{R}^n \setminus Z) = 0$.

Proof. If $v = w - u$, then v is a non-negative temperature on $\underline{R}^n \times$]0, a[. Furthermore, whenever $x \in \underline{R}^n \setminus Z$, it follows from (27) that

$$\liminf_{t \to 0^+} v(x,t) \leq \limsup_{t \to 0^+} w(x,t) - \limsup_{t \to 0^+} u(x,t) < \infty$$

Therefore $\{x \in \underline{R}^n : v(x, 0^+) = \infty\} \subseteq Z$. It now follows from Theorem 5.4 that

$$v(x,t) = \int_{\underline{R}^n} W(x - y, t)v(y, 0^+)\, dy + \int_{\underline{R}^n} W(x - y, t)\, d\sigma(y)$$

for all $(x,t) \in \underline{R}^n \times]0, a[$, where σ is a non-negative measure, singular with respect to m, such that

$$\sigma(\underline{R}^n \setminus Z) \leq \sigma(\underline{R}^n \setminus \{x \in \underline{R}^n : v(x, 0^+) = \infty\}) = 0$$

Furthermore, since w is a non-negative temperature on $\underline{R}^n \times]0, a[$ such that (27) holds, another application of Theorem 5.4 shows that

$$w(x,t) = \int_{\underline{R}^n} W(x - y, t)w(y, 0^+)\, dy$$

for all $(x,t) \in \underline{R}^n \times]0, a[$. Hence

$$\begin{aligned} u(x,t) &= w(x,t) - v(x,t) \\ &= \int_{\underline{R}^n} W(x - y, t)(w(y, 0^+) - v(y, 0^+))\, dy - \int_{\underline{R}^n} W(x - y, t)\, d\sigma(y) \\ &= \int_{\underline{R}^n} W(x - y, t)u(y, 0^+)\, dy - \int_{\underline{R}^n} W(x - y, t)\, d\sigma(y) \end{aligned}$$

for all $(x,t) \in \underline{R}^n \times]0, a[$.

Remark. Theorem 5.4 is a special case of Theorem 5.8, since if $u \geq 0$ then 0 is a thermic majorant of $(-u)^+$ which satisfies (27) for all $x \in \underline{R}^n$.

We now give a generalization of Theorem 5.5.

Theorem 5.9

Let $u \in R_a$, and let w be the least thermic majorant of u^+ on $\underline{R}^n \times$]0, a[. Suppose that

$$\limsup_{t\to 0^+} w(x,t) < \infty \tag{28}$$

and

$$\limsup_{t\to 0^+} u(x,t) > -\infty \tag{29}$$

for all $x \in \underline{R}^n$. Then u is the Gauss-Weierstrass integral of $u(\cdot,0^+)$ on $\underline{R}^n \times]0,a[$.

Proof. It follows from (28) and Theorem 5.8 that

$$u(x,t) = \int_{\underline{R}^n} W(x-y,t)u(y,0^+)\,dy - \int_{\underline{R}^n} W(x-y,t)\,d\sigma(y)$$

for all $(x,t) \in \underline{R}^n \times]0,a[$, where σ is non-negative and

$$\sigma(\underline{R}^n \setminus \{x \in \underline{R}^n : u(x,0^+) = -\infty\}) = 0$$

In view of (29), $u(x,0^+)$ is never $-\infty$, so that $\sigma(\underline{R}^n) = 0$. Since σ is non-negative, it follows that σ is null.

Corollary

Let $u \in R_a$, and let w be the least thermic majorant of u^+. Suppose that

$$\limsup_{t\to 0^+} w(x,t) < \infty$$

for all $x \in \underline{R}^n$, and

$$\limsup_{t\to 0^+} u(x,t) = 0$$

for m-almost all $x \in \underline{R}^n$. Then u = 0 on $\underline{R}^n \times]0,a[$.

Proof. By Theorem 5.9, u is the Gauss-Weierstrass integral of 0.

Remarks. The conclusion of Theorem 5.9 fails to hold if we relax the condition (28), even at a single point. For example, if u = -W then w = 0 (by Theorem 5.7), so that (28) holds for all $x \in \underline{R}^n$ and (29) holds for all non-zero x; but u is not the Gauss-Weierstrass integral of $u(\cdot, 0^+)$ since $W(\cdot, 0^+) = 0$ m-a. e. on $\underline{R}^n$.

We now give an example to show that the condition (28) on w is stronger than the same condition on u.

<u>Example</u>. Let f be defined on $\underline{R} \setminus \{0\}$ by

$$f(x) = \begin{cases} x^{-\frac{1}{2}} & \text{if} \quad x > 0 \\ -|x|^{-\frac{1}{2}} & \text{if} \quad x < 0 \end{cases}$$

Then

$$\int_{\underline{R}} \exp(-\alpha x^2) |f(x)| \ dx < \infty$$

for every positive constant α, so that the Gauss-Weierstrass integral u of f is defined and is a temperature on $\underline{R}^n \times]0, \infty[$, by Theorem 1.2, Corollary. Furthermore, $u(x, 0^+) = f(x)$ for all $x \in \underline{R} \setminus \{0\}$, by Theorem 1.3, Corollary. Also, since $f(-x) = -f(x)$ for all x, whenever $t > 0$ we have

$$u(0, t) = \int_{\underline{R}} W(y, t) f(y) \ dy = - \int_{\underline{R}} W(y, t) f(-y) \ dy$$

so that, if $z = -y$,

$$u(0, t) = - \int_{\underline{R}} W(z, t) f(z) \ dz = -u(0, t)$$

Hence $u(0, t) = 0$ for all $t > 0$, and in particular $u(0, 0^+) = 0$. Thus $u(x, 0^+)$ exists and is finite for all $x \in \underline{R}$.

By Theorem 5.7, the least thermic majorant v of u^+ on $\underline{R} \times]0, \infty[$ is the Gauss-Weierstrass integral of f^+. If ν is defined on $\underline{R}$ by $d\nu(y) = f^+(y) \ dy$, and $r > 0$, then

$$\frac{\nu(B(0, r))}{m(B(0, r))} = \frac{1}{2r} \int_0^r x^{-\frac{1}{2}} \ dx = r^{-\frac{1}{2}}$$

so that $(D\nu)(0) = \infty$. Therefore, by Theorem 5.1, $v(0, 0^+) = \infty$. Hence $u(x, 0^+)$ is finite for all x, but $v(0, 0^+) = \infty$.

The next result is slightly weaker than Theorem 5.9, but does not directly mention thermic majorization, and has a form which is closely related to that of the definition of the class R_a and Theorem 3.13.

Theorem 5.10

Let u be a temperature on $\underline{R}^n \times]0, a[$ such that, whenever $0 < c < b < a$, the function

$$t \longmapsto W(x, b - t) u^+(x, t) \tag{30}$$

is bounded on $]0, c[$. If

$$\limsup_{t \to 0^+} u(x, t) > -\infty$$

for all $x \in \underline{R}^n$, then u is the Gauss-Weierstrass integral of $u(\cdot, 0^+)$ on $\underline{R}^n \times]0, a[$.

Proof. Let $b \in]0, a[$. If v is defined on $\underline{R}^n \times [0, b[$ by

$$v(x, t) = (4\pi b^2)^{n/2} (b - t)^{-n/2} \exp\left(\frac{\|x\|^2}{4(b - t)}\right)$$

then v is continuous, and is a positive temperature on $\underline{R}^n \times]0, b[$. In fact, it follows from the example at the end of section 2 of Chapter I, that v is the Gauss-Weierstrass integral of $v(\cdot, 0)$. Let $c \in]0, b[$. Then the function in (30) is bounded on $]0, c[$, so that there is a positive constant κ such that

$$u^+(x, t) \leq \kappa(4\pi)^{n/2}(b - t)^{n/2} \exp\left(\frac{\|x\|^2}{4(b - t)}\right)$$

$$\leq \kappa(4\pi b^2)^{n/2}(b - t)^{-n/2} \exp\left(\frac{\|x\|^2}{4(b - t)}\right)$$

$$= \kappa v(x, t)$$

for all $(x, t) \in \underline{R}^n \times]0, c[$. Hence κv is a thermic majorant of u^+ on $\underline{R}^n \times]0, c[$, and $u \in R_c$. Therefore, by Theorem 5.7, u^+ has a least thermic majorant w on $\underline{R}^n \times]0, c[$, which satisfies

$$\limsup_{t \to 0^+} w(x, t) \leq \kappa v(x, 0) < \infty$$

for all $x \in \underline{R}^n$. It now follows from Theorem 5.9 that u is the Gauss-Weierstrass integral of $u(\cdot, 0^+)$ on $\underline{R}^n \times]0, c[$. Since b and c are any numbers such that $0 < c < b < a$, the result follows.

Remark. It follows from (30) that, whenever $0 < b < a$, there is a constant C such that

$$u(x, 0^+) \leq C \exp\left(\frac{\|x\|^2}{4b}\right)$$

for all $x \in \underline{R}^n$.

Corollary

Let u be a temperature on $\underline{R}^n \times]0, a[$, where $a < \infty$, and suppose that there are constants $\kappa > 0$ and $\alpha \in]0, 1/4a]$ such that

$$u(x, t) \leq \kappa \exp(\alpha \|x\|^2)$$

for all $(x, t) \in \underline{R}^n \times]0, a[$. If

$$\limsup_{t \to 0^+} u(x, t) > -\infty$$

for all $x \in \underline{R}^n$, then u is the Gauss-Weierstrass integral of $u(\cdot, 0^+)$ on $\underline{R}^n \times]0, a[$.

Proof. Since $\exp(\alpha \|x\|^2)$ increases with α, we can assume that $\alpha = 1/4a$. If $0 < t < c < b < a$, then

$$W(x, b - t) \leq (4\pi(b - c))^{-n/2} \exp(-\alpha \|x\|^2)$$

for all $x \in \underline{R}^n$, so that

$$W(x, b - t)u^+(x, t) \leq (4\pi(b - c))^{-n/2} \exp(-\alpha \|x\|^2)u^+(x, t)$$

$$\leq \kappa(4\pi(b - c))^{-n/2}$$

Hence the hypotheses of Theorem 5.10 are satisfied, and the result follows.

Remark. The condition that $\alpha \leq 1/4a$ in the above corollary cannot be removed. Shortly after Theorem 3.7, we gave an example in which we showed that the Gauss-Weierstrass integral of the function

$$x \longmapsto (1/2) \exp{(x^2/16)} \cos{(x^2/16)}$$

is defined on $\underline{R} \times]0,4[$, and can be extended to a temperature u on $\underline{R} \times]0,\infty[$, but is itself undefined on $\underline{R} \times]4,\infty[$. It follows easily from the explicit formula for u on $\underline{R} \times]0,4[$ that

$$|u(x,t)| \leq \kappa \ \exp{(x^2/16)} \tag{31}$$

for all $(x,t) \in \underline{R} \times]0,4[$, and it follows from the formula (9) of Chapter III that u is bounded on $\underline{R} \times]2,\infty[$, since $u(\cdot,2)$ is bounded. Hence (31) holds throughout $\underline{R} \times]0,\infty[$, and this implies that the hypotheses of the corollary are satisfied with any $\alpha \geq 1/16$. But the conclusion of the corollary does not hold on $\underline{R} \times]0,a[$ for any $a > 4$.

We now give a result which will assist us to extend Theorem 5.6, but which is also of independent interest.

Theorem 5.11

Let $u \in R_a$, and let w be the least thermic majorant of u^+ on $\underline{R}^n \times]0,a[$. If

$$\liminf_{t \to 0^+} w(x,t) < \infty \tag{32}$$

for all $x \in \underline{R}^n$, and

$$\liminf_{t \to 0^+} u(x,t) \leq 0 \tag{33}$$

for m-almost all $x \in \underline{R}^n$, then $u \leq 0$ on $\underline{R}^n \times]0,a[$.

Proof. Since $u \in R_a$, it is the Gauss-Weierstrass integral of a signed measure μ on $\underline{R}^n \times]0,a[$. Since w is a non-negative temperature, it follows from (32) and Theorem 5.5 that w is the Gauss-Weierstrass integral of $w(\cdot,0^+)$. Therefore, by Theorem 5.7 and the uniqueness of the Gauss-Weierstrass representation (Theorem 1.5), $d\mu^+(y) = w(y,0^+) \, dy$. Thus μ^+ has no singular part, so that u has the representation

$$u(x,t) = \int_{\underline{R}^n} W(x-y,t)u(y,0^+)\,dy - \int_{\underline{R}^n} W(x-y,t)\,d\sigma(y)$$

for all $(x,t) \in \underline{R}^n \times \,]0,a[$, where σ is a non-negative measure which is singular with respect to m. It therefore follows from (33) that

$$u(x,t) \le -\int_{\underline{R}^n} W(x-y,t)\,d\sigma(y) \le 0$$

for all $(x,t) \in \underline{R}^n \times \,]0,a[$.

Remark. It is interesting to compare Theorem 5.11 with the extended maximum principle of Theorem 3.1. The latter result has the weaker growth constraint on u^+, one which does not even imply that u^+ has a thermic majorant, but in Theorem 5.11 the initial condition involves only normal limits.

We now give a slightly weaker result, which does not directly mention thermic majorization, and has a form which is closely related to that of the definition of the class R_a and Theorem 3.13. Its relationship to Theorem 5.11 is similar to that of Theorem 5.10 to Theorem 5.9.

Theorem 5.12

Let u be a temperature on $\underline{R}^n \times \,]0,a[$ such that, whenever $0 < c < b < a$, the function

$$t \longmapsto W(x,b-t)u^+(x,t)$$

is bounded on $]0,c[$. If

$$\liminf_{t \to 0^+} u(x,t) \le 0$$

for m-almost all $x \in \underline{R}^n$, then $u \le 0$ on $\underline{R}^n \times \,]0,a[$.

Proof. If $0 < c < b < a$, and v is defined on $\underline{R}^n \times [0,b[$ by

$$v(x,t) = (4\pi b^2)^{n/2}(b-t)^{-n/2}\exp(\|x\|^2/4(b-t))$$

then there is a constant κ such that κv is a thermic majorant of u^+ on $\underline{R}^n \times \,]0,c[$, so that $u \in R_c$. The proof of this is given in the proof of

Theorem 5.10. Therefore, by Theorem 5.7, u^+ has a least thermic majorant w on $\underline{R}^n \times]0, c[$, which satisfies

$$\liminf_{t \to 0^+} w(x, t) \leq \kappa v(x, 0) < \infty$$

for all $x \in \underline{R}^n$. It now follows from Theorem 5.11 that $u \leq 0$ on $\underline{R}^n \times]0, c[$. Since b and c are any numbers such that $0 < c < b < a$, the result follows.

Corollary

Let u be a temperature on $\underline{R}^n \times]0, a[$, and suppose that there are positive constants κ and α such that

$$u(x, t) \leq \kappa \exp(\alpha \|x\|^2) \tag{34}$$

for all $(x, t) \in \underline{R}^n \times]0, a[$. If

$$\liminf_{t \to 0^+} u(x, t) \leq 0$$

for m-almost all $x \in \underline{R}^n$, then $u \leq 0$ on $\underline{R}^n \times]0, a[$.

Proof. Since $\exp(\alpha \|x\|^2)$ increases with α, we can assume that $\alpha = k/4a$ for some $k \in \underline{N}$. We begin by proving that $u \leq 0$ on $\underline{R}^n \times]0, a/k[$. If $0 < t < c < b < a/k$, then

$$W(x, b - t) \leq (4\pi(b - c))^{-n/2} \exp(-\alpha \|x\|^2)$$

for all $x \in \underline{R}^n$, so that

$$W(x, b - t) u^+(x, t) \leq (4\pi(b - c))^{-n/2} \kappa$$

by (34). Therefore the hypotheses of Theorem 5.12 are satisfied on $\underline{R}^n \times]0, a/k[$, which implies that $u \leq 0$ there. This proves the result in the case where $k = 1$. If $k > 1$, we apply Theorem 2.4 to the function

$$(x, t) \longmapsto u\left(x, t + \frac{a}{2k}\right)$$

on $\underline{R}^n \times [0, a - (a/2k)[$. Since $u(\cdot, a/2k) \leq 0$, we thus deduce that $u \leq 0$ on $\underline{R}^n \times]a/2k, a[$, which completes the proof.

Remark. In the above corollary, the positive constant α is unrestricted, and the case $a = \infty$ is included, in contrast to the corollary to Theorem 5.10.

It is now a simple matter to generalize Theorem 5.6, and thus characterize W within a wider class of temperatures.

Theorem 5.13

Let $u \in R_a$, let w be the least thermic majorant of u^+ on $\underline{R}^n \times]0, a[$, and let $\xi \in \underline{R}^n$. If

$$\liminf_{t \to 0^+} w(x, t) < \infty \tag{35}$$

for all $x \in \underline{R}^n$,

$$\limsup_{t \to 0^+} u(x, t) > -\infty \quad \text{whenever} \quad x \neq \xi \tag{36}$$

and

$$u(x, 0^+) = 0 \tag{37}$$

for m-almost all $x \in \underline{R}^n$, then there is a non-negative constant κ such that

$$u(x, t) = -\kappa W(x - \xi, t)$$

for all $(x, t) \in \underline{R}^n \times]0, a[$.

Proof. It follows from (35), (37), and Theorem 5.11, that $u \leq 0$ on $\underline{R}^n \times]0, a[$. Therefore, by (36), (37), and Theorem 5.6 applied to -u, there is a non-negative constant κ such that $-u(x, t) = \kappa W(x - \xi, t)$ for all $(x, t) \in \underline{R}^n \times]0, a[$.

Using Theorem 5.12 instead of Theorem 5.11, we can similarly obtain the following result.

Theorem 5.14

Let u be a temperature on $\underline{R}^n \times]0, a[$ such that, whenever $0 < c < b < a$, the function

$$t \longmapsto W(x, b - t) u^+(x, t)$$

is bounded on $]0, c[$, and let $\xi \in \underline{R}^n$. If

$$\limsup_{t \to 0^+} u(x, t) > -\infty \quad \text{whenever} \quad x \neq \xi$$

and

$$u(x, 0^+) = 0$$

for m-almost all $x \in \underline{R}^n$, then there is a constant $\kappa \geq 0$ such that

$$u(x, t) = -\kappa W(x - \xi, t)$$

for all $(x, t) \in \underline{R}^n \times]0, a[$.

Another version can similarly be obtained from the corollary to Theorem 5.12. We omit the details.

5. LINEAR PARABOLIC EQUATIONS

As usual, $W(x - y, t - s)$ must be replaced by $\Gamma(x, t; y, s)$ throughout, where Γ denotes the fundamental solution on $\underline{R}^n \times]0, a[$.

In the general case, the last inequality in Theorem 5.1 is replaced by

$$\limsup_{t \to 0^+} u(x, t) \leq \lambda(\bar{D}\mu)(x)$$

for some positive constant λ. This is proved by essentially the method given in the text, and requires the sharp lower estimate for Γ. The constant λ depends on n and the constants in both the upper and lower estimates for Γ. The first inequality in Theorem 5.1 can be extended without difficulty, and requires only that there is a constant $C > 0$ such that

$$\Gamma(x, t; y, 0) \geq C t^{-n/2} \tag{38}$$

whenever $\|x - y\| < \sqrt{t}$, and then κ depends on C. This inequality is a
consequence of others obtained during the parametrix construction of Γ,
and can therefore be obtained under mild conditions on the coefficients.
Specifically, the equation needs to be uniformly parabolic, with coefficients which are uniformly continuous, bounded, Hölder continuous in x
uniformly with respect to t, and Hölder continuous in t uniformly with
respect to x. Thus the first inequality in Theorem 5.1 does not require
such strong conditions on the coefficients as does the second, which is
important because only the first is used in the proofs of subsequent
results in this chapter. Theorems 5.4, 5.5 (and its corollary), and 5.6,
can be extended without alteration to statement or proof, under the conditions just described.

With regard to the remark following Theorem 5.6, see the com-
ments on Theorem 2.13 in Chapter II, section 8.

All of the results (and the example) in section 4 depend on earlier
theorems which require the sharp lower estimate for the fundamental
solution. Theorems 5.7, 5.8, 5.9, 5.11, 5.13, and 5.14 carry over to
the general case unchanged, using the given techniques.

In extending Theorem 5.10 the function in (30) is replaced by the
function $t \longmapsto \Gamma(0, b; x, t) u^+(x, t)$. If this is bounded on $]0, c[$, there are
positive constants C and γ such that

$$u(x, t) \leq C/\Gamma(0, b; x, t)$$

$$\leq C(b - t)^{n/2} \exp(\gamma \|x\|^2/(b - t))$$

$$\leq C \exp(\gamma \|x\|^2/(b - c))$$

for all $(x, t) \in \underline{R}^n \times]0, c[$; here we have used the sharp lower estimate
for Γ, and allowed C to change from line to line. Using this inequality,
it can be shown that u^+ has a thermic majorant on some substrip of
$\underline{R}^n \times]0, c[$, so that the conclusion of the theorem can be established
only on a substrip of $\underline{R}^n \times]0, a[$. The conclusion of the corollary must
be weakened in a similar way, but the weaker result does not require
that $\alpha \leq 1/4a$.

The above changes required to extend Theorem 5.10 affect the
proofs of Theorem 5.12 and its corollary, but their statements remain
essentially unchanged. As in the case $k > 1$ of the proof of the corollary,
the result is first obtained on some substrip of $\underline{R}^n \times]0, a[$, and then
carried across the whole strip using Theorem 2.4.

6. BIBLIOGRAPHICAL NOTES

For the heat equation, Watson [5] proved that the result of Theorem 5.1
holds with $\kappa = 1$, even if the measure μ is signed. In fact, the argument
used in the text to prove the inequality between the upper limits does
not require μ to be non-negative, and can therefore be applied to $-\mu$
also. However, for more general equations, we do not have an explicit
fundamental solution, and in order to use estimates we require that μ
is non-negative. The extension described in section 5 was proved by
Chabrowski and Watson [1, Theorem 2]. For the heat equation, Watson
[10] proved that, if u and v are the Gauss-Weierstrass integrals of μ
and ν respectively, where μ is a signed measure and ν a positive one
whose support is $\underline{R}^n$, then

$$\liminf_{r \to 0} \frac{\mu(B(x,r))}{\nu(B(x,r))} \leq \liminf_{t \to 0^+} \frac{u(x,t)}{v(x,t)} \leq \limsup_{t \to 0^+} \frac{u(x,t)}{v(x,t)} \leq \limsup_{r \to 0} \frac{\mu(B(x,r))}{\nu(B(x,r))}$$

Applications of these inequalities have been given by Watson [10, 12,
19]. Even with multiplicative constants and μ non-negative, this result
cannot easily be generalized, but Chabrowski [9, Section 6] has proved
some results in this direction, and extended some of the applications.

The results in section 2 are due to Besicovitch [1, 2], and Theorem 5.3 can be seen as a special case of his Theorem 3 in [1]. Some
of the proofs in the chapter are greatly modified versions of the originals.

Watson's paper [19] contains Theorem 5.4 in the general case, as
well as an extension to arbitrary members of R_a for the heat equation
only. This extension provides quick and easy proofs of sharper forms
of Theorems 5.8, 5.11, and 5.13, in which the conditions on the least
thermic majorant of u^+ are replaced by similar (or, in the case of Theorem 5.8, weaker) conditions on u itself.

Theorem 5.5 has evolved slowly. For the heat equation with $n = 1$,
Pollard [1] proved that a non-negative solution u, with finite normal limits everywhere on $\underline{R} \times \{0\}$, is the Gauss-Weierstrass integral of
$u(\cdot, 0^+)$. For the heat equation with n arbitrary, and the lower limits
replaced by upper limits, the result was proved by Kato [1] as part of
the proof of his Theorem 2. This result was extended to more general
parabolic equations by Chabrowski [2, 7]. Theorem 5.5 itself was
proved for the heat equation by Watson [9], and generalized by Chabrowski and Watson [1]. In addition, the corollary alone has been considered
in some papers. Widder [1] proved that a non-negative temperature on
$\underline{R} \times]0, a[$, with normal limits zero everywhere on $\underline{R} \times \{0\}$, is identically
zero, and Watson [5] proved the corollary itself, for the heat equation

with n arbitrary. The corollary can also be deduced from Theorem 5.6, which further complicates its history.

Krzyżański [6] proved Theorem 5.6, in the general case, with the hypotheses (24) and (25) strengthened to $u(x, 0^+) = 0$ for all $x \neq \xi$. Closely related results were then proved by Aronson [3], [5, p. 691]. A result equivalent to Theorem 5.6 was given by Watson [5, Theorem 5] for the heat equation, and by Chabrowski and Watson [1, Theorem 3] for the general case.

Theorem 5.7 is due to Watson [9] for the heat equation, and to Chabrowski and Watson [1, Theorem 8] for the general case, where the word 'thermic' is replaced by 'parabolic.'

Theorem 5.8 was given, for the general case, by Watson [19].

Theorem 5.9 was proved by Watson [9, Theorem 1] for the heat equation, and by Chabrowski and Watson [1, Theorem 7] for the general case, in which the result holds on the whole strip and not just a substrip as stated. Subsequently, Watson [19] proved that the hypothesis (28) can be replaced by

$$\liminf_{t \to 0^+} u(x, t) < \infty$$

for all $x \in \underline{R}^n$, in the case of the heat equation.

The precise form of Theorem 5.10 is new, but the corollary is given by Chabrowski and Watson [1, p. 487], along with details of the function v which appears in the proof of Theorem 5.10 in the general case [1, Lemma 2]. Earlier Dressel [1, Theorem 3] had proved the corollary for the heat equation with n = 1.

Theorem 5.11 is due to Watson [9] for the heat equation, and the extension to the general case is routine. Much earlier, for the heat equation with n = 1, Gehring [2, Theorem 10] proved a stronger result, in which the condition on w is replaced by a similar one on u. This has since been extended to the case of an arbitrary n by Watson [12, Theorem 7], [19].

The precise form of Theorem 5.12 is new, but the corollary was proved for the heat equation by Gehring [2, Theorem 11], in the case where n = 1, and by Watson [9] for arbitrary n.

Theorem 5.13 is new for the general case, but Watson [19] has given a sharper version for the heat equation. Theorem 5.14 is also new.

Further representation theorems involving normal limit hypotheses have been proved for the heat equation by Shapiro [1] and Watson [12, 19].

6

Hyperplane Conditions and
Representation Theorems

Theorems 5.5, 5.9, and 5.10 all give conditions under which a temperature u on $R^n \times]0, a[$ can be written as the Gauss-Weierstrass integral of $u(\cdot, 0^+)$. Each result combines a condition on u over the whole strip (such as $u \geq 0$), with a condition on the normal limits of u (such as $\liminf_{t \to 0^+} u(x, t) < \infty$ for all x). In the present chapter, we obtain the same conclusion from hypotheses which do not involve normal limits. Firstly, given any number $p > 1$, we characterize the Gauss-Weierstrass integrals u of functions f such that

$$\int_{R^n} W(x, b) | f(x) |^p \, dx < \infty \tag{1}$$

for all $b \in]0, a[$, in terms of the integral means

$$t \longmapsto \left(\int_{R^n} W(x, b - t) | u(x, t) |^p \, dx \right)^{1/p}$$

Of course, by Theorem 4.1, if u is the Gauss-Weierstrass integral of any function f, then $u(\cdot, 0^+) = f$ m-a. e. on $\underline{R}^n$. We also discuss those temperatures which satisfy a similar condition on u^+ rather than $|u|$, and obtain a result analogous to Theorem 5.8. Secondly, for non-negative temperatures, we derive a condition on the function

$$t \longmapsto m(\{x \in \underline{R}^n : u(x, t) > 1\})$$

as $t \to 0^+$ which again ensures that u is the Gauss-Weierstrass integral of $u(\cdot, 0^+)$.

1. CONVERGENT SEQUENCES OF INTEGRALS

Let μ be a non-negative measure on $\underline{R}^n$, and let $p \in]1, \infty[$. For any μ-measurable function f on $\underline{R}^n$, we put

$$\|f\|_{p, \mu} = \|f\|_p = \left(\int_{\underline{R}^n} |f|^p \, d\mu\right)^{1/p}$$

The class of all f such that $\|f\|_{p, \mu} < \infty$ is the Lebesgue class $L^p(\mu)$. We denote by q the Hölder conjugate of p, defined by

$$\frac{1}{p} + \frac{1}{q} = 1 \tag{2}$$

In this section, we prove a theorem on the convergence of sequences of integrals, under hypotheses which involve L^p conditions. This will be used in the next section to prove our characterization of the Gauss-Weierstrass integrals of functions f which satisfy (1).

Theorem 6.1

Let μ be a non-negative measure on $\underline{R}^n$, and let $p > 1$. Suppose that $\{f_j\}$ is a sequence of functions on $\underline{R}^n$ which converges pointwise to a function f μ-a. e., and that there is a real number κ such that

$$\|f_j\|_{p, \mu} \leq \kappa \tag{3}$$

for all j. Then

$$\lim_{j \to \infty} \int_{\underline{R}^n} f_j \, g \, d\mu = \int_{\underline{R}^n} fg \, d\mu \qquad (4)$$

for all $g \in L^q(\mu)$, where q is given by (2).

Proof. We first extend (3) to include f. By Fatou's lemma and (3),

$$\int_{\underline{R}^n} |f|^p \, d\mu = \int_{\underline{R}^n} \lim_{j \to \infty} |f_j|^p \, d\mu \le \liminf_{j \to \infty} \int_{\underline{R}^n} |f_j|^p \, d\mu \le \kappa^p$$

so that

$$\|f\|_p \le \kappa \qquad (5)$$

In order to prove (4), we must estimate

$$\left| \int_{\underline{R}^n} fg \, d\mu - \int_{\underline{R}^n} f_j \, g \, d\mu \right| \le \int_{\underline{R}^n} |f - f_j| \, |g| \, d\mu \qquad (6)$$

for an arbitrary $g \in L^q(\mu)$. Note that the finiteness of the integrals follows from Hölder's inequality and (5). We split up $\underline{R}^n$ into three subsets. Since the sequence $\{B(0, k)\}$ is expanding and has union $\underline{R}^n$,

$$\int_{B(0, k)} |g|^q \, d\mu \to \int_{\underline{R}^n} |g|^q \, d\mu$$

as $k \to \infty$. Therefore

$$\int_{\underline{R}^n \setminus B(0, k)} |g|^q \, d\mu \to 0$$

so that, given any $\epsilon > 0$, we can find k such that, if $A = B(0, k)$,

$$2\kappa \left(\int_{\underline{R}^n \setminus A} |g|^q \, d\mu \right)^{1/q} < \epsilon/3 \qquad (7)$$

Furthermore, $\mu(A) < \infty$ because A is bounded.

Next, by the absolute continuity of ν with respect to μ, where $d\nu = |g|^q \, d\mu$, we can find $\delta > 0$ such that

$$2\kappa \left(\int_E |g|^q \, d\mu \right)^{1/q} < \epsilon/3$$

for any set E with $\mu(E) < \delta$.

By Egoroff's theorem, there is a subset B of A such that $\mu(A \setminus B) < \delta$ and $f_j \to f$ uniformly on B. Since $\mu(A \setminus B) < \delta$, we have

$$2\kappa \left(\int_{A \setminus B} |g|^q \, d\mu \right)^{1/q} < \epsilon/3 \tag{8}$$

Since $|f - f_j| \to 0$ uniformly on B, we can find $N \in \underline{N}$ such that

$$\left(\int_B |f - f_j|^p \, d\mu \right)^{1/p} \|g\|_q < \epsilon/3 \tag{9}$$

whenever $j \geq N$. We now have our three subsets $\underline{R}^n \setminus A$, $A \setminus B$, and B, of $\underline{R}^n$, and can proceed to estimate the last integral in (6).

By Minkowski's inequality, (3), and (5),

$$\|f - f_j\|_p \leq \|f\|_p + \|f_j\|_p \leq 2\kappa$$

Therefore, by Hölder's inequality and (7),

$$\int_{\underline{R}^n \setminus A} |f - f_j| \, |g| \, d\mu \leq \|f - f_j\|_p \left(\int_{\underline{R}^n \setminus A} |g|^q \, d\mu \right)^{1/q} < \epsilon/3$$

and by Hölder's inequality and (8),

$$\int_{A \setminus B} |f - f_j| \, |g| \, d\mu \leq \|f - f_j\|_p \left(\int_{A \setminus B} |g|^q \, d\mu \right)^{1/q} < \epsilon/3$$

Finally, by Hölder's inequality and (9), whenever $j \geq N$ we have

$$\int_B |f - f_j| \, |g| \, d\mu \leq \left(\int_B |f - f_j|^p \, d\mu \right)^{1/p} \|g\|_q < \epsilon/3$$

Hence

$$\int_{\underline{R}^n} |f - f_j| \, |g| \, d\mu = \left(\int_{\underline{R}^n \setminus A} + \int_{A \setminus B} + \int_B \right) |f - f_j| \, |g| \, d\mu < \epsilon$$

for all $j \geq N$, so that the result follows from (6).

2. TEMPERATURES WHICH SATISFY L^p CONDITIONS

In Chapter III, we used the integral means M_b, $0 < b < a$, to characterize those temperatures which can be represented as the Gauss-Weierstrass integrals of signed measures on $\underline{R}^n \times]0, a[$. In this section, we consider the corresponding problem with the means M_b applied, not to the temperatures themselves, but to the p-th powers of their moduli, for each $p > 1$.

If $1 < p < \infty$, $0 \le t < b$, and v is a non-negative function on $\underline{R}^n \times \{t\}$, the integral mean $M_{b,p}(v; t)$ is defined by

$$M_{b,p}(v; t) = M_b(v^p; t)^{1/p} = \left(\int_{\underline{R}^n} W(x, b - t) v(x, t)^p \, dx \right)^{1/p}$$

Theorem 6.2

Let $p > 1$, let $r \in [0, a[$, and let f be a function on $\underline{R}^n \times \{r\}$ such that

$$M_{b,p}(|f|; r) < \infty \qquad (10)$$

for every $b \in]r, a[$. If

$$u(x, t) = \int_{\underline{R}^n} W(x - y, t - r) f(y, r) \, dy$$

then u is defined and is a temperature on $\underline{R}^n \times]r, a[$. Furthermore, for each $b \in]r, a[$, $M_{b,p}(|u|; \cdot)$ is non-increasing on $]r, b[$, and

$$M_{b,p}(|f|; r) = \lim_{t \to r^+} M_{b,p}(|u|; t) \qquad (11)$$

Proof. If $r < b < a$, it follows from Hölder's inequality that

$$\int_{\underline{R}^n} W(x, b - r) |f(x, r)| \, dx$$

$$= \int_{\underline{R}^n} W(x, b - r)^{1/q} (W(x, b - r)^{1/p} |f(x, r)|) \, dx$$

$$\le \left(\int_{\underline{R}^n} W(x, b - r) \, dx \right)^{1/q} M_{b,p}(|f|; r)$$

By Lemma 1.1 and (10), this is finite. Therefore, by Lemma 3.1, u is defined and is a temperature on $\underline{R}^n \times]r, a[$.

It follows from Theorem 1.4 that

$$u(x, t) = \int_{\underline{R}^n} W(x - y, t - s)u(y, s)\, dy$$

whenever $r < s < t < a$ and $x \in \underline{R}^n$. If we write $u(\cdot, r) = f(\cdot, r)$, then this holds when $s = r$ also. By Jensen's inequality and Lemma 1.1,

$$|u(x, t)|^p \leq \int_{\underline{R}^n} W(x - y, t - s)|u(y, s)|^p\, dy$$

Therefore, if $r \leq s < t < b < a$, it follows from Theorem 1.4 and Fubini's theorem that

$$M_{b,p}(|u|; t)^p \leq \int_{\underline{R}^n} W(x, b - t)\, dx \int_{\underline{R}^n} W(x - y, t - s)|u(y, s)|^p\, dy$$

$$= \int_{\underline{R}^n} |u(y, s)|^p\, dy \int_{\underline{R}^n} W(-x, b - t)W(x - y, t - s)\, dx$$

$$= \int_{\underline{R}^n} W(y, b - s)|u(y, s)|^p\, dy$$

Thus $M_{b,p}(|u|; \cdot)$ is non-increasing on $[r, b[$. Hence the limit in (11) exists and, by Theorem 4.1 and Fatou's lemma,

$$M_{b,p}(|f|; r) = \left(\int_{\underline{R}^n} \lim_{t \to r^+} (W(x, b - t)|u(x, t)|^p)\, dx \right)^{1/p}$$

$$\leq \lim_{t \to r^+} M_{b,p}(|u|; t)$$

$$\leq M_{b,p}(|f|; r)$$

which proves (11).

We seek a converse to Theorem 6.2. More precisely, if u is a temperature on $\underline{R}^n \times]0, a[$, we seek a condition on $M_{b,p}(|u|; t)$, whenever $0 < t < b < a$, which implies that u can be written as the Gauss-Weierstrass integral of a function f on $\underline{R}^n$ such that (10) holds with $r = 0$. Since (11) implies that $M_{b,p}(|u|; \cdot)$ is bounded on $]0, b[$ whenever b < a, we might expect that the boundedness of the means would be a sufficient condition for the converse result. This turns out to be the case, but we want as weak a sufficient condition as possible. By analogy with the first part of Theorem 3.13, we might expect that, if each $M_{b,p}(|u|; \cdot)$ is locally integrable on $]0, b[$ and satisfies

$$\liminf_{t \to 0^+} M_{b,p}(|u|; t) < \infty$$

then each $M_{b,p}(|u|; \cdot)$ is bounded. This, too, turns out to be true. However, closer analogy with the results of Chapter III cannot be obtained here, in that conditions on $M_{b,p}(u^+; \cdot)$ alone cannot imply that u is the Gauss-Weierstrass integral of any function. For example, if u = -W then $u^+ = 0$, but u is the Gauss-Weierstrass integral of minus the Dirac δ-measure concentrated at 0, and therefore not of any function (by Theorem 1.5). Despite this, we still give some results about $M_{b,p}(u^+; \cdot)$, as these are needed in section 3 and imply similar results about $M_{b,p}(|u|; \cdot)$.

Lemma 6.1

Let p > 1, and let u be a temperature on $R^n \times]0, a[$. If, for each b < a, $M_{b,p}(u^+; \cdot)$ is locally integrable on $]0, b[$ and

$$\liminf_{t \to 0^+} M_{b,p}(u^+; t) < \infty$$

then $u \in R_a$, so that u is the Gauss-Weierstrass integral on $R^n \times]0, a[$ of a signed measure on R^n.

Proof. By Hölder's inequality and Lemma 1.1, whenever $0 < t < b < a$ we have

$$\int_{R^n} W(x, b - t) u^+(x, t)\, dx$$

$$= \int_{R^n} W(x, b - t)^{1/q}(W(x, b - t)^{1/p} u^+(x, t))\, dx$$

$$\leq \left(\int_{R^n} W(x, b - t)\, dx\right)^{1/q} M_{b,p}(u^+; t)$$

$$= M_{b,p}(u^+; t)$$

Therefore $M_b(u^+; \cdot)$ is locally integrable on $]0, b[$ and

$$\liminf_{t \to 0^+} M_b(u^+; t) < \infty$$

for each $b < a$, so that $u \in R_a$ and u is the Gauss-Weierstrass integral on $\underline{R}^n \times]0, a[$ of a signed measure (by Theorem 3.15).

Theorem 6.3

Let $p > 1$, and let u be a temperature on $\underline{R}^n \times]0, a[$. If, for each $b < a$, $M_{b,p}(u^+; \cdot)$ is locally integrable on $]0, b[$ and

$$\liminf_{t \to 0^+} M_{b,p}(u^+; t) < \infty \tag{12}$$

then $M_{b,p}(u^+; \cdot)$ is non-increasing, bounded and continuous on $]0, b[$ for each $b < a$.

Proof. We first show that the means are non-increasing and bounded. By Lemma 6.1, $u \in R_a$. Since $R_a \subseteq S_a$, it follows from Theorem 3.5 that

$$u^+(x, t) \le \int_{\underline{R}^n} W(x - y, t - s) u^+(y, s) \, dy$$

whenever $x \in \underline{R}^n$ and $0 < s < t < a$. In view of Lemma 1.1, it now follows from Jensen's inequality that

$$u^+(x, t)^p \le \int_{\underline{R}^n} W(x - y, t - s) u^+(y, s)^p \, dy \tag{13}$$

Therefore, whenever $0 < s < t < b < a$, it follows from Fubini's theorem and Theorem 1.4 that

$$M_{b,p}(u^+; t)^p \le \int_{\underline{R}^n} W(x, b - t) \, dx \int_{\underline{R}^n} W(x - y, t - s) u^+(y, s)^p \, dy$$

$$= \int_{\underline{R}^n} u^+(y, s)^p \, dy \int_{\underline{R}^n} W(-x, b - t) W(x - y, t - s) \, dx$$

$$= \int_{\underline{R}^n} W(y, b - s) u^+(y, s)^p \, dy$$

$$= M_{b,p}(u^+; s)^p$$

Hence $M_{b,p}(u^+; \cdot)$ is non-increasing on $]0, b[$. Therefore $M_{b,p}(u^+; s)$ tends to a limit as $s \to 0^+$, and it follows from (12) that

$$M_{b,p}(u^+; t) \le \lim_{s \to 0^+} M_{b,p}(u^+; s) < \infty$$

for every $t \in]0, b[$. Thus $M_{b,p}(u^+; \cdot)$ is bounded on $]0, b[$.

We now prove that $M_{b,p}(u^+; \cdot)$ is continuous, using the argument with which we established the continuity of the means M_b in the proof of Theorem 3.8. Let $r \in]0, b[$, and let $\{r_j\}$ be an arbitrary sequence in $]r/2, b[$ with limit r. Since u^+ is continuous, it follows from Fatou's lemma that

$$\liminf_{j \to \infty} M_{b,p}(u^+; r_j)^p \ge \int_{\underline{R}^n} \lim_{j \to \infty} (W(x, b - r_j) u^+(x, r_j)^p) \, dx$$

$$= \int_{\underline{R}^n} W(x, b - r) u^+(x, r)^p \, dx$$

$$= M_{b,p}(u^+; r)^p \tag{14}$$

Next, if w is defined on $\underline{R}^n \times]r/2, a[$ by

$$w(x, t) = \int_{\underline{R}^n} W(x - y, t - (r/2)) u^+(y, r/2)^p \, dy$$

then w is a temperature and $M_b(w; t) = M_{b,p}(u^+; r/2)^p$ whenever $r/2 < t < b$, by Lemma 3.1 (using the finiteness of $M_{b,p}(u^+; r/2)$, which was established in the first part of this proof). Furthermore, $u^+(\cdot, r_j)^p \le w(\cdot, r_j)$ for all j, by (13). It therefore follows, from the continuity of $w - (u^+)^p$ and Fatou's lemma, that

$$M_{b,p}(u^+; r/2)^p - M_{b,p}(u^+; r)^p = M_b(w; r) - M_{b,p}(u^+; r)^p$$

$$= \int_{\underline{R}^n} W(x, b - r)(w(x, r) - u^+(x, r)^p) \, dx$$

$$= \int_{\underline{R}^n} \lim_{j\to\infty} (W(x, b - r_j)\{w(x, r_j) - u^+(x, r_j)^p\})\, dx$$

$$\leq \lim_{j\to\infty} \inf\ (M_b(w; r_j) - M_{b,p}(u^+; r_j)^p)$$

$$= M_{b,p}(u^+; r/2)^p - \lim_{j\to\infty} \sup M_{b,p}(u^+; r_j)^p$$

Hence

$$\lim_{j\to\infty} \sup M_{b,p}(u^+; r_j) \leq M_{b,p}(u^+; r)$$

which combines with (14) to show that $M_{b,p}(u^+; r_j) \to M_{b,p}(u^+; r)$ as $j \to \infty$, for any sequence $\{r_j\}$ which converges to r. Thus $M_{b,p}(u^+; \cdot)$ is continuous at r. Since r is any point of $]0, b[$, the theorem is proved.

Corollary

Let $p > 1$, and let u be a temperature on $\underline{R}^n \times]0, a[$. If, for each $b < a$, $M_{b,p}(|u|; \cdot)$ is locally integrable on $]0, b[$ and

$$\lim_{t\to 0^+} \inf M_{b,p}(|u|; t) < \infty$$

then $M_{b,p}(|u|; \cdot)$ is bounded on $]0, b[$ for each $b < a$.

Proof. Since $u^+ \leq |u|$ and $(-u)^+ = u^- \leq |u|$, our hypotheses on $|u|$ imply that similar conditions are satisfied by u^+ and $(-u)^+$. We can therefore apply Theorem 6.3 to both u and -u. Hence, because $(-u)^+ = u^-$, both $M_{b,p}(u^+; \cdot)$ and $M_{b,p}(u^-; \cdot)$ are bounded on $]0, b[$. Furthermore,

$$|u|^p = (\max\{u^+, u^-\})^p = \max\{(u^+)^p, (u^-)^p\} \leq (u^+)^p + (u^-)^p$$

so that

$$M_{b,p}(|u|; \cdot)^p \leq M_{b,p}(u^+; \cdot)^p + M_{b,p}(u^-; \cdot)^p$$

Therefore $M_{b,p}(|u|; \cdot)$ is bounded on $]0, b[$.

We can now prove the converse of Theorem 6.2.

Theorem 6.4

Let $p > 1$, and let u be a temperature on $\underline{R}^n \times]0, a[$. If, for each $b < a$, $M_{b,p}(|u|; \cdot)$ is locally integrable on $]0, \overline{b}[$ and

$$\liminf_{t \to 0^+} M_{b,p}(|u|; t) < \infty$$

then u is the Gauss-Weierstrass integral on $\underline{R}^n \times]0, a[$ of $u(\cdot, 0^+)$ and, for each $b < a$, $M_{b,p}(|u|; \cdot)$ is non-increasing on $]0, b[$ with

$$\lim_{t \to 0^+} M_{b,p}(|u|; t) = M_{b,p}(|u(\cdot, 0^+)|; 0) < \infty$$

Proof. We have only to prove that u is the Gauss-Weierstrass integral of $u(\cdot, 0^+)$ and $M_{b,p}(|u(\cdot, 0^+)|; 0) < \infty$, for the remaining conclusions will then follow from Theorem 6.2. Since $u^+ \leq |u|$, it follows from Lemma 6.1 that u is the Gauss-Weierstrass integral of a signed measure μ. Therefore $u(\cdot, 0^+)$ exists and is finite m-a. e. on $\underline{R}^n$, by Theorem 4.1.

Let $b \in]0, a[$. By the corollary to Theorem 6.3, there is a number κ such that $M_{b,p}(|u|; s) \leq \kappa$ for all $s \in]0, b[$. Therefore, if $0 < s < b/2$,

$$\int_{\underline{R}^n} \exp(-\|y\|^2/2b)\, |u(y, s)|^p\, dy$$

$$\leq (4\pi(b-s))^{n/2} \int_{\underline{R}^n} W(y, b-s)|u(y, s)|^p\, dy$$

$$\leq (4\pi b)^{n/2} \kappa^p$$

Furthermore, if $d\nu(y) = \exp(-\|y\|^2/2b)\, dy$, then $u(\cdot, 0^+)$ exists ν-a. e. on $\underline{R}^n$. Hence, by Theorem 6.1,

$$\lim_{s \to 0^+} \int_{\underline{R}^n} u(y, s) g(y)\, d\nu(y) = \int_{\underline{R}^n} u(y, 0^+) g(y)\, d\nu(y) \qquad (15)$$

whenever $g \in L^q(\nu)$.

We aim to apply (15) with

$$g(y) = W(x - y, t) \exp\left(\frac{\|y\|^2}{2b}\right)$$

for each fixed point $(x, t) \in R^n \times \,]0, b/4[$. Observe that, with this choice of g, we have $g(y) \, d\nu(y) = W(x - y, t) \, dy$. We must first verify that $g \in L^q(\nu)$. Since $\|y\|^2 \leq 2(\|y - x\|^2 + \|x\|^2)$, we have

$$- \|x - y\|^2 \leq \|x\|^2 - \left(\frac{\|y\|^2}{2}\right)$$

If $0 < t < b/4$, then $-1/8t < -1/2b$, so that

$$- \frac{\|x - y\|^2}{4t} \leq \frac{\|x\|^2}{4t} - \frac{\|y\|^2}{8t} \leq \frac{\|x\|^2}{4t} - \frac{\|y\|^2}{2b}$$

Hence, for all $(x, t) \in R^n \times \,]0, b/4[$,

$$\int_{R^n} g(y)^q \, d\nu(y) = \int_{R^n} W(x - y, t)^q \exp\left(\frac{\|y\|^2}{2b}\right)^{q-1} dy$$

$$= (4\pi t)^{-qn/2} \int_{R^n} \exp\left(- \frac{q\|x - y\|^2}{4t} + \frac{(q - 1)\|y\|^2}{2b}\right) dy$$

$$\leq (4\pi t)^{-qn/2} \exp\left(\frac{q\|x\|^2}{4t}\right) \int_{R^n} \exp\left(- \frac{\|y\|^2}{2b}\right) dy$$

$$< \infty$$

so that $g \in L^q(\nu)$. It therefore follows from (15) that

$$\lim_{s \to 0^+} \int_{R^n} W(x - y, t)u(y, s) \, dy = \int_{R^n} W(x - y, t)u(y, 0^+) \, dy \qquad (16)$$

whenever $(x, t) \in R^n \times \,]0, b/4[$. Since the integral on the right-hand side of (16) is the Gauss-Weierstrass integral of $u(\cdot, 0^+)$, we need to show that the left-hand side is just $u(x, t)$. We know that u is the Gauss-Weierstrass integral of some measure μ, so that Theorem 1.4 implies that u has the semigroup property on $R^n \times \,]0, a[$. Hence, whenever $(x, t) \in R^n \times \,]0, b/4[$,

$$\lim_{s \to 0^+} \int_{R^n} W(x - y, t)u(y, s) \, dy = \lim_{s \to 0^+} u(x, s + t) = u(x, t)$$

Therefore (16) shows that u is the Gauss-Weierstrass integral of $u(\cdot, 0^+)$ on $\underline{R}^n \times]0, b/4[$, so that $d\mu(y) = u(y, 0^+)\, dy$ by the uniqueness of the Gauss-Weierstrass representation on this strip (Theorem 1.5).

It only remains to prove that $M_{b,p}(|u(\cdot, 0^+)|; 0) < \infty$ for each $b < a$. By Fatou's lemma,

$$M_{b,p}(|u(\cdot, 0^+)|; 0) = \left(\int_{\underline{R}^n} \lim_{t \to 0^+} (W(x, b - t)|u(x,t)|^p)\, dx \right)^{1/p}$$

$$\leq \liminf_{t \to 0^+} M_{b,p}(|u|; t)$$

which is finite by hypothesis.

Corollary

Let $p > 1$, and let u be a temperature on $\underline{R}^n \times]0, a[$. If there is a constant $\alpha \in]0, 1/4a]$ such that the function

$$t \longmapsto \int_{\underline{R}^n} \exp(-\alpha\|x\|^2)|u(x,t)|^p\, dx \qquad (17)$$

is locally integrable on $]0, a[$, and

$$\liminf_{t \to 0^+} \int_{\underline{R}^n} \exp(-\alpha\|x\|^2)|u(x,t)|^p\, dx < \infty \qquad (18)$$

then u is the Gauss-Weierstrass integral on $\underline{R}^n \times]0, a[$ of $u(\cdot, 0^+)$, and

$$\int_{\underline{R}^n} \exp(-\alpha\|x\|^2)|u(x, 0^+)|^p\, dx < \infty$$

Proof. Whenever $0 < t < c < b < a \leq 1/4\alpha$, we have

$$W(x, b - t) \leq (4\pi(b - c))^{-n/2} \exp(-\alpha\|x\|^2)$$

for all $x \in \underline{R}^n$, so that

$$M_{b,p}(|u|; t)^p \leq (4\pi(b - c))^{-n/2} \int_{\underline{R}^n} \exp(-\alpha\|x\|^2)|u(x,t)|^p\, dx$$

It therefore follows from (17) and (18) that u satisfies the hypotheses of Theorem 6.4, so that u is the Gauss-Weierstrass integral on $\underline{R}^n \times]0, a[$ of $u(\cdot, 0^+)$. Finally, by Fatou's lemma and (18),

$$\int_{\underline{R}^n} \exp(-\alpha\|x\|^2) |u(x, 0^+)|^p \, dx \le \liminf_{t \to 0^+} \int_{\underline{R}^n} \exp(-\alpha\|x\|^2) |u(x, t)|^p \, dx$$

$$< \infty$$

Remark. The condition that $\alpha \le 1/4a$ in the above corollary cannot be removed. Shortly after Theorem 3.7, we gave an example in which we showed that the Gauss-Weierstrass integral of the function

$$x \longmapsto (1/2) \exp\left(\frac{x^2}{16}\right) \cos\left(\frac{x^2}{16}\right)$$

is defined on $\underline{R} \times]0, 4[$, and can be extended to a temperature u on $\underline{R} \times]0, \infty[$, but is itself undefined on $\underline{R} \times]4, \infty[$. As we showed in the remark at the end of section 6 in Chapter III,

$$|u(x, t)| \le C \exp\left(\frac{x^2}{16}\right)$$

for all $(x, t) \in \underline{R} \times]0, \infty[$. It follows that the conditions of the corollary are satisfied with any $\alpha > p/16$ and any $p > 1$, but the conclusion does not hold on $\underline{R} \times]0, a[$ for any $a > 4$.

3. TEMPERATURES WHICH SATISFY
   ONE-SIDED L^p CONDITIONS

We consider the Gauss-Weierstrass representation of temperatures which satisfy conditions on their positive parts similar to those applied to moduli in Theorem 6.4. As we remarked after Theorem 6.2, we cannot expect such one-sided conditions to imply that the temperatures are the Gauss-Weierstrass integrals of functions.

We use the concept of the least thermic majorant w of u^+, where u is the Gauss-Weierstrass integral of a signed measure μ. By Theorem 5.7, w is the Gauss-Weierstrass integral of μ^+, but we also require a different representation of w, which we now establish.

Theorem 6.5

Let u be the Gauss-Weierstrass integral on $\underline{R}^n \times]0, a[$ of a signed measure μ. Then the least thermic majorant w of u^+ on $\underline{R}^n \times]0, a[$ has the representation

$$w(x, t) = \lim_{r \to 0^+} \int_{\underline{R}^n} W(x - y, t - r) u^+(y, r) \, dy$$

for all $(x, t) \in \underline{R}^n \times]0, a[$.

Proof. We know, from Theorem 5.7, that w exists and is the Gauss-Weierstrass integral of μ^+. Therefore, by Theorem 1.4, w has the semi-group property on $\underline{R}^n \times]0, a[$. Hence, whenever $0 < s < t < a$ and $x \in \underline{R}^n$,

$$w(x, t) = \int_{\underline{R}^n} W(x - y, t - s) w(y, s) \, dy$$

$$\geq \int_{\underline{R}^n} W(x - y, t - s) u^+(y, s) \, dy$$

$$\geq u^+(x, t) \tag{19}$$

by Theorem 3.5, since $u \in R_a \subseteq S_a$. For each $r \in]0, a[$, we define v_r on $\underline{R}^n \times]r, a[$ by

$$v_r(x, t) = \int_{\underline{R}^n} W(x - y, t - r) u^+(y, r) \, dy$$

Then (19) shows that $u^+ \leq v_r \leq w$ on $\underline{R}^n \times]r, a[$. Furthermore, if $x \in \underline{R}^n$ and $0 < r < s < t < a$, it follows from (19), Fubini's theorem, and Theorem 1.4, that

$$v_s(x, t) \leq \int_{\underline{R}^n} W(x - y, t - s) \, dy \int_{\underline{R}^n} W(y - z, s - r) u^+(z, r) \, dz$$

$$= \int_{\underline{R}^n} u^+(z, r) \, dz \int_{\underline{R}^n} W(x - y, t - s) W(y - z, s - r) \, dy$$

$$= \int_{\underline{R}^n} W(x - z, t - r) u^+(z, r) \, dz$$

$$= v_r(x, t)$$

Hence, for each fixed $(x, t) \in \underline{R}^n \times {]}0, a{[}$, the function $r \mapsto v_r(x, t)$ is non-increasing on $]0, t[$. Therefore

$$v(x, t) = \lim_{r \to 0^+} v_r(x, t)$$

exists for all $(x, t) \in \underline{R}^n \times {]}0, a{[}$.

We must show that v is a temperature. If $0 < r < s < a$ and $(x, t) \in \underline{R}^n \times {]}s, a{[}$, then

$$v_r(x, t) = \int_{\underline{R}^n} W(x - y, t - s) v_r(y, s) \, dy$$

by Theorem 1.4. Since v_r increases to v as $r \to 0^+$, it follows from the monotone convergence theorem that

$$v(x, t) = \int_{\underline{R}^n} W(x - y, t - s) v(y, s) \, dy \tag{20}$$

for all $(x, t) \in \underline{R}^n \times {]}s, a{[}$. Furthermore, because $u^+ \leq v_r \leq w$ on $\underline{R}^n \times {]}r, a{[}$ for every r, $u^+ \leq v \leq w$ on $\underline{R}^n \times {]}0, a{[}$. Therefore the integrals in (20) are finite, so that v is a temperature on $\underline{R}^n \times {]}s, a{[}$, by Lemma 3.1. Since s is arbitrary in $]0, a[$, v is a temperature on the whole of $\underline{R}^n \times {]}0, a{[}$, so that the inequalities $u^+ \leq v \leq w$ imply that v is the least thermic majorant of u^+ on that strip. Thus $v = w$, which is the conclusion of the theorem.

As an application of Theorem 6.5, we can supplement our results in section 6 of Chapter III, on the class R_a.

Theorem 6.6

Let u be the Gauss-Weierstrass integral on $\underline{R}^n \times {]}0, a{[}$ of a signed measure μ. Then, for each $b < a$,

$$\lim_{t \to 0^+} M_b(u^+; t) = \int_{\underline{R}^n} W(y, b) \, d\mu^+(y) < \infty$$

Proof. Let $b < a$. Since $u \in R_a$, we know from Theorem 3.13 that $M_b(u^+; \cdot)$ is bounded on $]0, b[$, and from Theorem 3.8 that $M_b(u^+; \cdot)$ is non-increasing (because $R_a \subseteq S_a$). Therefore $M_b(u^+; t)$ tends to a finite limit as $t \to 0^+$. By Theorem 6.5, the least thermic majorant w of u^+ on $\underline{R}^n \times {]}0, a{[}$ has the representation

$$w(x, t) = \lim_{r \to 0^+} \int_{R^n} W(x - y, t - r)u^+(y, r) \, dy$$

for all $(x, t) \in R^n \times]0, a[$. Therefore

$$\lim_{r \to 0^+} M_b(u^+; r) = \lim_{r \to 0^+} \int_{R^n} W(y, b - r)u^+(y, r) \, dy$$

$$= w(0, b)$$

By Theorem 5.7, w is the Gauss-Weierstrass integral of μ^+, so that

$$w(0, b) = \int_{R^n} W(y, b) \, d\mu^+(y)$$

and the result follows.

We now come to the main result of this section, which it is interesting to compare with Theorem 5.8. It also improves upon Lemma 6.1.

Theorem 6.7

Let $p > 1$, and let u be a temperature on $R^n \times]0, a[$. Suppose that, for each $b < a$, $M_{b,p}(u^+; \cdot)$ is locally integrable on $]0, b[$ and

$$\liminf_{t \to 0^+} M_{b,p}(u^+; t) < \infty \qquad (21)$$

Then $u(\cdot, 0^+)$ exists m-a. e. on R^n, u has the representation

$$u(x, t) = \int_{R^n} W(x - y, t)u(y, 0^+) \, dy - \int_{R^n} W(x - y, t) \, d\sigma(y) \qquad (22)$$

for all $(x, t) \in R^n \times]0, a[$, where σ is a non-negative measure which is singular with respect to m, and

$$\lim_{t \to 0^+} M_{b,p}(u^+; t) = M_{b,p}(u^+(\cdot, 0^+); 0) < \infty \qquad (23)$$

for each $b < a$.

Proof. By Lemma 6.1, u is the Gauss-Weierstrass integral on $R^n \times]0, a[$ of a signed measure μ. Thus $u \in R_a$, so that, since $R_a \subseteq S_a$, Theorem 3.5 tells us that

$$u^+(x,t) \leq \int_{\underline{R}^n} W(x-y, t-r) u^+(y,r) \, dy < \infty \qquad (24)$$

whenever $0 < r < t < a$ and $x \in \underline{R}^n$. Let $v_r(x,t)$ denote the integral in (24). Since Theorem 6.3 tells us that $M_{b,p}(u^+; r) < \infty$ whenever $r < b < a$, it follows from Theorem 6.2 that v_r is a temperature on $\underline{R}^n \times]r, a[$, and

$$M_{b,p}(v_r; s) \leq M_{b,p}(u^+; r) \qquad (25)$$

whenever $r < s < b < a$. By Theorem 6.5, the least thermic majorant w of u^+ on $\underline{R}^n \times]0, a[$ is given by

$$w = \lim_{r \to 0^+} v_r$$

Therefore, by Fatou's lemma, (25), and (21),

$$M_{b,p}(w; s) = M_{b,p}(\lim_{r \to 0^+} v_r; s)$$

$$\leq \liminf_{r \to 0^+} M_{b,p}(v_r; s)$$

$$\leq \liminf_{r \to 0^+} M_{b,p}(u^+; r)$$

$$< \infty \qquad (26)$$

whenever $0 < s < b < a$. Thus, for each $b < a$, $M_{b,p}(w; \cdot)$ is bounded on $]0, b[$. It now follows from Theorem 6.4, and the non-negativity of w, that w is the Gauss-Weierstrass integral on $\underline{R}^n \times]0, a[$ of $w(\cdot, 0^+)$, and that

$$\lim_{t \to 0^+} M_{b,p}(w; t) = M_{b,p}(w(\cdot, 0^+); 0) < \infty \qquad (27)$$

for each $b < a$. But Theorem 5.7 tells us that w is the Gauss-Weierstrass integral of μ^+, so that $d\mu^+(y) = w(y, 0^+) \, dy$, in view of Theorem 1.5. However, by Theorem 4.1, $d\mu(y) = u(y, 0^+) \, dy + d\sigma(y)$ for some signed measure σ which is singular with respect to m, so that $d\mu^+(y) = u^+(y, 0^+) \, dy + d\sigma^+(y)$. Hence σ^+ is null, which establishes the representation (22), and $u^+(\cdot, 0^+) = w(\cdot, 0^+)$. Since $u^+ \leq w$, it therefore follows from (27) that

$$\limsup_{t\to0^+} M_{b,p}(u^+; t) \leq \lim_{t\to0^+} M_{b,p}(w; t) = M_{b,p}(u^+(\cdot,0^+); 0) < \infty$$

However, (26) implies that

$$\lim_{t\to0^+} M_{b,p}(w; t) \leq \liminf_{t\to0^+} M_{b,p}(u^+; t)$$

so that $M_{b,p}(u^+; t)$ tends to a limit as $t \to 0^+$, and

$$\lim_{t\to0^+} M_{b,p}(u^+; t) = \lim_{t\to0^+} M_{b,p}(w; t) = M_{b,p}(u^+(\cdot,0^+); 0) < \infty$$

which establishes (23).

Remark. Comparing Theorem 5.8 with Theorem 6.7, we see that the latter result gives more information about $u^+(\cdot,0^+)$ (namely (23)) but the former gives more about σ (namely that $\sigma(\underline{R}^n \setminus \{x \in \underline{R}^n \colon u(x,0^+) = -\infty\}) = 0$.

Corollary

Let $p > 1$, and let u be a temperature on $\underline{R}^n \times \,]0,a[$. If there is a constant $\alpha \in \,]0,1/4a]$ such that the function

$$t \longmapsto \int_{\underline{R}^n} \exp(-\alpha\|x\|^2) u^+(x,t)^p \, dx$$

is locally integrable on $]0,a[$, and

$$\liminf_{t\to0^+} \int_{\underline{R}^n} \exp(-\alpha\|x\|^2) u^+(x,t)^p \, dx < \infty$$

then $u(\cdot,0^+)$ exists m-a. e. on $\underline{R}^n$, u has the representation

$$u(x,t) = \int_{\underline{R}^n} W(x-y,t) u(y,0^+) \, dy - \int_{\underline{R}^n} W(x-y,t) \, d\sigma(y)$$

for all $(x,t) \in \underline{R}^n \times \,]0,a[$, where σ is a non-negative measure which is singular with respect to m, and

$$\int_{\underline{R}^n} \exp(-\alpha\|x\|^2) u^+(x,0^+)^p \, dx < \infty$$

Proof. The argument used to prove this result is the same as that for the corollary to Theorem 6.4, except that $|u|$ is replaced by u^+ throughout, and Theorem 6.7 is used instead of Theorem 6.4.

Remark. The condition that $\alpha \leq 1/4a$ in the above corollary cannot be removed, for the same reason that it could not be deleted from Theorem 6.4, Corollary.

Using Theorem 6.7, we can give new conditions under which $u \leq 0$, and hence new results related to Theorem 5.11 and the extended maximum principle of Theorem 3.1. One way is to add a normal limit hypothesis, such as

$$\liminf_{t \to 0^+} u(x, t) \leq 0$$

for m-almost all $x \in \underline{R}^n$, which, together with the representation (22), easily implies that $u \leq 0$. A slightly more interesting way is to impose an extra condition on one of the means $M_{b, p}$, as in the following result. Using Theorem 6.6 also, and writing $M_{b, 1}$ for M_b, we can include the case $p = 1$.

Theorem 6.8

Let $p \geq 1$, and let u be a temperature on $\underline{R}^n \times \,]0, a[$ such that, for each $b < a$, $M_{b, p}(u^+; \cdot)$ is locally integrable on $]0, b[$ and

$$\liminf_{t \to 0^+} M_{b, p}(u^+; t) < \infty$$

If there exists $c < a$ such that

$$\liminf_{t \to 0^+} M_{c, p}(u^+; t) = 0 \tag{28}$$

then $u \leq 0$ on $\underline{R}^n \times \,]0, a[$.

Proof. Consider first the case where $p = 1$. By Theorem 3.15, u is the Gauss-Weierstrass integral on $\underline{R}^n \times \,]0, a[$ of a signed measure μ. By Theorem 6.6 and (28),

$$\int_{\underline{R}^n} W(y, c) \, d\mu^+(y) = \lim_{t \to 0^+} M_c(u^+; t) = 0$$

Since $W(\cdot, c) > 0$, it follows that μ^+ is null, so that μ is non-positive and hence $u \le 0$.

Now consider the case where $p > 1$. By Theorem 6.7, u is the Gauss-Weierstrass integral on $\underline{R}^n \times]0, a[$ of a measure μ of the form

$$d\mu(y) = u(y, 0^+) \, dy - d\sigma(y)$$

for some non-negative measure σ. Furthermore, by Theorem 6.7 and (28),

$$M_{c,p}(u^+(\cdot, 0^+); 0) = \lim_{t \to 0^+} M_{c,p}(u^+; t) = 0$$

from which it follows that $u^+(\cdot, 0^+) = 0$ m-a. e on $\underline{R}^n$. Thus

$$d\mu(y) = -u^-(y, 0^+) \, dy - d\sigma(y)$$

so that μ is non-positive and hence $u \le 0$.

4. A RESULT LIKE EGOROFF'S THEOREM AND A COVERING LEMMA

We turn our attention from integral mean conditions to restrictions on $m(\{x \in \underline{R}^n : u(x, t) > 1\})$ as $t \to 0^+$, where u is a non-negative tempera-ture on $\underline{R}^n \times]0, a[$. This section contains two preliminary results of a measure-theoretic nature.

Let μ be a non-negative measure on $\underline{R}^n$ which is singular with respect to Lebesgue measure. By Theorem 5.3 and Lemma 5.1, $r^{-n}\mu(\bar{B}(x, r)) \to \infty$ as $r \to 0$, for μ-almost all $x \in \underline{R}^n$. We would like to apply Egoroff's theorem to the family of functions $\{f_r : r > 0\}$, given by

$$f_r(x) = r^{-n}\mu(\bar{B}(x, r))$$

for all x in some Borel set of finite μ-measure. However, Egoroff's theorem is applicable only to a sequence of functions, and so cannot be used here. We therefore prove the result we require.

Lemma 6.2

Let μ be a non-negative measure on $\underline{R}^n$ such that

$$r^{-n}\mu(\bar{B}(x, r)) \to \infty \quad \text{as} \quad r \to 0 \tag{29}$$

for μ-almost all x in a Borel set A of finite μ-measure. Then, given any positive ϵ, we can find a Borel subset E of A such that $\mu(A \setminus E) < \epsilon$ and $r^{-n}\mu(\bar{B}(\cdot, r)) \to \infty$ uniformly on E as $r \to 0$.

Proof. Arrange the rational numbers as a sequence $\{r_j\}$. For each $j, k, \ell \in \underline{N}$, put

$$S_j(k) = \{x \in A : r_j^{-n}\mu(\bar{B}(x, r_j)) \geq k\}$$

and

$$S(\ell, k) = \{x \in A : x \in S_j(k) \quad \text{for all j such that} \quad r_j < 1/\ell\}$$

We must show that each set $S(\ell, k)$ is a Borel set, so that it is μ-measurable. Since each $S(\ell, k)$ is the intersection of a sequence of the sets $S_j(k)$, this will follow if we prove that each $S_j(k)$ is closed. Fix j and k, and let $\{x_i\}$ be a sequence of points in $S_j(k)$ which has a limit x_0. We have to show that $x_0 \in S_j(k)$. If $R > r_j$, we can find $i_0 \in \underline{N}$ such that

$$\|x_0 - x_i\| \leq R - r_j \quad \text{whenever} \quad i \geq i_0$$

Therefore, if $i \geq i_0$ and $y \in \bar{B}(x_i, r_j)$,

$$\|y - x_0\| \leq \|y - x_i\| + \|x_i - x_0\| \leq r_j + (R - r_j) = R$$

Thus $\bar{B}(x_i, r_j) \subseteq \bar{B}(x_0, R)$ for all $i \geq i_0$. Therefore, because $x_i \in S_j(k)$ for all i,

$$\mu(\bar{B}(x_0, R)) \geq \mu(\bar{B}(x_i, r_j)) \geq k r_j^n$$

Thus $r_j^{-n}\mu(\bar{B}(x_0, R)) \geq k$ for every $R > r_j$, and it follows that

$$r_j^{-n}\mu(\bar{B}(x_0, r_j)) = \inf\{r_j^{-n}\mu(\bar{B}(x_0, R)) : R > r_j\} \geq k$$

Hence $x_0 \in S_j(k)$. It follows that each set $S_j(k)$ is closed, and hence that every $S(\ell, k)$ is a Borel set.

Note that, for each $k, \ell \in \underline{N}$, $S(\ell, k) \subseteq S(\ell + 1, k)$, because the latter set is the intersection of fewer of the sets $S_j(k)$ than is the former. Therefore $\mu(S(\cdot, k))$ is increasing, and

$$\lim_{\ell \to \infty} \mu(S(\ell, k)) = \mu\left(\bigcup_{\ell=1}^{\infty} S(\ell, k)\right) \tag{30}$$

for each k. If x is such that (29) holds, then given any $k \in \underline{N}$ we can find $\ell \in \underline{N}$ such that

$$r^{-n}\mu(\bar{B}(x, r)) \geq k \quad \text{for all} \quad r < 1/\ell$$

so that $x \in S_j(k)$ for all j such that $r_j < 1/\ell$, which means that $x \in S(\ell, k)$. Thus, if x satisfies (29), then for each $k \in \underline{N}$,

$$x \in \bigcup_{\ell=1}^{\infty} S(\ell, k)$$

Since (29) is true for μ-almost all $x \in A$, it follows that the right-hand side of (30) is equal to $\mu(A)$, so that

$$\lim_{\ell \to \infty} \mu(S(\ell, k)) = \mu(A)$$

for each k. Since $\mu(A) < \infty$ by hypothesis, it follows that to each k there corresponds ℓ_k such that

$$\mu(A \setminus S(\ell_k, k)) = \mu(A) - \mu(S(\ell_k, k)) < \epsilon 2^{-k}$$

Therefore, if $T_k = S(\ell_k, k)$ for each k, and

$$E = \bigcap_{k=1}^{\infty} T_k$$

we have

$$\mu(A \setminus E) = \mu\left(\bigcup_{k=1}^{\infty} (A \setminus T_k)\right) \leq \sum_{k=1}^{\infty} \mu(A \setminus T_k) < \epsilon \sum_{k=1}^{\infty} 2^{-k} = \epsilon \tag{31}$$

Given any $k \in \underline{N}$ and $x \in E$, we have

$x \in T_k = S(\ell_k, k) = \{x \in A : x \in S_j(k) \quad \text{for all } j \text{ such that} \quad r_j < 1/\ell_k\}$

Therefore, if $0 < \delta < 1/\ell_k$, for any $r_j \in \,]0, \delta]$ and $x \in E$ we have

$$\mu(\bar{B}(x, \delta)) \geq \mu(\bar{B}(x, r_j)) \geq k\, r_j^n$$

Hence, for any $k \in \underline{N}$ and $x \in E$,

$$\mu(\bar{B}(x, \delta)) \geq \sup\{k\, r_j^n : 0 < r_j \leq \delta\} = k\delta^n$$

whenever $0 < \delta < 1/\ell_k$. Thus $r^{-n}\mu(\bar{B}(\cdot, r)) \to \infty$ uniformly on E as $r \to 0$. In view of (31), the lemma is proved.

We shall also need the following covering lemma.

Lemma 6.3

If $\mathcal{F}$ is a finite collection of open balls, then there is a subcollection $\mathcal{G}$ of disjoint balls such that, for each ball $B \in \mathcal{F}$, there is a ball $B(x, r) \in \mathcal{G}$ such that $B \subseteq B(x, 3r)$.

Proof. Since $\mathcal{F}$ is finite, we can find a ball $B(x_1, r_1) \in \mathcal{F}$ such that $r_1 \geq \rho$ for every ball $B(y, \rho) \in \mathcal{F}$. Put

$$\mathcal{F}_1 = \{B(y, \rho) \in \mathcal{F} : B(y, \rho) \cap B(x_1, r_1) = \emptyset\}$$

If $\mathcal{F}_1$ is not empty, then since $\mathcal{F}_1$ is finite, we can find a ball $B(x_2, r_2) \in \mathcal{F}_1$ such that $r_2 \geq \rho$ for every ball $B(y, \rho) \in \mathcal{F}_1$. Put

$$\mathcal{F}_2 = \{B(y, \rho) \in \mathcal{F}_1 : B(y, \rho) \cap B(x_2, r_2) = \emptyset\}$$

If $\mathcal{F}_2$ is not empty, then we can choose a ball $B(x_3, r_3) \in \mathcal{F}_2$ such that $r_3 \geq \rho$ whenever $B(y, \rho) \in \mathcal{F}_2$. We continue in this way until we reach a class, say $\mathcal{F}_j$, which is empty. Put $\mathcal{G} = \{B(x_i, r_i) : i = 1, \ldots, j\}$. Let $B \in \mathcal{F}$. If $B \in \mathcal{G}$, there is nothing to prove, so suppose otherwise. Since $B \in \mathcal{F}$, we have $B \cap B(x_k, r_k) \neq \emptyset$ for some k, by our choice of the classes $\mathcal{F}_i$, and r_k is no smaller than the radius of B, by our choice of the balls $B(x_i, r_i)$. Hence $B \subseteq B(x_k, 3r_k)$, which proves the lemma.

5. NON-NEGATIVE TEMPERATURES AND A MEASURE GROWTH CONDITION

Every non-negative temperature u on $\underline{R}^n \times]0, a[$ is the Gauss-Weierstrass integral of a non-negative measure μ, by Theorem 2.10. We shall derive a lower estimate of $m(\{x \in \underline{R}^n : u(x,t) > 1\})$, for t near 0, which is satisfied whenever μ is not absolutely continuous with respect to m. If the lower estimate does not hold, then μ must be absolutely continuous, so that $d\mu(y) = u(y, 0^+) \, dy$, by Theorem 4.1. The estimate is first obtained in terms of an auxiliary function ρ_A, which is subsequently removed. In order to obtain the best estimate that our methods allow, we must keep track of the precise values of the various constants that occur.

LEMMA 6.4

Let $\chi > 1$, let u be the Gauss-Weierstrass integral on $\underline{R}^n \times]0, a[$ of a non-negative measure μ, let $z \in \underline{R}^n$ and let $r > 0$. Then, for all $(x,t) \in \bar{B}(z,r) \times]0, \chi r^2]$,

$$u(x,t) \geq (4\pi\chi)^{-n/2} \exp\left(-\frac{r^2}{t}\right) r^{-n} \mu(\bar{B}(z,r))$$

Proof. If $x, y \in \bar{B}(z,r)$, then

$$\|x - y\| \leq \|x - z\| + \|z - y\| \leq 2r$$

Therefore, if $0 < t \leq \chi r^2$ and $x \in \bar{B}(z,r)$,

$$u(x,t) = (4\pi t)^{-n/2} \int_{\underline{R}^n} \exp\left(-\frac{\|x-y\|^2}{4t}\right) d\mu(y)$$

$$\geq (4\pi\chi r^2)^{-n/2} \int_{\bar{B}(z,r)} \exp\left(-\frac{r^2}{t}\right) d\mu(y)$$

$$= (4\pi\chi)^{-n/2} \exp\left(-\frac{r^2}{t}\right) r^{-n} \mu(\bar{B}(z,r))$$

We proceed to describe the auxiliary function ρ_A mentioned above. Given a non-negative measure μ on $\underline{R}^n$, and a Borel subset A of $\underline{R}^n$, the modulus of continuity $\Omega_{A, \mu}$ of μ over A is defined by

$$\Omega_{A,\mu}(r) = \sup\{\mu(\bar{B}(x,r)) : x \in A\}$$

for all $r > 0$. Given any number $\chi > 1$, we define $\rho_A = \rho_{A,\mu,\chi}$ by

$$\rho_A(t) = \inf\{r : (4\pi\chi)^{-n/2} \exp(-r^2/t) r^{-n} \Omega_{A,\mu}(r) \leq 1\} \tag{32}$$

for all $t > 0$. As usual, the infimum of an empty set is defined to be ∞.

If μ is singular with respect to m, it follows from Theorem 5.3 and Lemma 5.1 that $r^{-n}\mu(\bar{B}(x,r)) \to \infty$ as $r \to 0$, for μ-almost all $x \in \underline{R}^n$. This illuminates one hypothesis in the following result.

Theorem 6.9

Let $\chi > 1$, let μ be a non-negative measure on $\underline{R}^n$ such that

$$r^{-n}\mu(\bar{B}(x,r)) \to \infty \quad \text{as} \quad r \to 0 \tag{33}$$

for μ-almost all x in a Borel set A with $\mu(A) < \infty$, and let u be the Gauss-Weierstrass integral of μ on $\underline{R}^n \times]0, a[$. Then there exists $t_0 > 0$ such that

$$m(\{x \in \underline{R}^n : u(x,t) > 1\}) \geq \omega_n 6^{-n} \pi^{-n/2} \chi^{-(n+2)/2} \mu(A) \exp(-\rho_A(t)^2/t)$$

whenever $0 < t \leq t_0$, where $\omega_n = m(B(0,1))$.

Proof. The result is trivial if $\mu(A) = 0$, so we suppose otherwise. By Lemma 6.2, there is a Borel set $E \subseteq A$ such that $\mu(A \setminus E) < (1 - \chi^{-1})\mu(A)$ and (33) holds uniformly for $x \in E$. Then

$$\mu(A) - \mu(E) < (1 - \chi^{-1})\mu(A)$$

so that $\mu(E) > \chi^{-1}\mu(A)$. Since μ is a regular measure, there is a compact set $K \subseteq E$ such that $\mu(K) > \chi^{-1}\mu(A)$. Then (33) holds uniformly for $x \in K$, so that we can find $t_0 > 0$ such that, whenever $0 < r \leq \sqrt{t_0}$,

$$(4\pi\chi)^{-n/2} r^{-n} \mu(\bar{B}(x,r)) \geq e \tag{34}$$

for all $x \in K$.

Fix $t \in]0, t_0]$. For every $x \in K$, put

$$\rho(x, t) = \inf\{r : (4\pi\chi)^{-n/2} \exp(-r^2/t) r^{-n} \mu(\bar{B}(x, r)) \leq 1\}$$

Since $\mu(\bar{B}(x, r)) \leq \Omega_A(r)$ for all $x \in A$ and all r,

$$\rho(x, t) \leq \rho_A(t) \tag{35}$$

for all $x \in K$. Also, for each fixed $x \in K$, there is a sequence $\{r_k\}$ which decreases to $\rho(x, t)$ and is such that

$$(4\pi\chi)^{-n/2} \exp\left(-\frac{r_k^2}{t}\right) r_k^{-n} \mu(\bar{B}(x, r_k)) \leq 1$$

for all k. Making $k \to \infty$, we deduce that

$$(4\pi\chi)^{-n/2} \exp(-\rho(x, t)^2/t) \rho(x, t)^{-n} \mu(\bar{B}(x, \rho(x, t))) \leq 1 \tag{36}$$

Next, if $0 < r < \sqrt{t}$, it follows from (34) that

$$(4\pi\chi)^{-n/2} \exp\left(-\frac{r^2}{t}\right) r^{-n} \mu(\bar{B}(x, r)) \geq \exp(-r^2/t) e > 1$$

Therefore

$$\rho(x, t) \geq \sqrt{t} \tag{37}$$

for all $x \in K$.

The family of open balls $\{B(x, \rho(x, t)/3) : x \in K\}$ covers the compact set K, so that there is a finite subfamily $\mathscr{F}$ which also covers K. By Lemma 6.3, $\mathscr{F}$ has a subfamily

$$\mathscr{G} = \left\{ B\left(x_\ell, \frac{\rho(x_\ell, t)}{3}\right) : \ell = 1, \ldots, i \right\}$$

of disjoint balls such that, for each ball $B \in \mathscr{F}$, there is a ball $B(x_\ell, \rho(x_\ell, t)/3)$ in $\mathscr{G}$ such that $B \subseteq B(x_\ell, \rho(x_\ell, t))$. Since $\mathscr{F}$ covers K, it follows that

$$K \subseteq \bigcup_{\ell=1}^{i} B(x_\ell, \rho(x_\ell, t))$$

so that, by our choice of K,

$$\mu(A) < \chi\mu(K) \le \chi \sum_{\ell=1}^{i} \mu(B(x_\ell, \rho(x_\ell, t)))$$

It now follows from (36) and (35) that, since $x_\ell \in K$ for every ℓ,

$$(4\pi\chi)^{-n/2}\mu(A) \le \chi \sum_{\ell=1}^{i} (4\pi\chi)^{-n/2}\mu(B(x_\ell, \rho(x_\ell, t)))$$

$$\le \chi \sum_{\ell=1}^{i} \exp\left(\frac{\rho(x_\ell, t)^2}{t}\right) \rho(x_\ell, t)^n$$

$$\le \chi \exp\left(\frac{\rho_A(t)^2}{t}\right) \sum_{\ell=1}^{i} \rho(x_\ell, t)^n$$

Therefore, because the balls in $\mathscr{G}$ are disjoint,

$$\omega_n 3^{-n}\chi^{-1}(4\pi\chi)^{-n/2}\mu(A) \exp\left(-\frac{\rho_A(t)^2}{t}\right) \le \sum_{\ell=1}^{i} \omega_n \left(\frac{\rho(x_\ell, t)}{3}\right)^n$$

$$= \sum_{\ell=1}^{i} m\left(B\left(x_\ell, \frac{\rho(x_\ell, t)}{3}\right)\right)$$

$$= m\left(\bigcup_{\ell=1}^{i} B\left(x_\ell, \frac{\rho(x_\ell, t)}{3}\right)\right)$$

$$\le m(Y) \tag{38}$$

where

$$Y = \bigcup_{\ell=1}^{i} B(x_\ell, \rho(x_\ell, t))$$

If $y \in Y$, there is $\ell \in \{1, \ldots, i\}$ such that $\|y - x_\ell\| < \rho(x_\ell, t)$. In view of (37), $\rho(x_\ell, t) > \sqrt{t/\chi}$, so that we can choose r such that $r > \sqrt{t/\chi}$ and $\|y - x_\ell\| < r < \rho(x_\ell, t)$. Then $(y, t) \in B(x_\ell, r) \times]0, \chi r^2]$, so that, by Lemma 6.4,

$$u(y, t) \ge (4\pi\chi)^{-n/2}\exp\left(-\frac{r^2}{t}\right) r^{-n}\mu(\bar{B}(x_\ell, r))$$

which exceeds 1 since $r < \rho(x_\ell, t)$. Thus $Y \subseteq \{y \in \underline{R}^n : u(y,t) > 1\}$. It therefore follows from (38) that

$$m(\{y \in \underline{R}^n : u(y,t) > 1\}) \geq \omega_n 6^{-n} \pi^{-n/2} \chi^{-(n+2)/2} \mu(A) \exp\left(-\frac{\rho_A(t)^2}{t}\right)$$

as required.

Even in the most simple case, where μ is the Dirac δ-measure concentrated at a point in A, we cannot evaluate ρ_A exactly. However, we can determine its behavior as $t \to 0$, and this is sufficient to enable us to prove a result, similar to Theorem 6.9, but which does not mention ρ_A explicitly.

Lemma 6.5

Let A be a Borel set, let $x_0 \in A$, let $\chi > 1$, and let ν be κ times the Dirac δ-measure concentrated at x_0, for some $\kappa > 0$. Then

$$\rho_{A,\nu,\chi}(t) \sim \left\{-\frac{nt}{2} \log(-\alpha t \log t)\right\}^{\frac{1}{2}}$$

as $t \to 0$, where $\alpha = 2n\pi\chi\kappa^{-2/n}$.

Proof. Since $x_0 \in A$, the modulus of continuity $\Omega_{A,\nu}$ of ν over A is given by

$$\Omega_{A,\nu}(r) = \sup\{\nu(\bar{B}(x,r)) : x \in A\} = \kappa$$

for all $r > 0$. Therefore, if $\rho = \rho_{A,\nu,\chi}$,

$$\rho(t) = \inf\{r : (4\pi\chi)^{-n/2} \exp(-r^2/t) r^{-n} \kappa \leq 1\}$$

so that the continuity of the function $r \longmapsto \exp(-r^2/t) r^{-n}$, for each fixed t, implies that

$$(4\pi\chi)^{-n/2} \exp(-\rho^2/t)\rho^{-n}\kappa = 1 \tag{39}$$

Consider the function σ, defined for all small $t > 0$ by

$$\sigma(t) = \left\{ - (nt/2) \log (-\alpha t \log t) \right\}^{\frac{1}{2}}$$

where $\alpha = 2n\pi\chi\kappa^{-2/n}$. Note that

$$\exp\left(-\frac{\sigma^2}{t}\right) = (-\alpha t \log t)^{n/2}$$

and $(2\alpha/n)^{-n/2} = (4\pi\chi)^{-n/2}\kappa$. Therefore, as $t \to 0$,

$$\exp(-\sigma^2/t)\sigma^{-n} = (-\alpha t \log t)^{n/2} \quad -\frac{nt}{2} \log (-\alpha t \log t)^{-n/2}$$

$$= \frac{(2\alpha/n) \log t}{\log (-\alpha t \log t)}^{n/2}$$

$$\to (2\alpha/n)^{n/2}$$

$$= (4\pi\chi)^{n/2}\kappa^{-1}$$

so that

$$(4\pi\chi)^{-n/2} \exp\left(-\frac{\sigma^2}{t}\right)\sigma^{-n}\kappa \to 1 \tag{40}$$

For all positive r and t, we put

$$f(r, t) = (4\pi\chi)^{-n/2} \exp\left(-\frac{r^2}{t}\right) r^{-n}\kappa$$

so that (39) and (40) can be written as $f(\rho(t), t) = 1$ and $f(\sigma(t), t) \to 1$ as $t \to 0$. Thus

$$f(\sigma(t), t) \sim f(\rho(t), t) \quad \text{as} \quad t \to 0 \tag{41}$$

We shall deduce from this that

$$\sigma(t) \sim \rho(t) \quad \text{as} \quad t \to 0 \tag{42}$$

Suppose that (42) is false. Then we can find $\eta > 0$ and a null sequence $\{t_k\}$ such that either $\sigma(t_k) > (1 + \eta)\rho(t_k)$ or $\rho(t_k) > (1 + \eta)\sigma(t_k)$ for all k. The proofs in the two cases are similar, so we give the details only for the former case. Since $f(\cdot, t)$ is strictly decreasing for each fixed t,

$$f(\sigma(t_k), t_k) < f((1 + \eta)\rho(t_k), t_k)$$

for all k. Therefore

$$\frac{f(\sigma(t_k), t_k)}{f(\rho(t_k), t_k)} < \frac{f((1 + \eta)\rho(t_k), t_k)}{f(\rho(t_k), t_k)}$$

$$= (1 + \eta)^{-n} \exp\left(-\frac{(1 + \eta)^2 \rho(t_k)^2}{t}\right) \exp\left(\frac{\rho(t_k)^2}{t}\right)$$

$$< (1 + \eta)^{-n}$$

for all k, which contradicts (41) since $(1 + \eta)^{-n} < 1$. It follows that (42) is true, and the lemma is proved.

Theorem 6.10

Let $\chi > 1$, let μ be a non-negative measure on $\underline{R}^n$ such that $r^{-n}\mu(\bar{B}(x, r)) \to \infty$ as $r \to 0$, for μ-almost all x in a Borel set A with $0 < \mu(A) < \infty$, and let u be the Gauss-Weierstrass integral of μ on $\underline{R}^n \times]0, a[$. Then there exists $t_0 > 0$ such that

$$m(\{x \in \underline{R}^n : u(x, t) > 1\}) \geq \omega_n\left(\frac{n}{18}\right)^{n/2} \chi^{-n-3} |t \log t|^{n/2}$$

whenever $0 < t \leq t_0$, where $\omega_n = m(B(0, 1))$.

Proof. Let $x_0 \in A$, and let ν be $\chi\mu(A)$ times the Dirac δ-measure concentrated at x_0. By Lemma 6.5,

$$\rho_{A, \nu, \chi}(t) \sim -\frac{nt}{2} \log(-\alpha t \log t)^{\frac{1}{2}} \tag{43}$$

as $t \to 0$, where $\alpha = 2n\pi\chi(\chi\mu(A))^{-2/n}$.

Since μ is a regular measure, we can find a compact subset K of A such that $\mu(K) \geq \chi^{-1}\mu(A)$, and an open superset V of K such that $\mu(V) \leq \chi\mu(A)$. Then, if r_0 denotes the distance between K and $\underline{R}^n \setminus V$, for all

$r < r_0$ the moduli of continuity satisfy

$$\Omega_{K,\mu}(r) = \sup\{\mu(\bar{B}(x,r)) : x \in K\} \leq \mu(V) \leq \chi\mu(A) = \Omega_{A,\nu}(r) \qquad (44)$$

Using (43), we see that $\rho_{A,\nu}(t) = \rho_{A,\nu,\chi}(t) \to 0$ as $t \to 0$, so that we can find $t_1 > 0$ such that $\rho_{A,\nu}(t) < r_0$ whenever $0 < t \leq t_1$. It follows from the definition (32) of $\rho_{A,\nu}$ that there are values of r less than r_0 such that

$$(4\pi\chi)^{-n/2} \exp\left(-\frac{r^2}{t}\right) r^{-n} \Omega_{A,\nu}(r) \leq 1$$

and hence that

$$(4\pi\chi)^{-n/2} \exp\left(-\frac{r^2}{t}\right) r^{-n} \Omega_{K,\mu}(r) \leq (4\pi\chi)^{-n/2} \exp\left(-\frac{r^2}{t}\right) r^{-n} \Omega_{A,\nu}(r) \leq 1$$

in view of (44). Therefore

$$\rho_{K,\mu}(t) = \inf\ r < r_0 : (4\pi\chi)^{-n/2} \exp\left(-\frac{r^2}{t}\right) r^{-n} \Omega_{K,\mu}(r) \leq 1$$

$$\leq \inf\ r < r_0 : (4\pi\chi)^{-n/2} \exp\left(-\frac{r^2}{t}\right) r^{-n} \Omega_{A,\nu}(r) \leq 1$$

$$= \rho_{A,\nu}(t)$$

It now follows from Theorem 6.9 that there exists $t_2 > 0$ such that, whenever $0 < t \leq t_2$,

$$m(\{x \in \underline{R}^n : u(x,t) > 1\}) \geq \omega_n 6^{-n} \pi^{-n/2} \chi^{-(n+2)/2} \mu(K) \exp\left(-\frac{\rho_{K,\mu}(t)^2}{t}\right)$$

$$\geq \omega_n 6^{-n} \pi^{-n/2} \chi^{-(n+4)/2} \mu(A) \exp\left(-\frac{\rho_{A,\nu}(t)^2}{t}\right)$$

by our choice of K. By (39), with $\kappa = \chi\mu(A)$,

$$\exp\left(-\frac{\rho_{A,\,\nu}(t)^2}{t}\right) = (4\pi\chi)^{n/2}(\chi\mu(A))^{-1}\rho_{A,\,\nu}(t)^n$$

so that

$$m(\{x \in \underline{R}^n : u(x,t) > 1\}) \geq \omega_n 3^{-n}\chi^{-3}\rho_{A,\,\nu}(t)^n \qquad (45)$$

In view of (43), for all small t we have

$$\frac{\rho_{A,\,\nu}(t)^2}{|t \log t|} \geq \frac{-(nt/2)\log(-\alpha t \log t)}{-\chi t \log t} = \frac{n \log(-\alpha t \log t)}{2\chi \log t}$$

The last expression tends to $n/2\chi$ as $t \to 0$, so that, for all sufficiently small t,

$$\rho_{A,\,\nu}(t)^2 \geq \frac{n}{2\chi^2}|t \log t|$$

Hence, by (45),

$$m(\{x \in \underline{R}^n : u(x,t) > 1\}) \geq \omega_n \left(\frac{n}{18}\right)^{n/2}\chi^{-n-3}|t \log t|^{n/2}$$

as required.

We can now prove our representation theorem.

Theorem 6.11

Let u be a non-negative temperature on $\underline{R}^n \times]0, a[$. If

$$\liminf_{t\to 0^+} |t \log t|^{-n/2} m(\{x \in \underline{R}^n : u(x,t) > 1\}) < \omega_n \left(\frac{n}{18}\right)^{n/2} \qquad (46)$$

then u is the Gauss-Weierstrass integral of $u(\cdot, 0^+)$ on $\underline{R}^n \times]0, a[$.

Proof. By Theorem 2.10, u is the Gauss-Weierstrass integral of a non-negative measure μ. The hypothesis (46) implies that there is a number $\phi < \omega_n (n/18)^{n/2}$ and a null sequence $\{t_i\}$ such that

$$|t_i \log t_i|^{-n/2} m(\{x \in \underline{R}^n : u(x, t_i) > 1\}) \rightarrow \phi$$

Choose χ_0 such that $\chi_0 > 1$ and $\phi < \chi_0^{-n-3} \omega_n (n/18)^{n/2}$. Then, given any $t_0 > 0$, we can find $k \in \underline{N}$ such that $t_k < t_0$ and

$$m(\{x \in \underline{R}^n : u(x, t_k) > 1\}) < \chi_0^{-n-3} \omega_n \left(\frac{n}{18}\right)^{n/2} |t_k \log t_k|^{n/2}$$

Thus the conclusion of Theorem 6.10 fails to hold, so that one of its hypotheses must also fail here. It follows that there is no Borel set A with $0 < \mu(A) < \infty$ such that

$$r^{-n} \mu(\bar{B}(x, r)) \rightarrow \infty \quad \text{as} \quad r \rightarrow 0 \tag{47}$$

for μ-almost all $x \in A$. Hence (47) holds only on a set E with $\mu(E) = 0$. In view of Theorem 4.1,

$$d\mu(y) = u(y, 0^+) \, dy + d\sigma(y)$$

for some non-negative measure σ which is singular with respect to m. By Theorem 5.3 and Lemma 5.1,

$$r^{-n} \sigma(\bar{B}(x, r)) \rightarrow \infty \quad \text{as} \quad r \rightarrow 0$$

for σ-almost all $x \in \underline{R}^n$. Since $u \geq 0$, it follows that (47) holds for σ-almost all $x \in \underline{R}^n$, so that $\sigma(\underline{R}^n \backslash E) = 0$. But $\mu(E) = 0$, so that $\sigma(E) = 0$. Hence $\sigma(\underline{R}^n) = 0$, so that σ is null and u is the Gauss-Weierstrass integral of $u(\cdot, 0^+)$.

Example. The sharpness of the above results can be illustrated by taking $u = W$. Since, for every $t > 0$,

$$\{x \in \underline{R}^n : W(x, t) > 1\} = B(0, \sqrt{2nt \log (1/4\pi t)})$$

we have

$$m(\{x \in \underline{R}^n : W(x, t) > 1\}) = \omega_n (2nt \log (1/4\pi t))^{n/2}$$

As $t \to 0^+$,

$$t \log(1/4\pi t) \sim |t \log t|$$

so that

$$m(\{x \in \underline{R}^n : W(x,t) > 1\}) \sim \omega_n (2n)^{n/2} |t \log t|^{n/2}$$

Thus Theorems 6.10 and 6.11 are sharp in respect of the term $|t \log t|^{n/2}$, but the constants may not be the best possible.

6. LINEAR PARABOLIC EQUATIONS

Consider the parabolic equation

$$\sum_{i,j=1}^{n} a_{ij}(x,t) D_i D_j u(x,t) + \sum_{j=1}^{n} b_j(x,t) D_j u(x,t) + c(x,t) u(x,t) - D_t u(x,t) = 0 \tag{48}$$

whose fundamental solution satisfies the estimates

$$C_1 (t-s)^{-n/2} \exp(-\gamma_1 \|x-y\|^2/(t-s))$$

$$\leq \Gamma(x,t;y,s)$$

$$\leq C_2 (t-s)^{-n/2} \exp(-\gamma_2 \|x-y\|^2/(t-s)) \tag{49}$$

for some positive constants C_1, C_2, γ_1, and γ_2, whenever $x,y \in \underline{R}^n$ and $0 \leq s < t < a$. It follows from (49) and Lemma 1.1 that there is a positive number λ such that

$$\int_{\underline{R}^n} \Gamma(x,t;y,s)\,dy \leq C_2 (t-s)^{-n/2} \int_{\underline{R}^n} \exp(-\gamma_2 \|x-y\|^2/(t-s))\,dy \leq \lambda$$

whenever $x \in \underline{R}^n$ and $0 \leq s < t < a$. Throughout this section, λ remains unchanged, and the integral mean $M_{b,p}(v;t)$ is defined by

$$M_{b,p}(v;t) = \left(\int_{\underline{R}^n} \Gamma(0,b;x,t) v(x,t)^p \, dx \right)^{1/p}$$

whenever $1 \leq p < \infty$, $0 \leq t < b$, and v is a non-negative function on $\underline{R}^n \times \{t\}$.

Theorem 6.2 does not carry over exactly to solutions of (48) unless $c(x,t) = 0$ for all $(x,t) \in \underline{R}^n \times]0,a[$. In general, all that can be proved is that u is a solution of (48) on $\underline{R}^n \times]r,a[$,

$$M_{b,p}(|u|;t) \leq \lambda^{1/q} M_{b,p}(|u|;s)$$

whenever $r \leq s < t < b < a$, and

$$M_{b,p}(|f|;r) \leq \liminf_{t \to r^+} M_{b,p}(|u|;t)$$

$$\leq \limsup_{t \to r^+} M_{b,p}(|u|;t)$$

$$\leq \lambda^{1/q} M_{b,p}(|f|;r)$$

Lemma 6.1 can be extended to solutions of (48). The constant λ appears in the proof, but does not create any difficulties.

Theorem 6.3 does not carry over exactly to solutions of (48) unless $c(x,t) = 0$ for all $(x,t) \in \underline{R}^n \times]0,a[$. In this case we have

$$\int_{\underline{R}^n} \Gamma(x,t;y,0) \, dy = 1 \tag{50}$$

for all $(x,t) \in \underline{R}^n \times]0,a[$, since the function $v = 1$ on $\underline{R}^n \times [0,a[$ is then the unique positive solution of the Cauchy problem for (48) with initial function 1 on $\underline{R}^n \times \{0\}$, and therefore has the representation

$$v(x,t) = \int_{\underline{R}^n} \Gamma(x,t;y,0) v(y,0) \, dy$$

on $\underline{R}^n \times]0,a[$, by the general form of Theorem 2.10. This is identical to (50), and when (50) holds the proof in the text carries over unchanged. In general, we cannot prove that each mean $M_{b,p}(u^+;\cdot)$ is monotone, only that

$$M_{b,p}(u^+;t) \leq \lambda^{1/q} M_{b,p}(u^+;s) \tag{51}$$

whenever $0 < s < t < b < a$. However, this is sufficient to imply that

$M_{b,p}(u^+; \cdot)$ is bounded on $]0,b[$ for each $b < a$, by the following argument. Since

$$\liminf_{t \to 0^+} M_{b,p}(u^+; t) < \infty$$

by hypothesis, there is a positive null sequence $\{s_j\}$ such that $\{M_{b,p}(u^+; s_j)\}$ converges to a finite limit, so that there is a real number K such that

$$M_{b,p}(u^+; s_j) \leq K$$

for all j. Applying (51) with $s = s_j$ we see that

$$M_{b,p}(u^+; t) \leq \lambda^{1/q} K$$

whenever $s_j < t < b$. Since $\{s_j\}$ is null, this holds whenever $0 < t < b$, so that $M_{b,p}(u^+; \cdot)$ is bounded on $]0,b[$.

The proof that each mean $M_{b,p}(u^+; \cdot)$ is continuous (in Theorem 6.3) carries over to solutions of (48), with the following minor changes. The function $u^+(\cdot, r_j)^p$ is dominated by $\lambda^{p-1} w(\cdot, r_j)$ rather than by $w(\cdot, r_j)$. However, the extra constant is eventually cancelled out, since we prove that

$$\lambda^{p-1} M_{b,p}(u^+; r/2)^p - M_{b,p}(u^+; r)^p$$

$$\leq \lambda^{p-1} M_{b,p}(u^+; r/2)^p - \limsup_{j \to \infty} M_{b,p}(u^+; r_j)^p$$

by the method in the text.

The corollary to Theorem 6.3 carries over to the general case without change to its statement or proof.

Theorem 6.4 does not carry over exactly to solutions of (48) unless $c(x,t) = 0$ for all (x,t). In general, all that can be proved is that

$$u(x,t) = \int_{\underline{R}^n} \Gamma(x,t; y, 0) u(y, 0^+) \, dy$$

for all $(x,t) \in \underline{R}^n \times]0,a[$,

$$M_{b,p}(|u|;t) \leq \lambda^{1/q} M_{b,p}(|u|;s)$$

whenever $0 < s < t < b < a$,

$$M_{b,p}(|u(\cdot,0^+)|;0) \leq \liminf_{t\to0^+} M_{b,p}(|u|;t)$$

and

$$\limsup_{t\to0^+} M_{b,p}(|u|;t) \leq \lambda^{1/q} M_{b,p}(|u(\cdot,0^+)|;0) < \infty$$

For the corollary, we must use the upper estimate for Γ in (49), and restrict α to the interval $]0,\gamma_2/a]$, but no other changes are necessary.

No changes need to be made to Theorems 6.5 and 6.6, or their proofs, to generalize them to solutions of (48).

If $c(x,t) = 0$ for all (x,t), the result and proof of Theorem 6.7 can be generalized without change. Otherwise the representation (22) can still be proved, by modifying the proof in the text, but instead of (23) our methods show only that

$$\lambda^{-1/q} M_{b,p}(u^+(\cdot,0^+);0) \leq \liminf_{t\to0^+} M_{b,p}(u^+;t) \tag{52}$$

and

$$\limsup_{t\to0^+} M_{b,p}(u^+;t) \leq \lambda^{1/q} M_{b,p}(u^+(\cdot,0^+);0) < \infty \tag{53}$$

The changes in the proof are as follows. Instead of (25), the generalization of Theorem 6.2 tells us only that

$$M_{b,p}(v_r;s) \leq \lambda^{1/q} M_{b,p}(u^+;r)$$

Therefore, instead of (26), all we get is

$$M_{b,p}(w;s) \leq \lambda^{1/q} \liminf_{r\to0^+} M_{b,p}(u^+;r)$$

and instead of (27) we get

$$M_{b,p}(w(\cdot,0^+);0) \leq \liminf_{t \to 0^+} M_{b,p}(w;t)$$

and

$$\limsup_{t \to 0^+} M_{b,p}(w;t) \leq \lambda^{1/q} M_{b,p}(w(\cdot,0^+);0) < \infty$$

The argument in the text now leads to the conclusions described above. For the corollary, we must use the upper estimate for Γ in (49), and restrict α to the interval $]0,\gamma_2/a]$, but no other changes are necessary.

The result of Theorem 6.8 can be generalized unchanged to solutions of (48), as can the proof in the case where $p = 1$. When $p > 1$, the extension of Theorem 6.7 described above gives only (52) and (53), but (52), together with the hypothesis (28) is sufficient for the given argument to work.

The results of section 5 carry over to solutions of (48) without any significant changes. Essential use is made of the lower estimate for Γ in (49), and the constants in the theorems depend on the constants C_1 and γ_1 therein.

7. BIBLIOGRAPHICAL NOTES

Most of the results of sections 2 and 3 are new, but have close connections with known results.

Consider Theorem 6.2. Gehring [1, Theorem 10] proved that, if u is the Gauss-Weierstrass integral on $\underline{R} \times]0,a[$ of a function f such that $M_{a,p}(|f|;0) < \infty$, then $M_{a,p}(|u|;t) \leq M_{a,p}(|f|;0)$ for all $t \in]0,a[$. Watson [13] proved that, if $M_{b,p}(|f|;0) < \infty$ for each $b < a$, and

$$u(x,t) = \int_{\underline{R}^n} \Gamma(x,t;y,0)f(y)\,dy$$

for all $(x,t) \in \underline{R}^n \times]0,a[$, then $M_{b,p}(|u|;\cdot)$ is bounded on $]0,b[$ for each $b < a$, for general parabolic equations. He also proved that

$M_{b,p}(|u|;t) \leq \lambda^{1/q} M_{b,p}(|u|;s)$ whenever $0 < s < t < b < a$, where λ is the same as in section 6, and considered whether λ could be replaced by 1. Flett [1, Theorem 2] studied the properties of the Gauss-Weierstrass integrals u of an arbitrary function $f \in L^p(m)$, and showed that u is a temperature on $\underline{R}^n \times]0,\infty[$, that the function

$$t \longmapsto \left(\int_{\underline{R}^n} |u(x,t)|^p \, dx \right)^{1/p} \tag{54}$$

is non-increasing on $]0, \infty[$, and that

$$\left(\int_{\underline{R}^n} |u(x,t)|^p \, dx \right)^{1/p} \to \left(\int_{\underline{R}^n} |f(x)|^p \, dx \right)^{1/p} \tag{55}$$

as $t \to 0$.

Lemma 6.1 and Theorem 6.3 are new, as is the associated material in section 6, but the corollary was given earlier by Watson [13].

Theorem 6.4 is also due to Watson [13], except for the last line, which is new. Earlier, Gehring [1, Theorem 10] proved that, if u is a temperature on $\underline{R} \times]0, a[$ such that $M_{a,p}(|u|; \cdot) \leq C$ on $]0, a[$, then u is the Gauss-Weierstrass integral of a function f such that $M_{a,p}(|f|; 0) \leq C$. Flett [1, Theorem 5] showed that, if u is a temperature on $\underline{R}^n \times]0, a[$ such that the mean in (54) is locally integrable on $]0, \infty[$ and bounded on $]0, c[$ for some $c > 0$, then u is the Gauss-Weierstrass integral of a function $f \in L^p(m)$, so that the mean in (54) is non-increasing and (55) holds. The corollary to Theorem 6.4 is also new, but similar results have been obtained, under the hypothesis that the mean in (17) is bounded, by Johnson [1] and Chabrowski [8]. Related work was done by Chabrowski and Johnson [1].

Theorem 6.5 was given by Watson [2, Theorem 19] for the heat equation, but the essence is contained in the proofs of two earlier results by Gehring [1, Theorems 4 and 6].

Theorems 6.6 and 6.7 are new, as are the associated remarks in section 6. The corollary to Theorem 6.7 is also new, but similar results were obtained, under the hypothesis that the mean is bounded, by Johnson [1, Corollary 2.4] and Chabrowski [8, Theorem 3, Corollary 1].

Theorem 6.8 is also new.

Sections 4 and 5 follow Watson's paper [18], except that there cubes were used instead of balls. The use of balls is preferable, since it gives a larger constant in (46) when $n > 1$.

Representation theorems for temperatures which satisfy special conditions, related to various spaces of functions, have also been proved by Flett [1], Johnson [2, 3], and Fabes and Neri [1]. Other results of this kind, for general parabolic equations, were given by Chabrowski in sections 2 and 5 of his article [9].

7

The Initial Measure of a
Gauss-Weierstrass Integral

In Theorem 3.15, we gave several necessary and sufficient conditions for a temperature u on $\underline{R}^n \times \,]0, a[$ to have the Gauss-Weierstrass representation. It is therefore possible to know that a particular temperature is the Gauss-Weierstrass integral of some signed measure, without knowing what the measure is. In this chapter, we show how the measure is determined by the temperature. Of course, in Chapters V and VI we did this in various ways, but only in special cases where the temperature satisfied certain additional conditions. We now treat the general case, where the temperature is assumed only to belong to the class R_a of Chapter III.

1. PRELIMINARY RESULTS

If u is the Gauss-Weierstrass integral of μ, then in order to obtain μ from u, we need to consider limits of the form

$$\lim_{t \to 0^+} \int_A u(x, t)\, dx$$

where A is an arbitrary member of some particular family of Borel
sets. In this section, we shall show that the limit exists and is equal
to $\mu(A)$, whenever A is a bounded Borel subset of $\underline{R}^n$ such that
$|\mu|(\partial A) = 0$. In section 2, we shall prove that the condition that
$|\mu|(\partial A) = 0$ presents no real difficulties. This will require some
results of a measure-theoretic nature, and the present section also contains those which have not appeared in earlier chapters.

Recall, from Chapter I, that our signed measures need not have
finite total variation. In fact, if μ is a signed measure such that

$$\int_{\underline{R}^n} W(x - y, t) \, d|\mu|(y) < \infty$$

for some $(x, t) \in \underline{R}^n \times]0, \infty[$, then there exist a positive number α and
a signed measure λ with $|\lambda|(\underline{R}^n) < \infty$, such that

$$\mu(E) = \int_E \exp(\alpha\|y\|^2) \, d\lambda(y)$$

for every bounded Borel set E. Furthermore, the positive and negative
variations of μ are defined by

$$d\mu^+(y) = \exp(\alpha\|y\|^2) \, d\lambda^+(y), \quad d\mu^-(y) = \exp(\alpha\|y\|^2) \, d\lambda^-(y)$$

By the Hahn decomposition theorem, there exist disjoint Borel sets M
and N, with $M \cup N = \underline{R}^n$, such that

$$\lambda^+(A) = \lambda(A \cap M), \quad \lambda^-(A) = -\lambda(A \cap N)$$

for every Borel set A. This implies a similar result for μ, at least
when the set is bounded. For, if E is a bounded Borel set, we have

$$\mu^+(E) = \int_E \exp(\alpha\|y\|^2) \, d\lambda^+(y) = \int_{E \cap M} \exp(\alpha\|y\|^2) \, d\lambda(y) = \mu(E \cap M)$$

and a similar result for μ^-.

Definitions. If μ and ν are signed measures on $\underline{R}^n$ such that

$$\int_{\underline{R}^n} W(x - y, t) \, d|\mu|(y) < \infty, \quad \int_{\underline{R}^n} W(x - y, t) \, d|\nu|(y) < \infty \qquad (1)$$

for some $(x, t) \in \underline{R}^n \times]0, \infty[$, the signed measure $\mu - \nu$ is defined by

$$\mu - \nu = (\overset{+}{\mu} + \overset{-}{\nu}) - (\overset{-}{\mu} + \overset{+}{\nu})$$

Then $\mu - \nu$ is said to be <u>non-negative</u> if $(\mu - \nu)^{-}$ is null.

Note that, if (1) holds, then

$$\int_{R^n} W(x - y, t) \, d|\mu - \nu|(y) < \infty \tag{2}$$

As in previous chapters, we always assume that our signed measures satisfy (1). Note also that, if μ and ν happen to have finite total variations, then our definition of $\mu - \nu$ coincides with the natural one.

We shall use the following measure-theoretic result, for which we recall a definition from Chapter V. Given a subset D of $\underline{R}^n$, and a family $\mathscr{F}$ of closed balls such that, given any $x \in D$ and $\epsilon > 0$, there is a ball $B(x, r) \in \mathscr{F}$ with $r < \epsilon$, we say that $\mathscr{F}$ covers D in the <u>narrow Vitali sense</u>.

Theorem 7.1

Let $\mathscr{F}$ be a family of closed balls which covers $\underline{R}^n$ in the narrow Vitali sense, and let μ and ν be signed measures on $\underline{R}^n$. If $\mu(B) \geq \nu(B)$ for all $B \in \mathscr{F}$, then $\mu - \nu$ is non-negative.

<u>Proof.</u> Let $\rho = \mu - \nu$, so that $\rho(B) \geq 0$ for all $B \in \mathscr{F}$. Let G be any bounded open set, and put

$$\mathscr{G} = \{B \in \mathscr{F} : B \subseteq G\}$$

Since G is open and $\mathscr{F}$ covers R^n in the narrow Vitali sense, $\mathscr{G}$ covers G in the same sense. Note that $|\rho|$ is a non-negative measure on R^n, so that we can use Theorem 5.2 to show that there is a sequence $\{B_i\}$ of disjoint balls in $\mathscr{G}$, whose union U satisfies $|\rho|(G \setminus U) = 0$. Since G is bounded and $U \subseteq G$, the $|\rho|$-measures of U and G are finite. Therefore, because $\rho(B_i) \geq 0$ for all i,

$$\rho(G) = \rho(G \setminus U) + \rho(U) = \rho(U) = \sum_{i=1}^{\infty} \rho(B_i) \geq 0$$

Thus $\rho(G) \geq 0$ for every bounded open set G.

If E is any bounded Borel set and $\epsilon > 0$, the regularity of $|\rho|$ implies that there is a bounded open superset H of E such that $|\rho|(H \setminus E) < \epsilon$. Therefore, because $\rho(H) \geq 0$ and $|\rho|(H) < \infty$,

$$\rho(E) \geq \rho(E) - \rho(H) = -\rho(H \setminus E) \geq -|\rho|(H \setminus E) > -\epsilon$$

Since ϵ is arbitrary, we deduce that $\rho(E) \geq 0$.

Now let M and N be disjoint Borel sets, with union $\underline{R}^n$, such that

$$\rho^+(E) = \rho(E \cap M), \quad \rho^-(E) = -\rho(E \cap N)$$

for every bounded Borel set E. The existence of these sets follows from (2) and the Hahn decomposition theorem, as described above. We have shown that $\rho(E) \geq 0$ for every bounded Borel set E, and it follows that $\rho(E \cap N) \geq 0$, and hence that

$$0 \leq \rho^-(E) = -\rho(E \cap N) \leq 0$$

for all such sets. Since ρ^- is a non-negative measure, we can deduce that, for an arbitrary Borel set A,

$$\rho^-(A) = \lim_{k \to \infty} \rho^-(A \cap B(0,k)) = 0$$

Hence ρ^- is null, and ρ is non-negative.

Theorem 7.1 will be used in conjunction with Theorem 7.2, which depends upon the following technical lemma about the Gauss-Weierstrass integrals of characteristic functions of Borel sets. Here, and subsequently, we use ω_E to denote the Gauss-Weierstrass integral on $\underline{R}^n \times]0,\infty[$ of the characteristic function χ_E of a bounded Borel set E. Thus

$$\omega_E(y,t) = \int_E W(y - x, t) \, dx \tag{3}$$

for all $(y,t) \in \underline{R}^n \times]0,\infty[$.

Lemma 7.1

Let E be a bounded Borel subset of $\underline{R}^n$, and let ω_E be defined by (3).

(i) If $b > 0$, there is a positive number κ, which depends only on E, b, and n, such that

$$\omega_E(y,t) \leq \kappa W(y,b)$$

for all $(y,t) \in \underline{R}^n \times]0,b/2[$.

(ii) For all $y \in \underline{R}^n \setminus \partial E$,

$$\lim_{t \to 0^+} \omega_E(y, t) = \chi_E(y) \qquad (4)$$

(iii) If E is compact, then

$$\lim_{t \to 0^+} \sup \omega_E(y, t) \leq \chi_E(y) \qquad (5)$$

for all $y \in \underline{R}^n$.

__Proof.__ (i) Choose r such that $E \subseteq B(0, r)$, and put $\delta = 1/2b$. Then

$$\omega_E(y, t) = (4\pi t)^{-n/2} \int_E \exp\left(- \frac{\|y - x\|^2}{4t}\right) dx$$

$$\leq (4\pi t)^{-n/2} \exp(\delta r^2) \int_E \exp\left(- \frac{\|y - x\|^2}{4t} - \delta \|x\|^2\right) dx \qquad (6)$$

Completing the square in the exponential, and writing

$$z = (1 + 4\delta t)^{\frac{1}{2}} x - (1 + 4\delta t)^{-\frac{1}{2}} y$$

we obtain

$$\frac{\|y - x\|^2}{4t} + \delta \|x\|^2 = \frac{1}{4t} \sum_{i=1}^{n} ((1 + 4\delta t) x_i^2 - 2x_i y_i + y_i^2)$$

$$= \frac{\|z\|^2}{4t} + \frac{\delta \|y\|^2}{1 + 4\delta t}$$

Putting this into (6), and noting that $dz = (1 + 4\delta t)^{n/2} dx$, we obtain

$$\omega_E(y, t) \leq (4\pi t)^{-n/2} (1 + 4\delta t)^{-n/2} \exp\left(\delta r^2 - \frac{\delta \|y\|^2}{1 + 4\delta t}\right) \int_E \exp\left(- \frac{\|z\|^2}{4t}\right) dz$$

Therefore, by Lemma 1.1,

$$\omega_E(y,t) \le (1 + 4\delta t)^{-n/2} \exp\left(\delta r^2 - \frac{\delta \|y\|^2}{1 + 4\delta t}\right)$$

Whenever $t \in {]0, b/2[}$, we have $(1 + 4\delta t)^{-n/2} \le 1$ and

$$\frac{\delta}{1 + 4\delta t} \ge \frac{\delta}{1 + 2\delta b} = \frac{\delta}{2} = \frac{1}{4b}$$

so that

$$\omega_E(y,t) \le \exp\left(\delta r^2 - \frac{\|y\|^2}{4b}\right) = \kappa W(y, b)$$

where $\kappa = \exp(\delta r^2)(4\pi b)^{n/2}$.

(ii) This follows immediately from Theorem 1.1, since χ_E is continuous at every point of $\underline{R}^n \setminus \partial E$.

(iii) Let V be a bounded open superset of E. Then $\chi_E \le \chi_V$, so that $\omega_E \le \omega_V$. Applying (ii) to V, we obtain

$$\limsup_{t\to 0^+} \omega_E(y,t) \le \lim_{t\to 0^+} \omega_V(y,t) = \chi_V(y) = \chi_E(y)$$

for all $y \in E$. Furthermore, since E is closed, it follows from (ii) that (4) holds for all $y \in \underline{R}^n \setminus E$. Hence (5) holds for all $y \in \underline{R}^n$.

We now come to the main result of this section.

Theorem 7.2

Let u be the Gauss-Weierstrass integral on $\underline{R}^n \times {]0, a[}$ of a signed measure μ.

(i) If E is a bounded Borel subset of $\underline{R}^n$, and $|\mu|(\partial E) = 0$, then

$$\lim_{t\to 0^+} \int_E u(x,t)\, dx = \mu(E)$$

(ii) If μ is non-negative, and E is a compact set in $\underline{R}^n$, then

$$\limsup_{t\to 0^+} \int_E u(x,t)\, dx \le \mu(E)$$

Proof. If μ is non-negative, it follows from Fubini's theorem that

$$\int_E u(x,t)\ dx = \int_E dx \int_{\underline{R}^n} W(x - y,t)\ d\mu(y) = \int_{\underline{R}^n} \omega_E(y,t)\ d\mu(y) \qquad (7)$$

where ω_E is defined by (3). By Lemma 7.1(i), given any $b \in\ ]0,a[$ we can find a positive constant κ such that $\omega_E(\cdot,t) \leq \kappa\,W(\cdot,b)$ for all $t \in\ ]0,b/2[$. Since $b < a$, it is one of our hypotheses that $W(\cdot,b)$ is μ-integrable over $\underline{R}^n$. We can therefore apply the dominated convergence theorem to the last integral in (7). It thus follows from Lemma 7.1(ii), and our hypothesis that $\mu(\partial E) = 0$, that

$$\lim_{t\to 0^+} \int_E u(x,t)\ dx = \lim_{t\to 0^+} \int_{\underline{R}^n} \omega_E(y,t)\ d\mu(y) = \int_{\underline{R}^n} \chi_E(y)\ d\mu(y) = \mu(E)$$

This proves (i) in the case where μ is non-negative. To prove (i) for a general signed measure μ, we apply the case just proved to the positive and negative variations of μ, noting that $\mu^+(\partial E) = \mu^-(\partial E) = 0$, by hypothesis. We thus obtain

$$\lim_{t\to 0^+} \int_E u(x,t)\ dx$$

$$= \lim_{t\to 0^+} \left(\int_E dx \int_{\underline{R}^n} W(x - y,t)\ d\mu^+(y) - \int_E dx \int_{\underline{R}^n} W(x - y,t)\ d\mu^-(y) \right)$$

$$= \mu^+(E) - \mu^-(E)$$

and this is equal to $\mu(E)$ since E is bounded.

To prove (ii), we apply Fatou's lemma to $-\omega_E$ in (7), noting that this is valid because $-\omega_E(\cdot,t) \geq -\kappa\,W(\cdot,b)$ whenever $0 < t < b/2$ and $b < a$, and $W(\cdot,b)$ is μ-integrable over $\underline{R}^n$. Thus, using Lemma 7.1(iii), we obtain

$$\limsup_{t\to 0^+} \int_E u(x,t)\ dx = \limsup_{t\to 0^+} \int_{\underline{R}^n} \omega_E(y,t)\ d\mu(y)$$

$$\leq \int_{\underline{R}^n} \limsup_{t\to 0^+} \omega_E(y,t)\ d\mu(y)$$

$$\leq \int_{\underline{R}^n} \chi_E(y)\ d\mu(y)$$

$$= \mu(E)$$

2. DETERMINATION OF THE INITIAL MEASURE
USING ABUNDANT VITALI COVERINGS

Before we come to use Theorems 7.1 and 7.2, we need to introduce the idea of a family of balls which effectively covers $\underline{R}^n$ in the narrow Vitali sense uncountably many times.

Definition. A family $\mathscr{F}$ of closed balls is called an <u>abundant Vitali covering of $\underline{R}^n$</u> if, given any $x \in \underline{R}^n$ and $\epsilon > 0$, $\mathscr{F}$ contains uncountably many balls centered at x with radius less than ϵ.

If $\mathscr{F}$ is an abundant Vitali covering, then it covers $\underline{R}^n$ in the narrow Vitali sense. The important point about $\mathscr{F}$ is that we can remove from it a sequence of balls centered at every point, and still be left with a family which covers $\underline{R}^n$ in the narrow Vitali sense.

The obvious example of an abundant Vitali covering is the family of all closed balls. The following lemma provides simple but non-trivial examples for the case where n = 1.

Lemma 7.2

If $A \subseteq \underline{R}$ and $\mathscr{F} = \{[a, b] : a, b \in A\}$, then $\mathscr{F}$ is an abundant Vitali covering of $\underline{R}$ if either

(i) $m(\underline{R} \setminus A) = 0$, or

(ii)   A is a dense open set.

Proof. In either case we must show that, given any $x \in \underline{R}$ and $\epsilon > 0$, there are uncountably many numbers $y, z \in A$ such that $x = (y + z)/2$ and $|y - z| < 2\epsilon$. The proofs in the two cases are similar. Let $\mathscr{C}_1$ and $\mathscr{C}_2$ denote the families of all subsets A of $\underline{R}^n$ which satisfy (i) and (ii) respectively. Let $i \in \{1, 2\}$, $x \in \underline{R}$, $\epsilon > 0$, and $A \in \mathscr{C}_i$. If

$$B = \{2x\} - A = \{\xi \in \underline{R} : \xi = 2x - \eta \quad \text{for some} \quad \eta \in A\}$$

then $B \in \mathscr{C}_i$ and hence $A \cap B \in \mathscr{C}_i$ (using Baire's theorem if i = 2). If $y \in A \cap B$, then $y \in A$ and there is $z \in A$ such that $y = 2x - z$, so that $x = (y + z)/2$ with $y, z \in A$. Since y is an arbitrary point of $A \cap B$, and $A \cap B \in \mathscr{C}_i$, the interval $]x - \epsilon, x + \epsilon[$ contains uncountably many choices of y. To each such y there corresponds a unique $z \in A$ such that $x = (y + z)/2$, and for this z we have

$$|y - z| = 2(|y - z| - (|y - z|/2)) = 2(|y - z| - |z - x|) \leq 2|y - x| < 2\epsilon$$

Remark. It is worth noting that a dense open set can have arbitrarily small Lebesgue measure, so that the conditions (i) and (ii) in Lemma 7.2 are quite independent of each other. For example, if we are given any $\epsilon > 0$ and arrange the rational numbers as a sequence $\{r_j\}$, then the union of all the open intervals

$$]r_j - \epsilon 2^{-j-1}, r_j + \epsilon 2^{-j-1}[, \quad j \in \underline{N}$$

is a dense open subset of $\underline{R}$ with Lebesgue measure at most

$$\sum_{j=1}^{\infty} \epsilon 2^{-j} = \epsilon$$

We now characterize the initial measure of a Gauss-Weierstrass integral using abundant Vitali coverings.

Theorem 7.3

Let u be the Gauss-Weierstrass integral on $\underline{R}^n \times]0, a[$ of a signed measure μ, and let $\mathscr{F}$ be an abundant Vitali covering of $\underline{R}^n$.

(i) There exists an abundant Vitali covering $\mathscr{F}_0 \subseteq \mathscr{F}$ such that

$$\lim_{t \to 0^+} \int_B u(x, t) \, dx = \mu(B) \tag{8}$$

and

$$|\mu|(\partial B) = 0 \tag{9}$$

for all $B \in \mathscr{F}_0$.

(ii) If there is a signed measure ν on $\underline{R}^n$ such that

$$\lim_{t \to 0^+} \int_B u(x, t) \, dx = \nu(B) \tag{10}$$

for all $B \in \mathscr{F}$, then $\nu = \mu$.

Proof. (i) If (9) holds for some ball B, then Theorem 7.2(i) shows that (8) also holds. It is therefore sufficient to prove that there is an abundant Vitali covering $\mathcal{F}_0 \subseteq \mathcal{F}$ such that (9) holds for all $B \in \mathcal{F}_0$.

Let $\mathcal{F}_0$ denote the family of all balls $B = \bar{B}(y, r) \in \mathcal{F}$ such that $r \leq 1$ and $|\mu|(\partial B) = 0$. The proof will be complete when we have shown that this $\mathcal{F}_0$ is an abundant Vitali covering. For each $x \in \underline{R}^n$, we denote by $\mathcal{C}_x$ the uncountable family of all balls $\bar{B}(x, r) \in \mathcal{F}$ with $r \leq 1$. If $\{B_j\}$ is any sequence of distinct elements of $\mathcal{C}_x$, then

$$\sum_{j=1}^{\infty} |\mu|(\partial B_j) = |\mu|\left(\bigcup_{j=1}^{\infty} \partial B_j\right) \leq |\mu|(\bar{B}(x, 1)) < \infty$$

Therefore, for each $k \in \underline{N}$, the family

$$\{B \in \mathcal{C}_x : |\mu|(\partial B) \geq 1/k\}$$

is finite, so that the family

$$\mathcal{C}_x \setminus \mathcal{F}_0 = \bigcup_{k=1}^{\infty} \{B \in \mathcal{C}_x : |\mu|(\partial B) \geq 1/k\}$$

is countable. Since $\mathcal{C}_x$ is uncountable, the family $\mathcal{F} \setminus (\mathcal{C}_x \setminus \mathcal{F}_0)$ is an abundant Vitali covering. Since x is arbitrary, the same is true of the family

$$\mathcal{F} \setminus \left(\bigcup_{x \in \underline{R}^n} (\mathcal{C}_x \setminus \mathcal{F}_0)\right)$$

and hence of $\mathcal{F}_0$, which consists of all balls in this family which have radius at most 1.

(ii) Suppose that (10) holds for all $B \in \mathcal{F}$. By (i), there is an abundant Vitali covering $\mathcal{F}_0 \subseteq \mathcal{F}$ such that (8) holds for all $B \in \mathcal{F}_0$. Therefore $\nu(B) = \mu(B)$ for all $B \in \mathcal{F}_0$. Since $\mathcal{F}_0$ covers $\underline{R}^n$ in the narrow Vitali sense, it follows from Theorem 7.1 that both $\nu - \mu$ and $\mu - \nu$ are non-negative, so that $\nu = \mu$.

Theorem 7.3(ii) has the following easy consequence.

Theorem 7.4

Let u be the Gauss–Weierstrass integral on $\underline{R}^n \times]0, a[$ of a signed measure μ, and let $\mathcal{F}$ be an abundant Vitali covering of $\underline{R}^n$. If there is a number κ such that

$$\lim_{t \to 0^+} \int_B u(x,t) \, dx = \kappa \, m(B)$$

for all $B \in \mathscr{F}$, then $u = \kappa$ on $\underline{R}^n \times]0, a[$.

Proof. If follows from Theorem 7.3(ii) that $\mu = \kappa m$. Therefore

$$u(x,t) = \kappa \int_{R^n} W(x - y, t) \, dy = \kappa$$

for all $(x,t) \in \underline{R}^n \times]0, a[$, by Lemma 1.1.

Corollary

Let u be the Gauss-Weierstrass integral on $\underline{R} \times]0, a[$ of a signed measure. If there is a number κ such that

$$\lim_{t \to 0^+} \int_b^c u(x,t) \, dx = \kappa(c - b) \tag{11}$$

holds either (i) for m-almost all $b, c \in \underline{R}$, or (ii) for all b, c in a dense open subset of $\underline{R}$, then $u = \kappa$ on $\underline{R} \times]0, a[$.

Proof. Put $\mathscr{F} = \{[b, c] : (11) \text{ holds}\}$. By Lemma 7.2, $\mathscr{F}$ is an abundant Vitali covering of $\underline{R}$ in both cases. Theorem 7.4 now shows that $u = \kappa$.

Remark. The use of abundant Vitali coverings has been necessary only to deal with measures μ which are not absolutely continuous with respect to m. For if $d\mu(y) = f(y) \, dy$, then $|\mu|(\partial B) = 0$ for every ball $B \subseteq \underline{R}^n$, so that Theorem 7.2(i) shows that

$$\lim_{t \to 0^+} \int_B u(x,t) \, dx = \mu(B)$$

3. DETERMINATION OF THE INITIAL MEASURE UNDER MILDER CONDITIONS

Our aim here is to obtain the conclusion of Theorem 7.3(ii) under weaker hypotheses. More specifically, we shall replace the abundant

Vitali covering by a possibly countable family of sets which need not be balls. The family must be such that an abundant Vitali covering can be manufactured from it using certain basic set-theoretic operations.

Definitions. Let $\mathscr{A}$ be a family of bounded Borel sets. Let $\mathscr{U}_{\mathscr{A}}$ denote the family of all closed sets which can be expressed as a union of finitely many disjoint sets from $\mathscr{A}$. We call $\mathscr{U}_{\mathscr{A}}$ the union family of $\mathscr{A}$. Let $\mathscr{F}$ be an abundant Vitali covering of $\underline{R}^n$. If each ball in $\mathscr{F}$ can be written as the intersection of some contracting sequence of sets in $\mathscr{U}_{\mathscr{A}}$, then $\mathscr{A}$ is called a generator of $\mathscr{F}$.

We shall replace the abundant Vitali covering $\mathscr{F}$ in Theorem 7.3(ii) by an arbitrary generator $\mathscr{A}$ of $\mathscr{F}$. We can show that $\mathscr{A}$ may be countable, even if $\mathscr{F}$ contains every closed ball in $\underline{R}^n$, using the following criterion. If, given any numbers r and s such that $0 < r < s$, and any $x \in \underline{R}^n$, there is a set F in the union family of $\mathscr{A}$ such that

$$\bar{B}(x, r) \subseteq F \subseteq \bar{B}(x, s) \tag{12}$$

then $\mathscr{A}$ generates the family of all closed balls. For, given any closed ball $\bar{B}(x, r)$, we can take a strictly decreasing sequence $\{r_j\}$ with limit r. Then, for each j, we can find a set $F_j \in \mathscr{U}_{\mathscr{A}}$ such that

$$\bar{B}(x, r_{j+1}) \subseteq F_j \subseteq \bar{B}(x, r_j)$$

Since the sequence $\{\bar{B}(x, r_j)\}$ contracts to $\bar{B}(x, r)$, so does $\{F_j\}$.

This criterion can be modified to suit a smaller abundant Vitali covering of $\underline{R}^n$. For example, suppose that to each $x \in \underline{R}^n$ there corresponds a positive number ϵ_x, and that

$$\mathscr{F} = \{\bar{B}(x, r) : x \in \underline{R}^n, \, 0 < r < \epsilon_x\}$$

Then $\mathscr{F}$ is generated by any family $\mathscr{A}$ of bounded Borel sets which has the following property. Whenever $x \in \underline{R}^n$ and $0 < r < s < \epsilon_x$, there is a set $F \in \mathscr{U}_{\mathscr{A}}$ such that (12) holds.

We now show that a generator $\mathscr{A}$ of an abundant Vitali covering may be countable. Let $\underline{Q}^n$ denote the set of all n-tuples of rational numbers, let $\underline{Q}^+ = \underline{Q} \cap \,]0, \infty[$, and let

$$\mathscr{A}_{\underline{Q}} = \{\bar{B}(y, q) : y \in \underline{Q}^n, \, q \in \underline{Q}^+\}$$

Then $\mathscr{A}_{\underline{Q}}$ is a countable family of compact sets. Given any numbers r and s such that $0 < r < s$, and any $x \in \underline{R}^n$, choose $y \in \underline{Q}^n$ such that

$$0 < \|y - x\| \le (s - r)/3$$

and then choose $q \in \underline{Q}^+$ such that

$$r + \|y - x\| \le q \le r + 2\|y - x\|$$

Then, if $z \in \bar{B}(x, r)$,

$$\|y - z\| \le \|y - x\| + \|x - z\| \le \|y - x\| + r \le q$$

so that $z \in \bar{B}(y, q)$. Also, if $\xi \in \bar{B}(y, q)$,

$$\|\xi - x\| \le \|\xi - y\| + \|y - x\| \le q + \|y - x\| \le r + 3\|y - x\| \le r + (s - r) = s$$

so that $\xi \in \bar{B}(x, s)$. Hence

$$\bar{B}(x, r) \subseteq \bar{B}(y, q) \subseteq \bar{B}(x, s)$$

so that (12) is satisfied with $F = \bar{B}(y, q) \in \mathscr{A}_{\underline{Q}}$. It follows that the countable family $\mathscr{A}_{\underline{Q}}$ generates the abundant Vitali covering which contains all closed balls in $\underline{R}^n$.

Lemma 7.3

Let u be the Gauss-Weierstrass integral on $\underline{R}^n \times \,]0, a[$ of a non-negative measure μ, and let $\mathscr{A}$ be a generator of an abundant Vitali covering $\mathscr{F}$ of $\underline{R}^n$. If ν is a non-negative measure on $\underline{R}^n$ such that

$$\limsup_{t \to 0^+} \int_F u(x, t)\, dx \le \nu(F) \tag{13}$$

for all F in the union family of $\mathscr{A}$, then $\nu - \mu$ is non-negative.

Proof. We begin by showing that (13) holds for all $F \in \mathscr{F}$. Let $B \in \mathscr{F}$, and let $\{F_j\}$ be a contracting sequence of sets in the union family of $\mathscr{A}$ whose intersection is B. Then, for each j, the non-negativity of u and (13) imply that

$$\limsup_{t \to 0^+} \int_B u(x, t)\, dx \le \limsup_{t \to 0^+} \int_{F_j} u(x, t)\, dx \le \nu(F_j)$$

Since $\nu(F_j) \to \nu(B)$ as $j \to \infty$, it follows that

$$\limsup_{t \to 0^+} \int_B u(x,t)\, dx \le \nu(B) \tag{14}$$

Thus we see that (13) holds for all sets in $\mathscr{F}$. By Theorem 7.3(i), there exists an abundant Vitali covering $\mathscr{F}_0 \subseteq \mathscr{F}$ such that

$$\lim_{t \to 0^+} \int_B u(x,t)\, dx = \mu(B)$$

for all $B \in \mathscr{F}_0$. This, together with (14), shows that $\mu(B) \le \nu(B)$ for all $B \in \mathscr{F}_0$. Since $\mathscr{F}_0$ covers $\underline{R}^n$ in the narrow Vitali sense, it follows from Theorem 7.1 that $\nu - \mu$ is non-negative.

Before we come to the main result of this section, we make a remark about its hypotheses and those of Theorem 7.3(ii). Let u be the Gauss-Weierstrass integral on $\underline{R}^n \times]0,a[$ of a signed measure μ, and let $\mathscr{F}$ be an abundant Vitali covering of $\underline{R}^n$. Theorem 7.3(ii) tell us that, if there is a signed measure ν such that

$$\lim_{t \to 0^+} \int_B u(x,t)\, dx = \nu(B) \tag{15}$$

for all $B \in \mathscr{F}$, then $\nu = \mu$. It therefore follows from Theorem 7.3(i) that, if (15) holds for all $B \in \mathscr{F}$, then there exists an abundant Vitali covering $\mathscr{F}_0 \subseteq \mathscr{F}$ such that both (15) and

$$|\nu|(\partial B) = 0 \tag{16}$$

hold for all $B \in \mathscr{F}_0$. Conversely, if both (15) and (16) hold for all balls B in some abundant Vitali covering of $\underline{R}^n$, then the hypotheses of Theorem 7.3(ii) are satisfied. Thus (15) and (16) together are equivalent to (15) alone. In the following theorem, we replace (15) and (16) by similar conditions on an arbitrary generator of an abundant Vitali covering.

Theorem 7.5

Let u be the Gauss-Weierstrass integral on $\underline{R}^n \times]0,a[$ of a signed measure μ, and let $\mathscr{A}$ be a generator of an abundant Vitali covering $\mathscr{F}$ of $\underline{R}^n$. If there is a signed measure ν on $\underline{R}^n$ such that

$$\lim_{t \to 0^+} \int_A u(x,t)\, dx = \nu(A) \tag{17}$$

and

$$|\nu|(\partial A) = 0 \tag{18}$$

for all $A \in \mathscr{A}$, then $\nu = \mu$.

Proof. We first use Theorems 7.2(ii) and 7.1 to show that $\nu^+(A) \leq \mu^+(A)$ and $\nu^-(A) \leq \mu^-(A)$ for every Borel set A. Let g and h be the Gauss-Weierstrass integrals on $R^n \times]0, a[$ of μ^+ and μ^- respectively, so that $u = g - h$. By Theorem 7.2(ii), for every compact set $F \subseteq R^n$,

$$\limsup_{t \to 0^+} \int_F g(x,t)\, dx \leq \mu^+(F) \tag{19}$$

In particular, (19) holds whenever F belongs to the union family $\mathscr{U}_{\mathscr{A}}$ of $\mathscr{A}$, since each F is a closed set which can be expressed as a union of finitely many disjoint sets from $\mathscr{A}$, and $\mathscr{A}$ contains only bounded Borel sets. Let $F \in \mathscr{U}_{\mathscr{A}}$ and let $A_1, \ldots, A_k$ be disjoint sets in $\mathscr{A}$ with union F. Using the fact that (17) holds for each A_i, the inequality $u \leq g$, and the fact that (19) holds for F, we obtain

$$\nu(F) = \sum_{i=1}^{k} \nu(A_i)$$

$$= \sum_{i=1}^{k} \left(\lim_{t \to 0^+} \int_{A_i} u(x,t)\, dx \right)$$

$$= \lim_{t \to 0^+} \left(\sum_{i=1}^{k} \int_{A_i} u(x,t)\, dx \right) \tag{20}$$

$$= \lim_{t \to 0^+} \int_F u(x,t)\, dx$$

$$\leq \limsup_{t \to 0^+} \int_F g(x,t)\, dx$$

$$\leq \mu^+(F)$$

Thus $\nu(F) \leq \mu^+(F)$ for all $F \in \mathscr{U}_{\mathscr{A}}$. Let $B \in \mathscr{F}$. Then there is a contracting sequence $\{F_j\}$, of sets in $\mathscr{U}_{\mathscr{A}}$, whose intersection is B. It follows that

$$\nu(B) = \lim_{j \to \infty} \nu(F_j) \leq \lim_{j \to \infty} \mu^+(F_j) = \mu^+(B)$$

Hence $\nu(B) \leq \mu^+(B)$ for all $B \in \mathcal{F}$. Since $\mathcal{F}$ covers $\underline{R}^n$ in the narrow Vitali sense, it follows from Theorem 7.1 that $\mu^+ - \nu$ is non-negative. Therefore, since ν is well-defined on bounded Borel sets, $\nu(A) \leq \mu^+(A)$ for every such set A. Let M be a Borel set such that $\nu^+(A) = \nu(A \cap M)$ for every bounded Borel set A; the existence of M follows from the Hahn decomposition theorem in the manner described in section 1. Then

$$\nu^+(A) = \nu(A \cap M) \leq \mu^+(A \cap M) \leq \mu^+(A)$$

whenever A is a bounded Borel set. Furthermore, if A is an unbounded Borel set,

$$\nu^+(A) = \lim_{\ell \to \infty} \nu^+(A \cap B(0, \ell)) \leq \lim_{\ell \to \infty} \mu^+(A \cap B(0, \ell)) = \mu^+(A)$$

so that $\nu^+(A) \leq \mu^+(A)$ for every Borel set A.

A similar argument, in which g, μ^+, and μ are replaced by h, μ^-, and $-\mu$ respectively, shows that $\nu^-(A) \leq \mu^-(A)$ for every Borel set A.

We now use the results just obtained, together with Theorem 7.2 and Lemma 7.3, to show first that $\nu(B) \geq \mu(B)$, then that $\mu(B) \geq \nu(B)$, for every closed ball $B \subseteq \underline{R}^n$.

Since the Gauss-Weierstrass integrals (g and h) of μ^+ and μ^- are both defined on $\underline{R}^n \times]0, a[$, and we have proved that $\nu^+(A) \leq \mu^+(A)$ and $\nu^-(A) \leq \mu^-(A)$ for every Borel set A, the Gauss-Weierstrass integrals of ν^+ and ν^- must also be defined on that strip. The same is therefore true of the Gauss-Weierstrass integrals of $\mu^+ - \nu^+$, $\mu^- - \nu^-$, and ν, which we denote by v, w, and $\bar{u}$, respectively. Note that $\mu^+ - \nu^+$ and $\mu^- - \nu^-$ are non-negative measures.

We have

$$\mu^+ - \nu^+ = \mu - \nu + (\mu^- - \nu^-)$$

so that

$$v = u - \bar{u} + w \tag{21}$$

Let $F \in \mathcal{U}_{\mathcal{A}}$, and let $A_1, \ldots, A_k$ be disjoint sets in $\mathcal{A}$ with union F. Then, by (20),

$$\nu(F) = \lim_{t \to 0^+} \int_F u(x, t) \, dx \tag{22}$$

Furthermore, since (18) holds for all $A \in \mathcal{A}$, we have

$$|\nu|(\partial F) \le |\nu|\left(\bigcup_{i=1}^{k} \partial A_i\right) \le \sum_{i=1}^{k} |\nu|(\partial A_i) = 0$$

so that $|\nu|(\partial F) = 0$. It therefore follows from Theorem 7.2(i) that

$$\lim_{t \to 0^+} \int_F \bar{u}(x, t)\, dx = \nu(F) \tag{23}$$

whenever $F \in \mathcal{U}_{\mathscr{A}}$. Next, because each $F \in \mathcal{U}_{\mathscr{A}}$ is compact and $\mu^- - \nu^-$ is non-negative, Theorem 7.2(ii) shows that

$$\limsup_{t \to 0^+} \int_F w(x, t)\, dx \le (\mu^- - \nu^-)(F)$$

It follows from this, together with (21), (22) and (23), that

$$\limsup_{t \to 0^+} \int_F v(x, t)\, dx$$
$$= \lim_{t \to 0^+} \int_F u(x, t)\, dx - \lim_{t \to 0^+} \int_F \bar{u}(x, t)\, dx + \limsup_{t \to 0^+} \int_F w(x, t)\, dx$$
$$\le \nu(F) - \nu(F) + (\mu^- - \nu^-)(F)$$
$$= (\mu^- - \nu^-)(F)$$

whenever $F \in \mathcal{U}_{\mathscr{A}}$. Hence, by Lemma 7.3, the measure $(\mu^- - \nu^-) - (\mu^+ - \nu^+)$ is non-negative. This measure coincides with $\nu - \mu$ on all bounded Borel sets, so that $\nu(B) \ge \mu(B)$ for every closed ball B.

We now interchange the roles of $\mu^+ - \nu^+$ and $\mu^- - \nu^-$, to obtain the reverse inequality. From (21) we have

$$w = \bar{u} - u + v$$

If $F \in \mathcal{U}_{\mathscr{A}}$ then F is compact. Therefore, by Theorem 7.2(ii),

$$\limsup_{t \to 0^+} \int_F v(x, t)\, dx \le (\mu^+ - \nu^+)(F)$$

It now follows, using (22) and (23), that

$$\lim_{t\to 0^+} \sup \int_F w(x,t)\ dx$$

$$= \lim_{t\to 0^+} \int_F \bar{u}(x,t)\ dx - \lim_{t\to 0^+} \int_F u(x,t)\ dx + \lim_{t\to 0^+} \sup \int_F v(x,t)\ dx$$

$$\le \nu(F) - \nu(F) + (\mu^+ - \nu^+)(F)$$

$$= (\mu^+ - \nu^+)(F)$$

whenever $F \in \mathcal{U}_{\mathscr{A}}$. Hence, by Lemma 7.3, the measure $(\mu^+ - \nu^+) - (\mu^- - \nu^-)$ is non-negative. Therefore $\mu(B) \ge \nu(B)$ for every closed ball B.

It follows that $\mu(B) = \nu(B)$ for every closed ball B. Hence $\mu = \nu$, by Theorem 7.1.

Corollary

Let u be the Gauss-Weierstrass integral on $\underline{R}^n \times]0, a[$ of a signed measure μ, and let $\mathscr{A}$ be a generator of an abundant Vitali covering of $\underline{R}^n$. If there is a function f on $\underline{R}^n$ such that

$$\lim_{t\to 0^+} \int_A u(x,t)\ dx = \int_A f(x)\ dx$$

for all $A \in \mathscr{A}$, and the Gauss-Weierstrass integral of f is defined on $\underline{R}^n \times]0, a[$, then $d\mu(x) = f(x)\ dx = u(x, 0^+)\ dx$.

Proof. Put $d\nu(x) = f(x)\ dx$. Then Theorem 7.5 shows that $\nu = \mu$, so that u is the Gauss-Weierstrass integral of f. By Theorem 4.1, $u(\cdot, 0^+)$ exists and is equal to f m-a. e. on $\underline{R}^n$, which completes the proof.

Another consequence of Theorem 7.5, which is essentially a special case of the above corollary, parallels Theorem 7.4. A comparison of the corollaries of this result and Theorem 7.4 illustrates the difference between using an abundant Vitali covering and a generator of one.

Theorem 7.6

Let u be the Gauss-Weierstrass integral on $\underline{R}^n \times]0, a[$ of a signed measure, and let $\mathscr{A}$ be a generator of an abundant Vitali covering of $\underline{R}^n$. If there is a number κ such that

$$\lim_{t\to 0^+} \int_A u(x,t)\ dx\ =\ \kappa m(A)$$

for all $A \in \mathscr{A}$, then $u = \kappa$ on $\underline{R}^n \times]0,a[$.

Proof. By the corollary of Theorem 7.5, u is the Gauss-Weierstrass integral of κ, so that $u = \kappa$ in view of Lemma 1.1.

Corollary

Let u be the Gauss-Weierstrass integral on $\underline{R} \times]0,a[$ of a signed measure. If there is a number κ such that

$$\lim_{t\to 0^+} \int_b^c u(x,t)\ dx\ =\ \kappa(c - b)$$

holds for all b, c in a countable dense subset of $\underline{R}$, then $u = \kappa$ on $\underline{R} \times]0,a[$.

Proof. Let C denote the countable dense set, and put $\mathscr{A} = \{ [b,c] : b,c \in C\}$. If $x \in \underline{R}$ and $0 < r < s$, then there exist points $b_0 \in C \cap]x - s, x - r[$ and $c_0 \in C \cap]x + r, x + s[$, since C is dense in $\underline{R}$. Then $[b_0, c_0] \in \mathscr{A}$ and

$$\bar{B}(x, r) = [x - r, x + r] \subseteq [b_0, c_0] \subseteq [x - s, x + s] = \bar{B}(x, s)$$

Therefore $\mathscr{A}$ generates the family of all closed balls in $\underline{R}$, by the criterion at the beginning of this section. The result now follows from Theorem 7.6.

Remarks. We have shown that a generator of an abundant Vitali covering may contain only countably many sets. But even when this is not the case, the conditions (17) and (18) in Theorem 7.5 may need to apply to only countably many sets A. Also, the sets need not be balls, as we now show. Let $\mathscr{K}_{\underline{Q}}$ denote the family of all half-open cubes in $\underline{R}^n$ of the form

$$K = \prod_{i=1}^n]a_i - r, a_i + r]$$

such that $r \in \underline{Q}$ and $a_i \in \underline{Q}$ for $i = 1, \ldots, n$. Let $\mathscr{A}$ denote the family

of all Borel sets A for which there is a cube $K \in \mathcal{K}_Q$ such that $K \subseteq A$
$\subseteq \bar{K}$. Thus each set in $\mathscr{A}$ is a cube with some portion of its boundary.
We use the criterion at the beginning of this section to show that $\mathscr{A}$
generates the abundant Vitali covering which consists of all closed balls
in $\underline{R}^n$. If $x \in \underline{R}^n$ and $0 < r < s$, we can cover the ball $\bar{B}(x, r)$ with disjoint cubes from $\mathcal{K}_Q$, of diameter $(r - s)/2$, whose closures lie within
$\bar{B}(x, s)$. The union P of these cubes then satisfies

$$\bar{B}(x, r) \subseteq P \subseteq \bar{P} \subseteq \bar{B}(x, s)$$

and $\bar{P}$ can be written as the union of finitely many sets in $\mathscr{A}$, namely
the cubes from $\mathcal{K}_Q$ some with extra parts of their boundaries included.
Thus $\bar{P}$ belongs to the union set of $\mathscr{A}$. It follows that $\mathscr{A}$ generates the
family of all closed balls.

Suppose that u is the Gauss-Weierstrass integral on $\underline{R}^n \times]0, a[$ of
a signed measure μ, and that there is a signed measure ν such that

$$\lim_{t \to 0^+} \int_K u(x, t) \, dx = \nu(K)$$

and

$$|\nu|(\partial K) = 0$$

for all $K \in \mathcal{K}_Q$. Since the boundary of any cube has zero Lebesgue
measure, and since, given any $A_0 \in \mathscr{A}$, there is $K_0 \in \mathcal{K}_Q$ such that
$\partial K_0 = \partial A_0$, we see that

$$\lim_{t \to 0^+} \int_{A_0} u(x, t) \, dx = \lim_{t \to 0^+} \int_{K_0} u(x, t) \, dx = \nu(K_0) = \nu(A_0)$$

and

$$|\nu|(\partial A_0) = |\nu|(\partial K_0) = 0$$

It follows that the hypotheses of Theorem 7.5 are satisfied, so that
$\nu = \mu$. Note that there are only countably many sets in $\mathcal{K}_Q$, but uncountably many in $\mathscr{A}$.

4. LINEAR PARABOLIC EQUATIONS

The form of Lemma 7.1 must be changed slightly, and its proof substantially, when general parabolic equations are considered. To begin
with, ω_E is defined by

$$\omega_E(y, t) = \int_E \Gamma(x, t; y, 0) \, dx \tag{24}$$

and since $\Gamma(x, t; y, 0)$ may not be symmetric in x and y, we cannot interchange them as we did for $W(x - y, t)$. Thus ω_E is not the direct analogue of the Gauss-Weierstrass integral of the characteristic function of E, and therefore the general form of Lemma 7.1(ii) is not a consequence of Theorem 1.1. However, for the general case of Theorem 1.1, the reader was referred to the details of the parametrix construction of Γ, and the method used there can be easily modified to prove a generalization of Lemma 7.1(ii). The general form of Lemma 7.1(iii) then follows by the method used for the heat equation, so it only remains to establish part (i) of the lemma.

Suppose that Γ satisfies the estimates

$$C_1 (t - s)^{-n/2} \exp(-\gamma_1 \|x - y\|^2 / (t - s))$$

$$\leq \Gamma(x, t; y, s)$$

$$\leq C_2 (t - s)^{-n/2} \exp(-\gamma_2 \|x - y\|^2 / (t - s)) \tag{25}$$

whenever $x, y \in \underline{R}^n$ and $0 \leq s < t < a$. Then ω_E, defined by (24), satisfies

$$\omega_E(y, t) \leq C_2 \int_E t^{-n/2} \exp(-\gamma_2 \|x - y\|^2 / t) \, dx$$

$$= C \int_E W(y - x, t/4\gamma_2) \, dx$$

for some positive constant C. By Lemma 7.1(i), the latter integral is dominated by $W(y, b/4\gamma_1)$ whenever $y \in \underline{R}^n$ and $0 < t < \gamma_2 b/2\gamma_1$. It therefore follows from the lower estimate for Γ in (25) that

$$\omega_E(y, t) \leq C W(y, b/4\gamma_1)$$

$$= C b^{-n/2} \exp(-\gamma_1 \|y\|^2 / b)$$

$$\leq C \Gamma(0, b; y, 0)$$

for all $(y, t) \in \underline{R}^n \times]0, \gamma_2 b/2\gamma_1[$. This is the required extension of Lemma 7.1(i).

The extensions of Theorems 7.2, 7.3, and 7.5 are all straight-forward, since their proofs depend only on the measure-theoretic results and Lemma 7.1. Those of Theorems 7.4 and 7.6 are also straight-forward, but if $\kappa \neq 0$ they require also that

$$\int_{\underline{R}^n} \Gamma(x,t;y,0) \, dy = 1 \tag{26}$$

for all $(x,t) \in \underline{R}^n \times]0,a[$. This occurs whenever the parabolic equation does not contain a term in the solution itself, only terms in its partial derivatives. For then $u = 1$ is the unique positive solution of the Cauchy problem with initial function 1, and has the representation

$$u(x,t) = \int_{\underline{R}^n} \Gamma(x,t;y,0)u(y,0) \, dy$$

on $\underline{R}^n \times]0,a[$, by the general form of Theorem 2.10. This is identical to (26).

5. BIBLIOGRAPHICAL NOTES

The material in the text largely follows Watson's paper [14], which includes the case of a general linear parabolic equation. However, there are several earlier works on this and related topics.

A generalization of Lemma 7.1(ii), far in excess of what is required here, was given by Il'in, Kalashnikov, and Oleinik [1, (4.71)]. Also, Aronson [5, p. 683] obtained a different generalization by different methods.

For the heat equation with $n = 1$, Widder [2, p. 69] proved that, if u is the Gauss-Weierstrass integral of a non-negative measure μ, then

$$\lim_{t \to 0^+} \int_b^c u(x,t) \, dx = \mu(]b,c[) + (\mu(\{b\}) + \mu(\{c\}))/2$$

which is closely related to Theorem 7.2(i). However, much earlier, Birkhoff and Kotik [1, Theorem 1] had proved a similar result under different conditions, which included temperatures that could take values of either sign. Widder's result was extended to arbitrary n by Aronson [7, Theorem B]. Wilcox [1, p. 184] proved a converse of Widder's theorem by showing that, if u is a non-negative temperature on $\underline{R} \times]0,a[$, and μ is a measure such that

$$\lim_{t\to 0^+} \int_b^c u(x,t)\ dx = \mu(]b,c[)$$

whenever $b < c$ and $\mu(\{b\}) = \mu(\{c\}) = 0$, then u is the Gauss-Weier-strass integral of μ. This was a forerunner of Theorem 7.3(ii). Aronson [7, Theorem C] proved an extension of Wilcox's result to the case of an arbitrary n.

Results which are closely related to the corollaries of Theorems 7.4 and 7.6 were obtained earlier, but only for the heat equation. Birkhoff and Kotik [1, Theorem 2] proved, in the case where n = 1, that if u is a temperature on $\underline{R} \times]0,a[$ such that

$$\left| \int_0^y u(x,t)\ dx \right| \le C \exp(\alpha y^2) \tag{27}$$

for all $y \in \underline{R}$ and all $t \in]0,a[$, and

$$\lim_{t\to 0^+} \int_0^y u(x,t)\ dx = 0 \tag{28}$$

for m-almost all y, then u = 0. Gehring [2] observed that (27) is equivalent to the condition that

$$\left| \int_{y_1}^{y_2} u(x,t)\ dx \right| \le C \exp(\alpha(y_1^2 + y_2^2))$$

for all $y_1, y_2 \in \underline{R}$ and all $t \in]0,a[$, and obviously (28) is equivalent to

$$\lim_{t\to 0^+} \int_{y_1}^{y_2} u(x,t)\ dx = 0$$

for m-almost all y_1 and y_2. He was then able to prove three stronger versions of Birkhoff and Kotik's result, in the case where n = 1. These are given in Gehring's paper [2, pp. 357-359]. His Theorem 12 states that, if u is a temperature on $\underline{R} \times]0,a[$ such that

$$\int_{y_1}^{y_2} u(x,t)\ dx \le C \exp(\alpha(y_1^2 + y_2^2)) \tag{29}$$

whenever $y_1 < y_2$ and $0 < t < a$, and there is a number κ such that

$$\liminf_{t \to 0^+} \int_{y_1}^{y_2} u(x, t) \, dx \le \kappa(y_2 - y_1) \tag{30}$$

for m-almost all y_1 and y_2 such that $y_1 < y_2$, then $u \le \kappa$ on $\underline{R} \times]0, a[$. His Theorem 13 is similar, except that (30) is replaced by

$$\lim_{t \to 0^+} \int_{y_1}^{y_2} u(x, t) \, dx = \kappa(y_2 - y_1)$$

and the conclusion is that $u = \kappa$. His Corollary 9 states that, if u has the Gauss-Weierstrass representation on $\underline{R} \times]0, a[$, and

$$\lim_{t \to 0^+} \int_{y_1}^{y_2} u(x, t) \, dx \le \kappa(y_2 - y_1)$$

for all y_1 and y_2 in a dense subset of $\underline{R}$ such that $y_1 < y_2$, then $u \le \kappa$ on $\underline{R} \times]0, a[$. Finally, Watson [9, p. 486] extended Gehring's Theorem 12, and hence the result of Birkhoff and Kotik, to the case of a general n, in the following form. Let u be a temperature on $\underline{R}^n \times]0, a[$ and let $\{h_k\}$ be a positive null sequence. If

$$\int_{B(y, h_k)} u(x, t) \, dx \le C \exp(\alpha \|y\|^2)$$

for all $(y, t) \in \underline{R}^n \times]0, a[$ and all $k \in \underline{N}$, and there is a number κ such that

$$\liminf_{t \to 0^+} \int_{B(y, h_k)} u(x, t) \, dx \le \kappa \, m(B(y, h_k))$$

for almost all $y \in \underline{R}^n$ and all $k \in \underline{N}$, then $u \le \kappa$ on $\underline{R}^n \times]0, a[$.

We have concerned ourselves with the convergence of $u(\cdot, t)$ to μ in the sense that

$$\lim_{t \to 0^+} \int_A u(x, t) \, dx = \mu(A)$$

for all A in some family of Borel sets. This can be written as

$$\lim_{t \to 0^+} \int_{\underline{R}^n} \chi_A(x) u(x, t) \, dx = \int_{\underline{R}^n} \chi_A(x) \, d\mu(x)$$

and in this form the parallel with "weak" convergence, namely

$$\lim_{t \to 0^+} \int_{\underline{R}^n} \psi(x) u(x, t) \, dx = \int_{\underline{R}^n} \psi(x) \, d\mu(x)$$

for all ψ in some specified class of functions, is apparent. Theorems involving this kind of convergence have been proved by several authors, but the class of functions varies, and is not always given explicitly. Such results were obtained, for the heat equation with $n = 1$, by Cooper [1] and Gehring [2, Theorem 14], and for linear parabolic equations with n arbitrary by Aronson [6] and Johnson [1, pp. 344-345].

8

Maximum Principles and Initial Limits

In Theorem 4.5, we proved that any Gauss-Weierstrass integral of a function has certain initial limits m-a. e. on $\underline{R}^n$. We shall use this result to prove a similar theorem for temperatures on $\underline{R}^n \times]0, a[$ which are not necessarily the Gauss-Weierstrass integrals of functions. This will imply that Theorem 4.5 can be extended to all Gauss-Weierstrass integrals of measures. In order to do this, we need a maximum principle for more general bounded open sets than was given in Theorem 2.1. We shall, in fact, prove two forms of the maximum principle, both of which are stronger and more generally applicable than is necessary for our application. The first, called the strong maximum principle, asserts that a temperature u on a bounded open set E can have a maximum at a point of E only if u is constant on a particular subset of E. The second, called the weak maximum principle, asserts that, if the values of u(x, t), as (x, t) approaches ∂E in a particular way, have an upper bound, then the values of u(x, t) for all (x, t) $\in$ E have the same upper bound.

1. THE STRONG MAXIMUM PRINCIPLE

Since the result is not more difficult to prove for solutions of a differential inequality, rather than just for temperatures, we do so. We use
θ to denote the heat operator on $\underline{R}^{n+1}$, so that

$$\theta u = \sum_{i=1}^{n} D_i^2 u - D_t u$$

for any function u which possesses the appropriate derivatives.

We need to introduce some notation and terminology. By a polygonal
path in $\underline{R}^{n+1}$, we mean a path whose trace can be written as the union
of finitely many closed line segments. We identify a path with its trace.

Let E be an open subset of $\underline{R}^{n+1}$, and let $(x_0, t_0) \in E$. We denote
by $\Lambda(x_0, t_0)$ the set of all points $(x, t) \in E$, with $t < t_0$, which can be
connected to (x_0, t_0) by a polygonal path in E along which the last coordinate is strictly increasing.

Let $(y, s), (z, r) \in \underline{R}^{n+1}$, with $r < s$, and let $\delta > 0$. The set of all
points $(x, t) \in \underline{R}^{n+1}$ such that

$$r < t < s, \quad \left| (x_i - z_i) - \left(\frac{t - r}{s - r} \right) (y_i - z_i) \right| < \delta \quad \text{for } i = 1, \ldots, n$$

(where the subscript i indicates the i-th coordinate) will be called an
open parallelogram and denoted by $P = P(y, s; z, r; \delta)$. The boundary
of P is divided up into three subsets, namely the open upper face

$$U = \{ (x, s) : |x_i - y_i| < \delta \quad \text{for} \quad i = 1, \ldots, n \}$$

the closed lower face

$$L = \{ (x, r) : |x_i - z_i| \le \delta \quad \text{for} \quad i = 1, \ldots, n \}$$

and the lateral surface $\partial P \setminus (U \cup L)$.

We now introduce an auxiliary function which is needed in the proof
of the strong maximum principle.

Lemma 8.1

Given any open parallelogram $P = P(y, s; z, r; \delta)$, there is a continuous function w on $\underline{R}^{n+1}$ such that

(i) $w(y, s) > 0$,

(ii) $w = 0$ on the lateral surface of P,

(iii) $\theta w > 0$ on $P \cup U$, where U is the open upper face of P.

Proof. Put

$$a = \frac{y - z}{s - r}, \quad b = \frac{ry - sz}{s - r}$$

so that $a, b \in \underline{R}^n$ and

$$x - at + b = x - t\left(\frac{y - z}{s - r}\right) + r\left(\frac{y - z}{s - r}\right) - z$$

$$= (x - z) - \left(\frac{t - r}{s - r}\right)(y - z) \tag{1}$$

Now we define v and w on $\underline{R}^{n+1}$ by putting

$$v(x, t) = \exp(2nc^2 t) \prod_{i=1}^{n} \cos(cx_i)$$

where $c = \pi/2\delta$, and

$$w(x, t) = \exp\left(-\sum_{i=1}^{n} x_i\left(\frac{a_i}{2}\right) + \left\|\frac{a}{2}\right\|^2 t\right) v(x - at + b, t)$$

where $x = (x_1, \ldots, x_n)$ and $a = (a_1, \ldots, a_n)$. Note that, by (1),

$$y - as + b = (y - z) - (y - z) = 0$$

so that

$$w(y, s) = v(0, s) = \exp(2nc^2 s) > 0$$

Note also that $v(x, t) = 0$ for all (x, t) such that $\cos(\pi x_i/2\delta) = 0$ for some i, and hence whenever $|x_i| = \delta$ for some i. Therefore $w(x, t) = 0$ for all (x, t) such that $|x_i - a_i t + b_i| = \delta$, and hence $w = 0$ on the lateral surface of P, in view of (1). Furthermore, routine calculations show that

$\theta v(x, t) = nc^2 v(x, t)$

and

$$\theta w(x, t) = \exp\left(-\sum_{i=1}^{n} x_i\left(\frac{a_i}{2}\right) + \left\|\frac{a}{2}\right\|^2 t\right) \theta v(x - at + b, t)$$

so that $\theta w(x, t) > 0$ if and only if $v(x - at + b, t) > 0$. Hence, in particular, $\theta w > 0$ on $P \cup U$.

We now come to the strong maximum principle, in which we do not assume that the open set E is bounded.

Theorem 8.1

Let E be an open subset of $\underline{R}^{n+1}$, and let $(x_0, t_0) \in E$. Suppose that u is continuous and satisfies $\theta u \geq 0$ on E, and that

$u(x, t) \leq u(x_0, t_0)$ whenever $(x, t) \in \Lambda(x_0, t_0)$

Then

$u(x, t) = u(x_0, t_0)$ for all $(x, t) \in \Lambda(x_0, t_0)$

Proof. Let $M = u(x_0, t_0)$, and suppose that there is a point $(x_1, t_1) \in \Lambda(x_0, t_0)$ such that $u(x_1, t_1) < M$. Choose a polygonal path γ in E which connects (x_1, t_1) to (x_0, t_0) in such a way that the t-coordinate is strictly monotonic along γ. Since u is continuous on E, and γ is closed, the set

$S = \{(x, t) \in \gamma : u(x, t) = M\}$

is also closed, so that there is a point $(y_0, s_0) \in S$ such that

$s_0 = \min\{t : (x, t) \in S\}$

Then

$u(y_0, s_0) = M$ \hfill (2)

and $u(x, t) < M$ whenever $(x, t) \in \gamma$ and $t < s_0$. We can therefore choose a point $(z_0, r_0) \in \gamma$ such that $r_0 < s_0$, $u(z_0, r_0) < M$, and the set

$$\lambda = \gamma \cap (\underline{R}^n \times [r_0, s_0])$$

is a closed line segment.

Since λ is compact and $\underline{R}^n \setminus E$ is closed, the distance η_1 between them is positive. Furthermore, because u is continuous and $u(z_0, r_0) < M$, we can find $\eta_2 \in]0, \eta_1[$ such that $u(x, r_0) < M$ whenever $\|x - z_0\| < \eta_2$. Therefore, if $0 < \delta < n^{-1/2}\eta_2$, the open parallelogram

$$P = P(y_0, s_0; z_0, r_0; \delta)$$

has its closure within E and

$$u < M \quad \text{on its closed lower face L} \tag{3}$$

Now let w be a continuous function on $\underline{R}^{n+1}$ such that $w(y_0, s_0) > 0$, $w = 0$ on the lateral surface of P, and $\theta w > 0$ on $P \cup U$. (The existence of w is assured by Lemma 8.1.) Since u is continuous on the compact set L, it attains a maximum over L. Therefore, in view of (3), there is a number κ such that

$$u \leq \kappa < M \quad \text{on} \quad L \tag{4}$$

Furthermore, w is continuous and is therefore bounded above on L. Therefore we can find $\epsilon > 0$ such that

$$\epsilon w \leq M - \kappa \quad \text{on} \quad L \tag{5}$$

Put

$$v = u + \epsilon w$$

Then, on $P \cup U$,

$$\theta v = \theta u + \epsilon \theta w \geq \epsilon \theta w > 0 \tag{6}$$

Since v is continuous on the closure of P, it has a maximum over $\bar{P}$. The inequality (6) implies that this maximum is not attained at a point of $P \cup U$, since at a point where it is attained we have

$$D_i^2 v \leq 0 \quad \text{for} \quad i = 1, \ldots, n, \quad D_t v \geq 0$$

and hence $\theta v \leq 0$. Hence v attains its maximum over $\bar{P}$ either on the lateral surface of P, or on L. On the lateral surface $w = 0$, so that

$$v = u \leq M$$

and on L we have (4) and (5), so that

$$v \leq \kappa + (M - \kappa) = M$$

It follows that $v \leq M$ on $\bar{P}$. In particular, $v(y_0, s_0) \leq M$. This, together with (2) and the inequality $w(y_0, s_0) > 0$, yields the contradiction

$$M \geq v(y_0, s_0) = u(y_0, s_0) + \epsilon w(y_0, s_0) = M + \epsilon w(y_0, s_0) > M$$

Therefore our supposition that there is a point $(x_1, t_1) \in \Lambda(x_0, t_0)$ such that $u(x_1, t_1) < M$ is untenable. Since $u \leq M$ on $\Lambda(x_0, t_0)$ by hypothesis, we deduce that $u = M$ there.

Corollary

Let E be a bounded open subset of $\underline{R}^{n+1}$, and suppose that u is a continuous function which satisfies $\theta u \geq 0$ on E. If there is a number M such that

$$\lim_{k \to \infty} \sup u(x_k, t_k) \leq M \tag{7}$$

for every sequence $\{(x_k, t_k)\}$ in E which tends to some point of ∂E, then $u \leq M$ on E.

Proof. If K is a compact subset of E, the continuity of u implies that u attains a maximum over K. It follows that either the supremum σ (possibly ∞) of u on E is attained in E, or there is a sequence $\{(y_j, s_j)\}$ of points in E, which converges to a point in ∂E, such that $u(y_j, s_j) \to \sigma$. Theorem 8.1 reduces the former case to the latter. In the latter case, we apply condition (7) to $u(y_j, s_j)$ and deduce that $\sigma \leq M$. Hence $u \leq M$ on E.

Remark. The above corollary is a naive form of the weak maximum principle. Thus the strong maximum principle easily implies the weak one, which explains the names. However, the form of Theorem 8.1 suggests that a sharper version of the weak maximum principle might be obtained by making use of the subdomains $\Lambda(x, t)$ of E, and this idea is pursued in the next section.

2. THE WEAK MAXIMUM PRINCIPLE

In this section we extend the weak maximum principle, given for temperatures on a circular cylinder in Theorem 2.1, to solutions of $\theta u \geq 0$ on an arbitrary bounded open set, in a sharper form than that of the corollary to Theorem 8.1. We first recall Theorem 2.1.

Let B be an open ball in $\underline{R}^n$, and let u be a temperature on the cylinder $C = B \times]a, b[$. If

$$\limsup u(x, t) \leq 0 \qquad\qquad\qquad\qquad (8)$$

whenever (x, t) approaches any point of $\partial C \setminus (B \times \{b\})$ from within C, then $u \leq 0$ on C.

Note that (8) is not required to hold as (x, t) approaches any point of $B \times \{b\}$. For an arbitrary bounded open set, we would expect that some subset of the boundary would similarly be free of the condition (8), but it is not so easy to see which. We shall describe the negligible subset, and also show that (8) can be weakened at other parts of the boundary.

We use the subsets $\Lambda(x, t)$, of the arbitrary open set E, which were described in section 1, and which featured in the statement of the strong maximum principle (Theorem 8.1). Instead of requiring condition (8) at some points of ∂E, and no condition at others, we require that

$$\limsup_{k \to \infty} u(x_k, t_k) \leq 0$$

for every sequence $\{(x_k, t_k)\}$ in E which tends to a point of ∂E and satisfies

$$(x_{k+1}, t_{k+1}) \in \Lambda(x_k, t_k) \quad \text{for all} \quad k \qquad\qquad (9)$$

This automatically determines the negligible subset of ∂E as that for which no such approach is possible. The obvious example is the subset $\bar{B} \times \{b\}$ of the boundary of the cylinder C in Theorem 2.1, but there are other more interesting ones. Indeed, there may be sequences $\{(x_k, t_k)\}$ which tend to points of ∂E and satisfy

$$t_{k+1} < t_k \quad \text{for all} \quad k \qquad\qquad\qquad (10)$$

but which do not satisfy (9). For example, if

$$E = \{(x,t) \in \underline{R}^2 : \sin(1/x) < t < 1 + \sin(1/x), \quad 0 < x < 1\}$$

then the subset $Z = \{0\} \times [-1, 2]$ of ∂E consists entirely of points which cannot be written as the limit of any sequence in E which satisfies (9). But every point of $Z \setminus \{(0,2)\}$ is the limit of some sequence in E which satisfies (10).

We need Hausdorff's maximality theorem, so we recall the result and the necessary definitions.

A set S is said to be <u>partially ordered</u> by a binary relation $<<$ if the following conditions are satisfied:

(i) $\xi << \xi$ for every $\xi \in S$

(ii) $\xi << \eta$ and $\eta << \zeta$ together imply that $\xi << \zeta$

(iii) $\xi << \eta$ and $\eta << \xi$ together imply that $\xi = \eta$

If S is partially ordered by $<<$, and $T \subseteq S$, then T is said to be <u>totally ordered</u> by $<<$ if every pair ξ, η in T satisfies either $\xi << \eta$ or $\eta << \xi$. In this situation, T is called a <u>maximal totally ordered subset</u> of S if there is no point $\zeta \in S \setminus T$ such that $T \cup \{\zeta\}$ is totally ordered by $<<$.

<u>Hausdorff's maximality theorem</u> states that every non-empty partially ordered set contains a maximal totally ordered subset.

We are now ready to prove the weak maximum principle.

Theorem 8.2

Let E be a bounded open subset of $\underline{R}^{n+1}$, and suppose that u is continuous and satisfies $\theta u \geq 0$ on E. If there is a number M such that

$$\limsup_{k \to \infty} u(x_k, t_k) \leq M \tag{11}$$

for every sequence $\{(x_k, t_k)\}$ in E which satisfies

$$(x_{k+1}, t_{k+1}) \in \Lambda(x_k, t_k) \quad \text{for all} \quad k \tag{12}$$

and tends to some point of ∂E, then $u \leq M$ on E.

<u>Proof.</u> For each $\alpha \in \underline{R}$, put

$$S_\alpha = \{(x,t) \in E : u(x,t) \geq \alpha\}$$

We shall prove that $S_\alpha \neq \emptyset$ only if $\alpha \leq M$, which implies that $S_\alpha = \emptyset$ whenever $\alpha > M$, and hence that $u \leq M$ on E.

Suppose that $S_\alpha \neq \emptyset$, and define a binary operation $<<$ on S_α by putting

$$(x, t) << (y, s) \quad \text{whenever} \quad (x, t) \in \Lambda(y, s) \cup \{(y, s)\}$$

Then (i) $(x, t) << (x, t)$ for every $(x, t) \in S_\alpha$, (ii) $(x, t) << (y, s)$ and $(y, s) << (z, r)$ together imply that $(x, t) << (z, r)$, and (iii) $(x, t) << (y, s)$ and $(y, s) << (x, t)$ together imply that $(x, t) = (y, s)$. Hence S_α is partially ordered by $<<$. It therefore follows from Hausdorff's maximality theorem that S_α contains a maximal totally ordered subset T. Put

$$\tau = \inf\{t : (x, t) \in T\} \tag{13}$$

and let

$$U = \bar{T} \cap (\underline{R}^n \times \{\tau\})$$

which is non-empty.

Suppose that $U \cap \partial E = \emptyset$. We show that this cannot occur, and the first step is to prove that

$$T \cap (\underline{R}^n \times \{\tau\}) \neq \emptyset \tag{14}$$

Since $U \subseteq \bar{T} \subseteq E$ and $U \cap \partial E = \emptyset$, we see that $U \subseteq E$. Furthermore, U is closed, and hence compact, so that the distance ρ between U and ∂E is positive. Therefore, for each point $(\eta, \tau) \in U$, the $(n + 1)$-dimensional open ball B_η, with center (η, τ) and radius ρ, is contained in E. Because $(\eta, \tau) \in \bar{T}$, (13) implies that either (i) there is a sequence $\{(\eta_i, \sigma_i)\}$ in $T \cap B_\eta$ such that $\{\sigma_i\}$ is strictly decreasing and $(\eta_i, \sigma_i) \to (\eta, \tau)$, or (ii) $(\eta, \tau) \in T$. In case (ii), (14) is obvious. In case (i), the fact that $(\eta_i, \sigma_i) \in B_\eta$ for all i implies that

$$(\eta, \tau) \in \Lambda(\eta_i, \sigma_i) \quad \text{for all} \quad i \tag{15}$$

Since T is totally ordered by $<<$ and $\{\sigma_i\}$ decreases to τ, given any point $(z, r) \in T$ with $r > \tau$ we can find i such that $\sigma_i < r$, so that $(\eta_i, \sigma_i) \in \Lambda(z, r)$. Hence (15) implies that $(\eta, \tau) \in \Lambda(z, r)$, so that $(\eta, \tau) << (z, r)$ whenever $(z, r) \in T$ and $r > \tau$. There are therefore two possibilities. Either there is a point $(y, \tau) \in T$ with $y \neq \eta$, so that (14) holds, or there is no such point and hence $T \cup \{(\eta, \tau)\}$ is totally ordered by $<<$, in which case the maximality of T implies that $(\eta, \tau) \in T$, and again (14) holds. Thus there is some point (ζ, τ) in T. We now observe that $\Lambda(\zeta, \tau) \cap S_\alpha$ is empty, since if it contained a point (y, s) we would have

$$(y, s) \ll (\zeta, \tau) \ll (z, r)$$

for every $(z, r) \in T$, which implies that $(y, s) \in T$ since T is maximal, and this is impossible because $s < \tau$. Hence $u < \alpha$ on $\Lambda(\zeta, \tau)$, whereas $u(\zeta, \tau) \geq \alpha$ since $(\zeta, \tau) \in T \subseteq S_\alpha$. Theorem 8.1 shows that this cannot happen, and it follows that our supposition that $U \cap \partial E = \emptyset$ is untenable.

Therefore $U \cap \partial E \neq \emptyset$, so that there is a point $(\xi, \tau) \in \bar{T} \cap \partial E$. Since $(\xi, \tau) \in \partial E$ and T is a subset of the open set E, $(\xi, \tau) \notin T$. It therefore follows from (13) that there is a sequence $\{(x_i, t_i)\}$ in T such that $\{t_i\}$ is strictly decreasing and $(x_i, t_i) \to (\xi, \tau)$. Since T is totally ordered by $\ll$, it follows that

$$(x_{i+1}, t_{i+1}) \in \Lambda(x_i, t_i) \quad \text{for all} \quad i$$

Therefore (11) and (12) imply that

$$\limsup_{i \to \infty} u(x_i, t_i) \leq M$$

whereas the fact that $(x_i, t_i) \in T \subseteq S_\alpha$ for all i implies that

$$\limsup_{i \to \infty} u(x_i, t_i) \geq \alpha$$

Hence $\alpha \leq M$, and the result follows.

The standard application of the weak maximum principle is to the question of uniqueness of solution of a boundary value problem.

Theorem 8.3

Let E be a bounded open subset of $\underline{R}^{n+1}$. Suppose that v and w are temperatures on E such that

$$\lim_{k \to \infty} v(x_k, t_k) = \lim_{k \to \infty} w(x_k, t_k) \in \underline{R}$$

for every sequence $\{(x_k, t_k)\}$ in E which satisfies $(x_{k+1}, t_{k+1}) \in \Lambda(x_k, t_k)$ for all k, and tends to some point of ∂E. Then $v = w$ on E.

Proof. Put u = v - w, so that $u(x_k, t_k) \to 0$ for every sequence
$\{(x_k, t_k)\}$ of the type described in the hypothesis. Applying Theorem
8.2, first to u and then to -u, we see first that $u \leq 0$ and then that
$-u \leq 0$ on E. Hence $u = 0$ and $v = w$ on E.

3. CONVERGENCE OF SEQUENCES OF INTEGRALS

The principal result of this section, Theorem 8.4, will be used in the
next section during the proof of the parabolic limits theorem. Although
we shall use it only in the case where $p = 2$ and $\mu = m$, we prove it for
$1 < p < \infty$ and an arbitrary non-negative measure μ, since the extra
generality is sometimes useful. For example, it could be used instead
of Theorem 6.1 in the proof of Theorem 6.4, although the approach we
took in Chapter VI is slightly neater.

The proof of Theorem 8.4 depends on the following auxiliary result,
in which the case $p = 1$ is included for completeness.

Lemma 8.2

If μ is a non-negative measure on $\underline{R}^n$ and $1 \leq p < \infty$, then $L^p(\mu)$ has
a countable dense subset.

Proof. If $y = (y_1, \cdots, y_n) \in \underline{R}^n$ and $\eta > 0$, the set

$$Q(y, \eta) = \coprod_{i=1}^{n} [y_i, y_i + \eta[$$

will be called the η-box with corner at y. For each $k \in \underline{N}$, we denote
by Δ_k the set of all points in $\underline{R}^n$ whose coordinates are integer multiples
of 2^{-k}, and by $\mathscr{B}_k$ the family of all 2^{-k}-boxes with corners at points of
Δ_k. We also put

$$\mathscr{B} = \bigcup_{k=1}^{\infty} \mathscr{B}_k$$

Then each set Δ_k is countable, so that each $\mathscr{B}_k$ is also countable, and
so $\mathscr{B}$ is too. Furthermore, each set in $\mathscr{B}$ is a bounded Borel set, so
that its characteristic function belongs to $L^p(\mu)$. Therefore the same
is true of any finite linear combination of such functions. We denote by
D the collection of all finite linear combinations, with rational coeffi-

cients, of characteristic functions of sets in $\mathcal{B}$. Thus D is a countable subset of $L^p(\mu)$.

We shall show that D is dense in $L^p(\mu)$ by using the functions in D to approximate, in the L^p-metric, successively larger subclasses of $L^p(\mu)$. The first is the class of all characteristic functions of open sets with finite μ-measure. Let V be any non-empty open subset of $\underline{R}^n$ such that $\mu(V) < \infty$, and let χ_V denote its characteristic function. Since V is open, it can be expressed as the union of a sequence of disjoint boxes $\{Q_j\}$ in $\mathcal{B}$ (Rudin [1, p. 50]). Therefore, because $\mu(V) < \infty$, to each positive number ϵ there corresponds $J \in \underline{N}$ such that

$$\mu\left(V \setminus \left(\bigcup_{j=1}^{J} Q_j\right)\right) < \epsilon^p$$

Note that the characteristic function f_J of $\bigcup_{j=1}^{J} Q_j$ belongs to D. Now

$$\int_{\underline{R}^n} |\chi_V - f_J|^p \, d\mu = \mu\left(V \setminus \left(\bigcup_{j=1}^{J} Q_j\right)\right) < \epsilon^p$$

so that $\|\chi_V - f_J\|_p < \epsilon$. Since V is arbitrary, we see that the closure $\bar{D}$ of D in $L^p(\mu)$ contains the characteristic functions of all open sets with finite μ-measure.

The next subclass consists of all characteristic functions of G_δ-sets with finite μ-measure, where we call a subset of $\underline{R}^n$ a G_δ-set if it can be written as the intersection of a contracting sequence of open sets. Let A be a G_δ-set with $\mu(A) < \infty$, and let $\{V_j\}$ be a contracting sequence of open sets with intersection A and $\mu(V_j) < \infty$ for all j. Then to each $\epsilon > 0$ there corresponds $K \in \underline{N}$ such that

$$\mu(V_K \setminus A) < \epsilon^p$$

If g_K and χ_A denote, respectively, the characteristic functions of V_K and A, then $g_K \in \bar{D}$ and

$$\|g_K - \chi_A\|_p = \mu(V_K \setminus A)^{1/p} < \epsilon$$

Hence χ_A belongs to the closure of $\bar{D}$, namely $\bar{D}$. Thus $\bar{D}$ contains the characteristic functions of all G_δ-sets with finite μ-measure.

We now extend the subclass under consideration by removing the restriction that the sets therein are G_δ-sets. Let F be any set with

$\mu(F) < \infty$. Then there is a G_δ-set C such that $F \subseteq C$ and $\mu(C \setminus F) = 0$ (Rudin [1, p. 47]), so that the characteristic functions of F and C are equal μ-a. e. But any two functions which are equal μ-a. e. are indistinguishable in $L^p(\mu)$, so it follows that $\bar{D}$ contains the characteristic functions of all sets with finite μ-measure.

The next class we consider consists of all real-valued μ-measurable simple functions s on $\underline{R}^n$ such that

$$\mu(\{x : s(x) \neq 0\}) < \infty \tag{16}$$

Any such function s can be expressed in the form

$$s = \sum_{\ell=1}^{r} \alpha_\ell \chi_\ell$$

where the α_ℓ are distinct non-zero real numbers, each χ_ℓ is the characteristic function of a μ-measurable set B_ℓ, and the sets B_ℓ are disjoint. Condition (16) implies that $\mu(B_\ell) < \infty$ for each ℓ, so that each χ_ℓ belongs to $\bar{D}$. Fix ℓ, and choose a rational number β_ℓ, which is an integer multiple of 2^{-k} for some $k \in \underline{N}$, such that

$$|\alpha_\ell - \beta_\ell| < \epsilon \tag{17}$$

Next, choose a function $h_\ell \in D$ such that

$$\|\chi_\ell - h_\ell\|_p < \epsilon \tag{18}$$

and put

$$\sigma = \sum_{\ell=1}^{r} \beta_\ell h_\ell$$

Then $\sigma \in D$ since $h_\ell \in D$ for every ℓ. Furthermore, by Minkowski's inequality,

$$\|s - \sigma\|_p = \left\| \sum_{\ell=1}^{r} (\alpha_\ell \chi_\ell - \beta_\ell h_\ell) \right\|_p$$

$$\leq \sum_{\ell=1}^{r} \|\alpha_\ell \chi_\ell - \beta_\ell h_\ell\|_p$$

$$= \sum_{\ell=1}^{r} \| (\alpha_\ell - \beta_\ell) \chi_\ell + \beta_\ell (\chi_\ell - h_\ell) \|_p$$

$$\leq \sum_{\ell=1}^{r} (|\alpha_\ell - \beta_\ell| \|\chi_\ell\|_p + |\beta_\ell| \|\chi_\ell - h_\ell\|_p)$$

The inequalities (17) and (18) now imply that

$$\| s - \sigma \|_p \leq \sum_{\ell=1}^{r} (\epsilon \|\chi_\ell\|_p + |\beta_\ell| \epsilon) \leq \epsilon \sum_{\ell=1}^{r} (\mu(B_\ell)^{1/p} + |\alpha_\ell| + \epsilon)$$

Since each $\mu(B_\ell)$ and $|\alpha_\ell|$ is fixed in advance, and ϵ is arbitrary, we can make $\| s - \sigma \|_p$ arbitrarily small. Hence $s \in \bar{D}$.

Finally, the class of all real-valued μ-measurable simple functions s on $\underline{R}^n$, which satisfy (16), is dense in $L^p(\mu)$ (Rudin [1, p. 67]). Since $\bar{D}$ contains this class it contains its closure, and hence $\bar{D} = L^p(\mu)$.

We now come to the main result of this section. Observe the similarity between this theorem and Theorem 2.8.

Theorem 8.4

Let μ be a non-negative measure on $\underline{R}^n$, let $1 < p < \infty$, let κ be a fixed real number, and let $\{f_i\}$ be a sequence of functions on $\underline{R}^n$ such that

$$\|f_i\|_{p,\mu} \leq \kappa \quad \text{for all} \quad i \tag{19}$$

Then there is a function f such that $\|f\|_{p,\mu} \leq \kappa$, and a subsequence $\{f_{i(j)}\}$ such that

$$\lim_{j \to \infty} \int_{\underline{R}^n} f_{i(j)} g \, d\mu = \int_{\underline{R}^n} fg \, d\mu \tag{20}$$

for all $g \in L^q(\mu)$, where $q = p/(p-1)$.

Proof. By Lemma 8.2, $L^q(\mu)$ contains a countable dense subset D. Arrange the members of D as a sequence $\{\phi_k\}$. For each i and k, it follows from Hölder's inequality and (19) that

$$\left| \int_{\underline{R}^n} f_i \phi_k \, d\mu \right| \leq \|f_i\|_p \|\phi_k\|_q \leq \kappa \|\phi_k\|_q$$

Thus the sequence of real numbers

$$\left\{ \int_{\underline{R}^n} f_i \phi_1 \, d\mu \right\}$$

is bounded by $\kappa \|\phi_1\|_q$, and therefore has a convergent subsequence. Hence there is a subsequence $\{f_i^{(1)}\}$ of $\{f_i\}$ such that the sequence

$$\left\{ \int_{\underline{R}^n} f_i^{(1)} \phi_1 \, d\mu \right\}$$

is convergent. Next, the real sequence

$$\left\{ \int_{\underline{R}^n} f_i^{(1)} \phi_2 \, d\mu \right\}$$

is bounded by $\kappa \|\phi_2\|_q$, so that it too has a convergent subsequence. Therefore there is a subsequence $\{f_i^{(2)}\}$ of $\{f_i^{(1)}\}$ such that

$$\left\{ \int_{\underline{R}^n} f_i^{(2)} \phi_2 \, d\mu \right\}$$

is convergent. We successively choose subsequences in this way, so that for each fixed j, $\{f_i^{(j)}\}$ is a subsequence of $\{f_i^{(j-1)}\}$ and

$$\left\{ \int_{\underline{R}^n} f_i^{(j)} \phi_j \, d\mu \right\}$$

is convergent. The sequence $\{f_i^{(i)}\}$ is a subsequence of $\{f_i^{(j)}\}$ for each fixed j, and therefore

$$\left\{ \int_{\underline{R}^n} f_i^{(i)} \phi_j \, d\mu \right\}$$

converges for each fixed j.

Given any $h \in L^q(\mu)$ and $\epsilon > 0$, choose $\phi_\ell \in D$ such that

$$\| h - \phi_\ell \|_q < \epsilon / 3\kappa$$

Then, for each i, k,

$$\left| \int_{\underline{R}^n} f_i^{(i)} h \, d\mu - \int_{\underline{R}^n} f_k^{(k)} h \, d\mu \right|$$

$$\leq \left| \int_{\underline{R}^n} f_i^{(i)} (h - \phi_\ell) \, d\mu \right| + \left| \int_{\underline{R}^n} (f_i^{(i)} - f_k^{(k)}) \phi_\ell \, d\mu \right| + \left| \int_{\underline{R}^n} f_k^{(k)} (\phi_\ell - h) \, d\mu \right|$$

Now, by Hölder's inequality and (19),

$$\left| \int_{\underline{R}^n} f_i^{(i)} (h - \phi_\ell) \, d\mu \right| \leq \| f_i^{(i)} \|_p \| h - \phi_\ell \|_q < \epsilon / 3$$

and similarly for k. Furthermore, since $\phi_\ell \in D$, the real sequence

$$\left\{ \int_{\underline{R}^n} f_i^{(i)} \phi_\ell \, d\mu \right\}$$

is convergent, and therefore we can find M such that

$$\left| \int_{\underline{R}^n} f_i^{(i)} \phi_\ell \, d\mu - \int_{\underline{R}^n} f_k^{(k)} \phi_\ell \, d\mu \right| < \epsilon / 3$$

whenever $i, k \geq M$. It follows that

$$\left| \int_{\underline{R}^n} f_i^{(i)} h \, d\mu - \int_{\underline{R}^n} f_k^{(k)} h \, d\mu \right| < \epsilon$$

for all $i, k \geq M$. Therefore the sequence of real numbers

$$\left\{ \int_{\underline{R}^n} f_i^{(i)} h \, d\mu \right\}$$

is convergent.

Put

$$\Lambda(h) = \lim_{i\to\infty} \int_{R^n} f_i^{(i)} h \, d\mu$$

for all $h \in L^q(\mu)$. By Hölder's inequality and (19),

$$\left| \int_{R^n} f_i^{(i)} h \, d\mu \right| \le \kappa \|h\|_q$$

for all i, so that

$$|\Lambda(h)| \le \kappa \|h\|_q$$

for every $h \in L^q(\mu)$. Therefore Λ is a bounded linear functional on $L^q(\mu)$. Hence there is a function $f \in L^p(\mu)$ such that

$$\Lambda(h) = \int_{R^n} fh \, d\mu$$

for all $h \in L^q(\mu)$, (Rudin [1, p. 128]) and furthermore

$$\|f\|_p = \sup \left\{ \left| \int_{R^n} fg \, d\mu \right| \|g\|_q^{-1} : g \in L^q, \, g \ne 0 \right\} \tag{21}$$

(Rudin [1, p. 128]). Equating the two expressions for $\Lambda(h)$, we obtain

$$\lim_{i\to\infty} \int_{R^n} f_i^{(i)} h \, d\mu = \int_{R^n} fh \, d\mu$$

for all $h \in L^q(\mu)$, and (20) is established.

It only remains to prove that $\|f\|_p \le \kappa$, and for this we use (21) and the fact that (21) also holds if f is replaced by $f_i^{(i)}$ on both sides. Thus, for all non-zero $h \in L^q(\mu)$,

$$\left| \int_{R^n} fh \, d\mu \right| \|h\|_q^{-1} = \lim_{i\to\infty} \left| \int_{R^n} f_i^{(i)} h \, d\mu \right| \|h\|_q^{-1}$$

$$\le \limsup_{i\to\infty} \|f_i^{(i)}\|_p$$

$$\le \kappa$$

by (19), and hence

$$\|f\|_p = \sup \left\{ \left| \int_{\underline{R}^n} fh\, d\mu \right| \|h\|_q^{-1} : h \in L^q, h \neq 0 \right\} \leq \kappa$$

Remark. We cannot expect the result of Theorem 8.4 to hold if $p = 1$. For example, if $f_i = W(\cdot, i^{-1})$ for all i, and $\mu = m$, then $\|f_i\|_1 = 1$ for all i, by Lemma 1.1, but

$$\int_{\underline{R}^n} f_i g\, dm = \int_{\underline{R}^n} W(x, i^{-1}) g(x)\, dx \to g(0) = \int_{\underline{R}^n} g(x)\, d\delta_0(x)$$

whenever g is bounded on $\underline{R}^n$ and continuous at 0, by Theorem 1.3, Corollary. (Here δ_0 denotes the Dirac δ-measure concentrated at 0.) Thus Theorem 2.8 (with $d\mu_i = f_i\, dm$) gives the correct conclusion.

4. EXISTENCE OF INITIAL LIMITS

In this section we prove two extensions of Theorem 4.5. We begin by establishing some notation which is needed to describe these results.

If $\xi \in \underline{R}^n$, $\gamma > 0$, and $b > 0$, we denote by $\Pi_\gamma^b(\xi)$ the truncated open paraboloid in $\underline{R}^n \times]0, b[$ with vertex $(\xi, 0)$ and aperture γ, that is

$$\Pi_\gamma^b(\xi) = \{(x, t) : \|x - \xi\| < \gamma\sqrt{t}, \ 0 < t < b\}$$

Let Ω be an open subset of $\underline{R}^n \times]0, \infty[$ with the properties that (i) there exists $\alpha > 0$ such that, for every $(x, t) \in \Omega$,

$$\{(y, s) : \|y - x\| < \alpha(\sqrt{s} - \sqrt{t}), \ t < s\} \subseteq \Omega$$

(ii) there exists $\beta > 0$ such that $m(\{x \in \underline{R}^n : (x, t) \in \Omega\}) \leq \beta t^{n/2}$ for every $t > 0$, (iii) the point $(0, 0)$ is a limit point of Ω. In Theorem 4.5, we proved that if w is the Gauss-Weierstrass integral on $\underline{R}^n \times]0, a[$ of a function f on $\underline{R}^n$, then $w(x, t) \to f(\xi)$ as $(x, t) \to (\xi, 0)$ through the set

$$\Omega_\xi = (\Omega + \{(\xi, 0)\}) \cap (\underline{R}^n \times]0, a[)$$

for m-almost all $\xi \in \underline{R}^n$. Theorem 8.5 below asserts that, if u is a temperature on $\underline{R}^n \times]0, a[$ and E is the set of all $\xi \in \underline{R}^n$ such that u

is bounded on $\Pi_\gamma^b(\xi)$ for some γ and b, then u(x, t) tends to a finite limit as $(x, t) \to (\xi, 0)$ through Ω_ξ for m-almost all $\xi \in E$. Here γ, b, and the bound for u on $\Pi_\gamma^b(\xi)$, may all depend on ξ. Note that the boundedness hypothesis involves only one paraboloid with vertex at a given point ξ, whereas the conclusion is that a limit exists through Ω_ξ for any Ω which satisfies the above conditions. Examples of such sets Ω were given in section 4 of Chapter IV.

Theorem 8.6 below asserts that, if v is the Gauss-Weierstrass integral on $\underline{R}^n \times]0, a[$ of a signed measure on $\underline{R}^n$, then v(x, t) tends to a finite limit as $(x, t) \to (\xi, 0)$ through Ω_ξ for m-almost all $\xi \in \underline{R}^n$. This is achieved by combining the results of Theorems 4.5 and 8.5.

We require a lemma which implies the measurability of the set E mentioned above.

Lemma 8.3

Let S be a subset of $\underline{R}^n \times]0, b[$, and let $\gamma > 0$. Then the set

$$E = \{\xi \in \underline{R}^n : \Pi_\gamma^b(\xi) \subseteq S\}$$

is closed.

Proof. Let $\eta \in \underline{R}^n$, and suppose that there is a sequence $\{\xi_j\}$ in E which converges to η. We must show that $\eta \in E$, that is, that $\Pi_\gamma^b(\eta) \subseteq S$. Let $(x, t) \in \Pi_\gamma^b(\eta)$. Since $\Pi_\gamma^b(\eta)$ is open, we can choose r > 0 such that

$$B(x, r) \times \{t\} \subseteq \Pi_\gamma^b(\eta)$$

Then, for each j,

$$B(x + \xi_j - \eta, r) \times \{t\} \subseteq \Pi_\gamma^b(\xi_j)$$

Since $\xi_j \to \eta$, we can find j_0 such that $\|\xi_j - \eta\| < r$ whenever $j > j_0$. Then $\|(x + \xi_j - \eta) - x\| < r$ whenever $j > j_0$, so that

$$(x, t) \in B(x + \xi_j - \eta, r) \times \{t\} \subseteq \Pi_\gamma^b(\xi_j) \subseteq S$$

It follows that $\Pi_\gamma^b(\eta) \subseteq S$, as required.

Theorem 8.5

Let u be a temperature on $\underline{R}^n \times]0, a[$, and let E be the set of all $\xi \in \underline{R}^n$ with the property that there exist numbers $\gamma > 0$, $M > 0$, and $b \in]0, a[$, such that

$$|u(x, t)| \leq M \quad \text{for all} \quad (x, t) \in \Pi_\gamma^b(\xi) \tag{22}$$

Let Ω be an open subset of $\underline{R}^n \times]0, \infty[$ with the following properties:

(i) there exists $\alpha > 0$ such that, whenever $(x, t) \in \Omega$,

$$\{(y, s) : \|y - x\| < \alpha(\sqrt{s} - \sqrt{t})\} \subseteq \Omega$$

(ii) there exists $\beta > 0$ such that $m(\{x \in \underline{R}^n : (x, t) \in \Omega\}) \leq \beta t^{n/2}$ for all $t > 0$;

(iii) the point $(0, 0)$ is a limit point of Ω.

Then $u(x, t)$ tends to a finite limit as $(x, t) \to (\xi, 0)$ through Ω_ξ for m-almost all $\xi \in E$, where $\Omega_\xi = (\Omega + \{(\xi, 0)\}) \cap (\underline{R}^n \times]0, a[)$.

Proof. We first note that there is no loss of generality in assuming that b is independent of ξ. For, if $0 < d < c < a$, the closure of $\Pi_\gamma^c(\xi) \setminus \Pi_\gamma^d(\xi)$ is a compact subset of $\underline{R}^n \times]0, a[$ on which u is bounded since u is continuous. Therefore, if (22) holds with $b = d$, there is a number $N \geq M$ such that

$$|u(x, t)| \leq N \quad \text{for all} \quad (x, t) \in \Pi_\gamma^c(\xi) \tag{23}$$

Conversely, if (23) holds, then obviously (22) holds with $b = d$ and $M = N$. Hence b can be chosen arbitrarily in $]0, a[$, and can therefore be chosen to be the same for all ξ.

We now consider γ and M. By reducing γ and increasing M in (22), we can make both of them rational. Then there are only countably many pairs (γ, M), so that E is the union of the countably many sets

$$E_{\gamma, M} = \{\xi \in \underline{R}^n : |u(x, t)| \leq M \quad \text{for all} \quad (x, t) \in \Pi_\gamma^b(\xi)\}$$

If we prove that u has limits through Ω_ξ m-a. e. on each set $E_{\gamma, M}$, the result of the theorem will follow from the fact that the union of countably many sets of measure zero itself has measure zero. Furthermore, since

$$E_{\gamma, M} = \bigcup_{\ell=1}^{\infty} (E_{\gamma, M} \cap B(0, \ell))$$

a similar argument shows that it is enough to prove that u has parabolic limits m-a. e. on each set $E_{\gamma, M} \cap B(0, \ell)$. Note also that, if S denotes the subset of $\underline{R}^n \times]0, a[$ on which $|u| \leq M$, then

$$E_{\gamma, M} = \{\xi \in \underline{R}^n : \Pi_\gamma^b(\xi) \subseteq S\}$$

so that Lemma 8.3 implies that $E_{\gamma, M}$ is closed, and hence $E_{\gamma, M} \cap B(0, \ell)$ is Borel measurable.

Given γ, M, and ℓ, we put

$$F = E_{\gamma, M} \cap B(0, \ell)$$

and

$$J = \bigcup_{\xi \in F} \Pi_\gamma^b(\xi)$$

so that $|u| \leq M$ on J. Let χ denote the characteristic function of J, and for each integer $k > 1/b$ define v_k on $\underline{R}^n \times]0, a[$ by

$$v_k(x, t) = \int_{\underline{R}^n} W(x - y, t) u(y, k^{-1}) \chi(y, k^{-1}) \, dy$$

Since u is bounded on J, $u\chi$ is bounded on $\underline{R}^n \times]0, a[$, so that each v_k is a temperature, by Theorem 1.2, Corollary. The set J is bounded because F is bounded, and so we can choose r such that

$$J \subseteq B(0, r) \times]0, a[$$

Then because $|u| \leq M$ on J and $\chi = 0$ outside $B(0, r) \times]0, a[$,

$$\|u(\cdot, k^{-1})\chi(\cdot, k^{-1})\|_{2, m} \leq \left(\int_{B(0, r)} M^2 \, dy\right)^{\frac{1}{2}}$$

for all k. Therefore, by Theorem 8.4, there is a function $f \in L^2(m)$ and a subsequence $\{u(\cdot, k(j)^{-1})\chi(\cdot, k(j)^{-1})\}$ such that

$$\lim_{j\to\infty} \int_{\underline{R}^n} u(y,k(j)^{-1})\chi(y,k(j)^{-1})g(y)\ dy = \int_{\underline{R}^n} f(y)g(y)\ dy$$

for all $g \in L^2(m)$. For each fixed point $(x,t) \in \underline{R}^n \times]0,a[$, the function

$$y \longmapsto W(x-y,t)$$

belongs to $L^2(m)$, so that

$$\lim_{j\to\infty} \int_{\underline{R}^n} W(x-y,t)u(y,k(j)^{-1})\chi(y,k(j)^{-1})\ dy = \int_{\underline{R}^n} W(x-y,t)f(y)\ dy$$

that is,

$$\lim_{j\to\infty} v_{k(j)}(x,t) = \int_{\underline{R}^n} W(x-y,t)f(y)\ dy$$

If $v(x,t)$ denotes this last integral, then by Hölder's inequality

$$|v(x,t)| \le \left(\int_{\underline{R}^n} W(x-y,t)^2\ dy \right)^{\frac{1}{2}} \|f\|_2 < \infty$$

for all $(x,t) \in \underline{R}^n \times]0,a[$, so that v is a temperature, by Theorem 1.2. Thus there is a subsequence $\{v_{k(j)}\}$ of $\{v_k\}$ which converges point-wise to a temperature v which is the Gauss-Weierstrass integral of a function.

For each integer $k > 1/b$, put

$$w_k(x,t) = u(x,t+k^{-1}) - v_k(x,t)$$

for all $(x,t) \in \underline{R}^n \times]0, a - k^{-1}[$, and let

$$w(x,t) = u(x,t) - v(x,t)$$

whenever $(x,t) \in \underline{R}^n \times]0,a[$, so that $w_{k(j)} \to w$ pointwise as $j \to \infty$.

Since v is the Gauss-Weierstrass integral of f, $v(x,t) \to f(\xi)$ as $(x,t) \to (\xi,0)$ through Ω_ξ for m-almost all $\xi \in \underline{R}^n$, by Theorem 4.5. We shall prove that w has the same type of limit 0 m-a. e. on F, and this will imply the result of the theorem. We introduce an auxiliary temperature h which dominates w on J and has the appropriate limit 0 m-a. e. on F. Put

$$\delta = 2M/(\gamma^2/4\pi)^{n/2} \omega_n \exp(-\gamma^2/4) \qquad (24)$$

where $\omega_n = m(B(0,1))$, and for each $(x,t) \in \underline{R}^n \times]0,a[$ let

$$h(x,t) = \delta \int_{\underline{R}^n \setminus F} W(x-y,t)\, dy$$

By Theorem 1.2, Corollary, h is a temperature. By Theorem 4.5, $h(x,t) \to 0$ as $(x,t) \to (\xi,0)$ through Ω_ξ for m-almost all $\xi \in F$, because h is the Gauss-Weierstrass integral of δ times the characteristic function of $\underline{R}^n \setminus F$.

   In order to show that h dominates w on J, we need a lower estimate for h. In fact, we prove that $h \geq 2M$ on the set

$$P = (\partial J) \cap (\underline{R}^n \times]0,b[)$$

If $(x,t) \in \underline{R}^n \times]0,b[$, and there is a point $\zeta \in B(x,\gamma\sqrt{t}) \cap F$, then $\zeta \in F$ and $x \in B(\zeta,\gamma\sqrt{t})$, so that $\zeta \in F$ and $(x,t) \in \Pi_\gamma^b(\zeta)$, so that (x,t) belongs to J and not to P, since J is open. Therefore, if $(x,t) \in P$ there is no such ζ, and hence $B(x,\gamma\sqrt{t}) \subseteq \underline{R}^n \setminus F$. It follows that, whenever $(x,t) \in P$,

$$h(x,t) = \delta \int_{\underline{R}^n \setminus F} W(x-y,t)\, dy$$

$$\geq \delta \int_{B(x,\gamma\sqrt{t})} W(x-y,t)\, dy$$

$$\geq \delta (4\pi t)^{-n/2} \exp(-\gamma^2/4)\, m(B(x,\gamma\sqrt{t}))$$

$$= \delta \omega_n (\gamma^2/4\pi)^{n/2} \exp(-\gamma^2/4)$$

$$= 2M$$

by (24). Thus

$$h \geq 2M \quad \text{on} \quad (\partial J) \cap (\underline{R}^n \times]0,b[) \qquad (25)$$

   We now use this lower bound for h, together with the weak maximum principle, to show that $|w| \leq h$ on J. For each $k \in \underline{N}$, the set

$$J_k = \{x : (x,k^{-1}) \in J\}$$

is relatively open in $\underline{R}^n$, so that $u(\cdot,k^{-1})\chi(\cdot,k^{-1})$ is continuous on J_k. Hence, as $(x,t) \to (\eta,0^+)$ for each $\eta \in J_k$, the corollary of Theorem 1.3 shows that $v_k(x,t) \to u(\eta,k^{-1})$, and hence that $w_k(x,t) \to 0$. Therefore,

$$\liminf (h(x,t) - |w_k(x,t)|) = \liminf h(x,t) \geq 0 \qquad (26)$$

as $(x,t) \to (\eta,0^+)$ for every $\eta \in J_k$, and in particular whenever $\eta \in (\partial J) \cap (\underline{R}^n \times \{0\})$. Since $|u| \leq M$ on J, and u is continuous, $|u| \leq M$ on $\bar{J} \cap (\underline{R}^n \times]0,b[)$. Therefore, for every $(x,t) \in \bar{J} \cap (\underline{R}^n \times]0,b-k^{-1}[)$,

$$|w_k(x,t)| \leq |u(x,t+k^{-1})| + |v_k(x,t)|$$

$$\leq M + \int_{\underline{R}^n} W(x-y,t)|u(y,k^{-1})\chi(y,k^{-1})|\,dy$$

$$\leq M + \int_{\underline{R}^n} W(x-y,t)M\,dy$$

$$\leq 2M$$

by Lemma 1.1. It now follows from (25) that $|w_k| \leq h$ on $(\partial J) \cap (\underline{R}^n \times]0,b-k^{-1}[)$. Therefore the continuity of w_k and h on $\underline{R}^n \times]0,b-k^{-1}[$ implies that

$$\lim (h(x,t) - |w_k(x,t)|) \geq 0$$

as $(x,t) \to (y,s)$ for every $(y,s) \in (\partial J) \cap (\underline{R}^n \times]0,b-k^{-1}[)$. This, combined with (26), shows that

$$\liminf (h(x,t) - |w_k(x,t)|) \geq 0$$

as $(x,t) \to (y,s)$ from inside J, for every $(y,s) \in (\partial J) \cap (\underline{R}^n \times [0,b-k^{-1}[)$. By applying Theorem 8.2 to both $-h + w_k$ and $-h - w_k$, we deduce that $|w_k| \leq h$ on $J \cap (\underline{R}^n \times]0,b-k^{-1}[)$. Since $w_{k(j)} \to w$ pointwise as $j \to \infty$, it follows that $|w| \leq h$ on $J \cap (\underline{R}^n \times]0,b[)$.

We can now quickly complete the proof. We know that h has limit 0 through Ω_η for m-almost all $\eta \in F$, and also that $h \geq 2M > 0$ on $(\partial J) \cap (\underline{R}^n \times]0,b[)$, by (25). Therefore, for m-almost all $\eta \in F$, the set $\Omega_\eta \cap J$ is of the form $\Omega_\eta \cap G$ for some neighborhood G of $(\eta,0)$. Since $|w| \leq h$ on $\Omega_\eta \cap J$, it follows that $w(x,t) \to 0$ as $(x,t) \to (\eta,0)$ through Ω_η. Finally, $u = w + v$ on $\underline{R}^n \times]0,a[$, and we now know that both v and w have the desired limits m-a. e. on F, so the result is proved.

Remark. Theorem 8.5 cannot be replaced by a similar result at each individual point. Immediately after the proof of Theorem 4.1, we gave an example of a bounded function on R whose Gauss-Weierstrass integral u does not have a parabolic limit at 0. Since u itself is bounded, the hypothesis (22) of Theorem 8.5 holds at 0, where no parabolic limit exists.

As an application of Theorem 8.5, we now extend the result of Theorem 4.5 to the Gauss-Weierstrass integrals of arbitrary signed measures.

Theorem 8.6

Let u be the Gauss-Weierstrass integral on $\underline{R}^n \times]0, a[$ of a signed measure μ, and let $d\mu(y) = f(y) \, dy + d\sigma(y)$, where σ is singular with respect to m. Let Ω be an open subset of $\underline{R}^n \times]0, \infty[$ with the following properties:

(i) there exists $\alpha > 0$ such that, whenever $(x, t) \in \Omega$,

$$\{(y, s) : \|y - x\| < \alpha(\sqrt{s} - \sqrt{t})\} \subseteq \Omega$$

(ii) there exists $\beta > 0$ such that $m(\{x \in \underline{R}^n : (x, t) \in \Omega\}) \leq \beta t^{n/2}$ for all $t > 0$;

(iii) the point $(0, 0)$ is a limit point of Ω.

Then $u(x, t) \to f(\xi)$ as $(x, t) \to (\xi, 0)$ through Ω_ξ for m-almost all $\xi \in \underline{R}^n$, where $\Omega_\xi = (\Omega + \{(\xi, 0)\}) \cap (\underline{R}^n \times]0, a[)$.

Proof. By Theorem 4.1, u has parabolic limits m-a. e. on $\underline{R}^n$. Therefore, for m-almost every $\xi \in \underline{R}^n$, we can find numbers $\gamma > 0$, $M > 0$, and $b \in]0, a[$, such that

$$|u(x, t)| \leq M \quad \text{for all} \quad (x, t) \in \Pi_\gamma^b(\xi)$$

(In fact, we can do this for any γ and b.) It now follows from Theorem 8.5 that $u(x, t)$ tends to a finite limit as $(x, t) \to (\xi, 0)$ through Ω_ξ for m-almost all $\xi \in \underline{R}^n$. By the remark following Theorem 4.5, $\Pi_\alpha^a(\xi) \subseteq \Omega_\xi$ for every ξ, so that the values of these limits are equal m-a. e. to those of the parabolic limits guaranteed by Theorem 4.1, and hence $u(x,t) \to f(\xi)$ through Ω_ξ for m-almost all $\xi \in \underline{R}^n$.

5. LINEAR PARABOLIC EQUATIONS

Theorem 8.1 can be generalized to a wide class of parabolic equations, with essentially the same proof, although sometimes the form of the result is rather different. Let

$$Lu(x,t)$$

$$= \sum_{i,j=1}^{n} a_{ij}(x,t) D_i D_j u(x,t) + \sum_{i=1}^{n} b_i(x,t) D_i u(x,t) + c(x,t)u(x,t) - D_t u(x,t)$$

If $c(x,t) = 0$ for all $(x,t) \in \underline{R}^n \times]0,a[$, the result is unchanged. If $c(x,t) \leq 0$ for all (x,t), the result still holds, but with the extra hypothesis that $u(x_0,t_0) > 0$. The underlying reason for this is the necessity for the constant $\kappa = u(x_0,t_0)$ to satisfy $L\kappa < 0$, that is $\kappa c(x,t) < 0$, whenever $c(x,t) < 0$. The following form of the strong maximum principle, which is suitable for generalization to non-linear equations, sheds some light on this matter.

Let E be an open subset of $\underline{R}^{n+1}$, and let $(x_0,t_0) \in E$. Suppose that u and v are continuous and satisfy $Lu \geq 0$, $Lv \leq 0$ on E, $u \leq v$ on $\Lambda(x_0,t_0)$, and that

$$u(x_0,t_0) = v(x_0,t_0)$$

Then $u = v$ on $\Lambda(x_0,t_0)$.

This result can be easily proved by applying the generalized version of Theorem 8.1 to u - v. Conversely, if the constant $\kappa = u(x_0,t_0)$ satisfies $L\kappa \leq 0$ on E, then the opposite implication is obtained by taking $v = \kappa$.

In extending Theorem 8.1, only minor changes in the proof are required. The auxiliary function w is different, but has similar properties, and the parallelogram P is usually replaced by an oblique cylinder

$$\{(x,t) : r < t < s, \, \|(x-z) - (t-r)(s-r)^{-1}(y-z)\| < \delta\}$$

Theorems 8.2 and 8.3 can be generalized without change provided that $c(x,t) = 0$ for all $(x,t) \in \underline{R}^n \times]0,a[$. If $c(x,t) \leq 0$ for all (x,t), the result of Theorem 8.2 remains valid if the condition that $M > 0$ is added to the hypotheses.

The generalization of Theorem 8.5 depends on Eidel'man's estimates and the sharp lower estimate for the fundamental solution Γ. The function v_k is defined by

$$v_k(x,t) = \int_{\underline{R}^n} \Gamma(x,t; y, k^{-1}) u(y, k^{-1}) \chi(y, k^{-1}) \, dy$$

for all $(x,t) \in \underline{R}^n \times]k^{-1}, a[$, and so is not the exact analogue of the one in section 4. The function v has the natural definition

$$v(x,t) = \int_{\underline{R}^n} \Gamma(x,t; y, 0) f(y) \, dy$$

where f is the limit in $L^2(m)$ of a subsequence of $\{u(\cdot, k^{-1}) \chi(\cdot, k^{-1})\}$. The proof that $v_{k(j)} \rightarrow v$ pointwise is now more difficult, because the kernel $\Gamma(x,t; y, k(j)^{-1})$ depends on j, whereas that in section 4 did not. These changes are necessary only if the coefficients of the equation depend on t, for then the function

$$(x,t) \mapsto \int_{\underline{R}^n} \Gamma(x, t - k^{-1}; y, 0) u(y, k^{-1}) \chi(y, k^{-1}) \, dy$$

is not necessarily a solution. The remainder of the proof is essentially unchanged, although the constant δ must be modified.

Theorem 8.6 can be extended without change.

6. BIBLIOGRAPHICAL NOTES

Maximum principles for parabolic equations have been proved by many authors, of which we mention only a few. The strong maximum principle was first formulated by Nirenberg [1]. The method of proof of Theorem 8.1 was used earlier by Il'in, Kalashnikov, and Oleinik [1], Walter [1], and Besala [4]. The last two authors give generalizations to non-linear equations, which extend the form of the result given in section 5. For certain relatively simple equations, such as the heat equation, a quite different approach is possible. Variations of this technique have been used by Fulks [1], Watson [1, Theorem 7], and Colton [1], for the heat equation, and by Kuptsov [1] for certain equations whose coefficients depend only on t. The weak maximum principle has a longer history, and can be applied on unbounded sets if the condition at the boundary is extended to include unbounded sequences. The form of the result given in Theorem 8.2 is due to Watson [16]. Earlier versions which involved some one-sided limiting conditions can be found in the works of Redheffer [1, Theorem 2] and Watson [4, Theorem 2]. The probabilistic approach to parabolic equations yields a closely related maximum principle (Doob [3]). Note that, despite Theorem 8.2, the maximum of a continuous function on $\bar{E}$ which is a temperature on the bounded open set E, can occur at a point where the largest value of t over $\bar{E}$ is attained. Indeed, if C is the cone

$$\{(x,t) : 0 < t < 1 - \|x\|\}$$

and f is an arbitrary continuous real-valued function on ∂c, then there exists a continuous function u on $\bar{C}$ such that u is a temperature on C and u = f on ∂C. Here f can be chosen so that it has a strict maximum at $(0,1)$, and then Theorem 8.2 shows that this is also the maximum of u. This is widely known, but a convenient reference is Watson's paper [4].

Theorem 8.3 has not previously been given in quite this form, but is an obvious consequence of Theorem 8.2.

The result of Lemma 8.2 is often referred to by saying that $L^p(\mu)$ is separable.

Theorem 8.4 is often referred to as a weak or weak* compactness theorem. See the bibliographical notes on Theorem 2.8 for an explanation of this expression in a parallel context. The fact that (21) holds for every $f \in L^p(\mu)$ is sometimes called the converse of Hölder's inequality. The use of Theorem 8.4, or more general results, in the proof of Theorem 6.4 and its analogues, is more common than the use of Theorem 6.1, which is neater but requires Theorem 4.1 whereas the use of Theorem 8.4 does not. For this alternative approach to Theorem 6.4, see the works of Gehring [1, Theorem 10], Flett [1, Theorem 5], Johnson [1], Chabrowski [8], and Chabrowski and Johnson [1].

Theorem 8.5 is new, but the special case where $\Omega_\xi = \Pi_\alpha^a(\xi)$ for an arbitrary $\alpha > 0$, which has essentially the same proof, was given by Hattemer [1] for the heat equation. The extension to equations with coefficients that depend only on x was given by Chabrowski [9, Theorem 8.1], and the general case was proved by Watson [15]. The result is sometimes called the local Fatou theorem. The boundedness hypothesis in Theorem 8.5 can be weakened to one-sided boundedness in the case where $\Omega_\xi = \Pi_\alpha^a(\xi)$, but substantially different techniques must be used. This was done by Kemper [1, 2] for the heat equation, and by Mair [1] for more general equations. The methods used by these authors require a good deal of general potential theory, and the necessary background can be found in Doob's book [3].

Theorem 8.6 is new.

Further parabolic limit theorems for temperatures were given by Hattemer [1, 2].

References

D. G. Aronson
1. The fundamental solution of a linear parabolic equation containing a small parameter, <u>Illinois J. Math.</u>, 3(1959)580-619.
2. Uniqueness of positive weak solutions of second order parabolic equations, <u>Ann. Polon. Math.</u>, 16(1965)285-303.
3. Isolated singularities of solutions of second order parabolic equations, <u>Arch. Rational Mech. Anal.</u>, 19(1965)231-238.
4. Bounds for the fundamental solution of a parabolic equation, <u>Bull. Amer. Math. Soc.</u>, 73(1967)890-896.
5. Non-negative solutions of linear parabolic equations, <u>Ann. Scuola Norm. Sup. Pisa Cl. Sci.</u>, 22(1968)607-694.
6. Non-negative solutions of linear parabolic equations: an addendum, <u>Ann. Scuola Norm. Sup. Pisa Cl. Sci.</u>, 25(1971)221-228.
7. Widder's inversion theorem and the initial distribution problem, <u>S.I.A.M. J. Math. Anal.</u>, 12(1981)639-651.

D. G. Aronson and P. Besala
1. Uniqueness of solutions of the Cauchy problem for parabolic

equations, J. Math. Anal. Appl., 13(1966)516-526. Correction, ibid., 17(1967)194-196.

2. Parabolic equations with unbounded coefficients, J. Differential Equations, 3(1967)1-14.

3. Uniqueness of positive solutions of parabolic equations with unbounded coefficients, Colloq. Math., 18(1967)125-135.

P. Besala

1. A remark on a problem of M. Krzyżański concerning second order parabolic equations, Colloq. Math., 10(1963)161-164.

2. Evaluations of solutions of a second order parabolic equation, Colloq. Math., 10(1963)165-171.

3. On a certain property of the fundamental solution of a linear parabolic equation the last coefficient of which is unbounded, Bull. Acad. Polon. Sci. Sér. Sci. Math. Astr. Phys., 11(1963)155-158.

4. An extension of the strong maximum principle for parabolic equations, Bull. Acad. Polon. Sci. Sér. Sci. Math. Astr. Phys., 19(1971)1003-1006.

5. Function classes pertaining to differential inequalities of parabolic type in unbounded regions, Ann. Polon. Math., 25(1972)281-291.

6. On the existence of a fundamental solution for a parabolic differential equation with unbounded coefficients, Ann. Polon. Math., 29(1975)403-409.

P. Besala and M. Krzyżański

1. Un théorème d'unicité de la solution du problème de Cauchy pour l'équation linéaire normale parabolique du second ordre, Atti Accad. Naz. Lincei Rend. Cl. Sci. Fis. Mat. Natur., 33(1962) 230-236.

P. Besala and H. Ugowski

1. Some uniqueness theorems for solutions of parabolic and elliptic partial differential equations in unbounded regions, Colloq. Math., 20(1969)127-141.

A. S. Besicovitch

1. A general form of the covering principle and relative differentiation of additive functions, Proc. Cambridge Philos. Soc., 41 (1945)103-110.

2. A general form of the covering principle and relative differentiation of additive functions II, Proc. Cambridge Philos. Soc., 42(1946)1-10.

G. Birkhoff and J. Kotik

1. Note on the heat equation, Proc. Amer. Math. Soc., 5(1954) 162-167.

J. Blackman

1. The inversion of solutions of the heat equation for the infinite rod, Duke Math. J., 19(1952)671-682.

W. Bodanko
1. Sur le problème de Cauchy et les problèmes de Fourier pour les équations paraboliques dans un domaine non borné, Ann. Polon. Math., 18(1966)79-94.
2. Les propriétés des solutions non négatives de l'équation linéaire normale parabolique, Ann. Polon. Math., 20(1968)107-117.

P. L. Butzer and H. Berens
1. Semi-groups of operators and approximation, Springer, 1967.

S. Cąkała
1. Solution fondamentale spéciale de l'équation de la chaleur et son application à la construction de la solution fondamentale spéciale d'un système parabolique, Demonstratio Math., 5(1973)147-164.

L. Carleson
1. On the existence of boundary values for harmonic functions in several variables, Ark. Mat., 4(1961)393-399.

J. Chabrowski
1. Sur l'unicité de la solution du problème de Cauchy pour l'équation linéaire du type parabolique, Ann. Scuola Norm. Sup. Pisa Cl. Sci., 23(1969)547-552.
2. Propriétés des solutions faibles non négatives de l'équation parabolique, Nagoya Math. J., 39(1970)119-125.
3. Remarques sue l'unicité du problème de Cauchy, Colloq. Math., 21(1970)133-139.
4. Sur la construction de la solution fondamentale de l'équation parabolique aux coefficients non bornés, Colloq. Math., 21(1970) 141-148.
5. Sur la mesure parabolique, Colloq. Math., 21(1970)291-301.
6. Sur l'unicité du problème de Cauchy dans une classe de fonctions non bornées, Ann. Polon. Math., 24(1971)127-135.
7. Certaines propriétés des solutions non négatives d'un système parabolique d'équations, Ann. Polon. Math., 24(1971)137-143.
8. Representation theorems and Fatou theorems for parabolic systems in the sense of Petrovskiĭ, Colloq. Math., 31(1974)301-315.
9. Representation theorems for parabolic systems, J. Austral. Math. Soc. Ser. A, 32(1982)246-288.
10. On maximal functions and parabolic limits, Math. Z., 184(1983) 271-282.

J. Chabrowski and R. Johnson
1. An axiomatic approach to the problem of representation of solutions of parabolic equations and systems, Bull. Acad. Polon. Sci. Sér. Sci. Math. Astr. Phys., 26(1978)887-893.

J. Chabrowski and N. A. Watson
1. Properties of solutions of weakly coupled parabolic systems, J. London Math. Soc. (2), 23(1981)475-495.

L.-S. Chen
1. On the behavior of solutions for large |x| of parabolic equations with unbounded coefficients, Tôhoku Math. J., 20(1968)585-595.

L.-S. Chen and T. Kuroda
1. On the behavior of solutions of parabolic equations with unbounded coefficients, Ann. Polon. Math., 23(1970)57-64.

L.-S. Chen, T. Kuroda and T. Kusano
1. Some parabolic equations with unbounded coefficients, Funkcialaj Ekvacioj, 16(1973)1-28.
2. Weakly coupled parabolic systems with unbounded coefficients, Hiroshima Math. J., 3(1973)1-14.

L.-S. Chen, J.-S. Lin and C.-C. Yeh
1. Asymptotic behavior of solutions for large |x| of weakly coupled parabolic systems with unbounded coefficients, Hiroshima Math. J., 4(1974)477-490.

D. Colton
1. The strong maximum principle for the heat equation, Proc. Edinburgh Math. Soc., 27(1984)297-299.

J. L. B. Cooper
1. The uniqueness of the solution of the equation of heat conduction, J. London Math. Soc., 25(1950)173-180.

C. Cosner
1. Asymptotic behavior of solutions of second order parabolic partial differential equations with unbounded coefficients, J. Differential Equations, 35(1980)407-428.

J. L. Doob
1. A relative limit theorem for parabolic functions, Trans. 2nd Prague Conference Information Theory, Stat. Decision Functions, Random Processes, (1959)61-70.
2. Relative limit theorems in analysis, J. Anal. Math., 8(1960/61) 281-306.
3. Classical potential theory and its probabilistic counterpart, Springer, 1984.

F. G. Dressel
1. Uniqueness theorems for the heat equation, Trans. 19th Conference Army Mathematicians, (1973)1181-1187.

S. D. Eidel'man
1. Fundamental matrices of the solutions of general parabolic systems, Dokl. Akad. Nauk S. S. S. R., 120(1958)980-983.
2. On the Cauchy problem for parabolic systems with increasing coefficients, Dokl. Akad. Nauk S. S. S. R., 127(1959)760-763.
3. On fundamental solutions of parabolic systems II, Mat. Sb., 53(1961)73-136. Corrections, ibid., 58(1962)128.

S. D. Eidel'man and I. M. Petrushko

1. Solvability of the Cauchy problem for second-order parabolic equations in the class of arbitrarily rising functions, Ukrainsk. Mat. Zh., 19 No. 1 (1967)108-113; Ukr. Math. J., 19 No. 1 (1967)93-97.

S. D. Eidel'man and F. O. Porper

1. Properties of solutions of second order parabolic equations with dissipation, Differentsial'nye Uravneniya, 7(1971)1684-1695; Differential Equations, 7(1971)1280-1288.

2. Two-sided estimates of fundamental solutions of second order parabolic equations, and some applications, Uspekhi Mat. Nauk, 39 No. 3 (1984)107-156; Russian Math. Surveys, 39 No. 3 (1984) 119-178.

E. B. Fabes and C. E. Kenig

1. Examples of singular parabolic measures and singular transition probability densities, Duke Math. J., 48(1981)845-856.

E. B. Fabes and U. Neri

1. Characterization of temperatures with initial data in BMO, Duke Math. J., 42(1975)725-734.

T. M. Flett

1. Temperatures, Bessel potentials and Lipschitz spaces, Proc. London Math. Soc. (3),22(1971)385-451.

C. Foias and M. Nicolescu

1. Représentation de Poisson et problème de Cauchy pour l'équation de la chaleur I, Atti Accad. Naz. Lincei Ren. Cl. Sci. Fis. Mat. Natur., 38(1965)466-476.

2. Représentation de Poisson et problème de Cauchy pour l'équation de la chaleur II, Atti Accad. Naz. Lincei Rend. Cl. Sci. Fis. Mat. Natur., 38(1965)621-626.

3. Sur l'unicité du problème de Cauchy pour l'équation de la chaleur, Atti Accad. Naz. Lincei Rend. Cl. Sci. Fis. Mat. Natur., 40 (1966)785-791.

A. Friedman

1. On the uniqueness of the Cauchy problem for parabolic equations, Amer. J. Math., 81(1959)503-511.

2. Partial differential equations of parabolic type, Prentice-Hall, 1964.

W. Fulks

1. A mean value theorem for the heat equation, Proc. Amer. Math. Soc., 17(1966)6-11.

F. W. Gehring

1. On solutions of the equation of heat conduction, Michigan Math. J., 5(1958)191-202.

2. The boundary behavior and uniqueness of solutions of the heat equation, Trans. Amer. Math. Soc., 94(1960)337-364.

R. Guenther

1. Representation theorems for linear second-order parabolic partial differential equations, J. Math. Anal. Appl., 17(1967)488-501.
2. A lower bound for fundamental solutions for parabolic equations, Portugal Math., 26(1967)93-100.

A. L. Gusarov

1. On a sharp Liouville theorem for solutions of a parabolic equation on a characteristic, Mat. Sb., 97(1975)379-394; Math. U. S. S. R. Sb., 26(1975)349-364.

J. R. Hattemer

1. Boundary behavior of temperatures I, Studia Math., 25(1964) 111-155.
2. Boundary behavior of temperatures II, Illinois J. Math., 10(1966) 466-469.

R. M. Hayne

1. Uniqueness in the Cauchy problem for parabolic equations, Trans. Amer. Math. Soc., 241(1978)373-399.

A. M. Il'in

1. On the fundamental solution for a parabolic equation, Dokl. Akad. Nauk S. S. S. R., 147(1962)768-771; Soviet Math. Dokl., 3(1962) 1697-1700.

A. M. Il'in, A. S. Kalashnikov and O. A. Oleinik

1. Linear equations of the second order of parabolic type, Uspekhi Mat. Nauk, 17 No. 3 (1962)3-147; Russian Math. Surveys, 17 No. 3 (1962)1-143.

R. Johnson

1. Representation theorems and Fatou theorems for second-order linear parabolic partial differential equations, Proc. London Math. Soc. (3), 23(1971)325-347. Corrigendum, ibid., 25(1972) 192.
2. Representation theorems for the heat equation, Proc. London Math. Soc. (3), 24(1972)367-384.
3. Temperatures, Riesz potentials, and the Lipschitz spaces of Herz, Proc. London Math. Soc. (3), 27(1973)290-316.

B. F. Jones, Jr.

1. A fundamental solution for the heat equation which is supported in a strip, J. Math. Anal. Appl., 60(1977)314-324.

L. I. Kamynin and B. N. Khimchenko

1. On uniqueness of the solution of the Cauchy problem for a second order parabolic equation with non-negative characteristic form, Dokl. Akad. Nauk S. S. S. R., 248 No. 2 (1979)290-293; Soviet Math. Dokl., 20(1979)1004-1008.

2. On Tikhonov-Täcklind classes of uniqueness for degenerate parabolic equations of second order, Dokl. Akad. Nauk S. S. S. R., 252 No. 4 (1980)784-788; Soviet Math. Dokl., 21(1980)815-819.
3. Tikhonov-Petrovskiĭ problem for second-order parabolic equations, Sibirsk. Mat. Zh., 22 No. 5 (1981)78-109; Siberian Math. J., 22(1981)709-734.
4. On an aspect of the uniqueness problem for second-order parabolic equations, Dokl. Akad. Nauk S. S. S. R., 270 No. 2 (1983)274-277; Soviet Math. Dokl., 27(1983)584-587.

M. Kato
1. On positive solutions of the heat equation, Nagoya Math. J., 30 (1967)203-207.

J. T. Kemper
1. Kernel functions and parabolic limits for the heat equation, Bull. Amer. Math. Soc., 76(1970)1319-1320.
2. Temperatures in several variables: kernel functions, representations, and parabolic boundary values, Trans. Amer. Math. Soc., 167(1972)243-262.

A. Korányi and J. C. Taylor
1. Fine convergence and parabolic convergence for the Helmholtz equation and the heat equation, Illinois J. Math., 27(1983)77-93.

M. Krzyżański
1. Certaines inégalités relatives aux solutions de l'équation parabolique linéaire normale, Bull. Acad. Polon. Sci. Sér. Sci. Math. Astr. Phys., 7(1959)131-135.
2. Sur la solution fondamentale de l'équation linéaire normale du type parabolique dont le dernier coefficient est non borné I, Atti Accad. Naz. Lincei Rend. Cl. Sci. Fis. Mat. Natur., 32(1962) 326-330.
3. Sur la solution fondamentale de l'équation linéaire normale du type parabolique dont le dernier coefficient est non borné II, Atti Accad. Naz. Lincei Rend. Cl. Sci. Fis. Mat. Natur., 32(1962) 471-476.
4. Évaluations des solutions de l'équation linéaire du type parabolique à coefficients non bornés, Ann. Polon. Math., 11(1962)253-260.
5. Une propriété des solutions de l'équation linéaire du type parabolique aux coefficients non bornés, Ann. Polon. Math., 12(1962) 209-211.
6. Sur les solutions non négatives de l'équation linéaire normale parbolique, Rev. Roumaine Math. Pures Appl., 9(1964)393-408.
7. La mesure parabolique et le problème de Cauchy pour l'équation linéaire parabolique normale, Colloq. Math., 16(1967)123-131.

M. Krzyżański and A. Szybiak

1. Construction et étude de la solution fondamentale de l'équation
   linéaire du type parabolique dont le dernier coefficient est non
   borné I, Atti Accad. Naz. Lincei Rend. Cl. Sci. Fis. Mat. Natur.,
   27(1959)26-30.
2. Construction et étude de la solution fondamentale de l'équation
   linéaire du type parabolique dont le dernier coefficient est non
   borné II, Atti Accad. Naz. Lincei Rend. Cl. Sci. Fis. Mat.
   Natur., 27(1959)113-117.

L. P. Kuptsov (Kupcov)

1. The mean property and the maximum principle for parabolic
   equations of second order, Dokl. Akad. Nauk S. S. S. R., 242
   No. 3 (1978)529-532; Soviet Math. Dokl., 19(1978)1140-1144.

T. Kusano

1. On the decay for large |x| of solutions of parabolic equations with
   unbounded coefficients, Publ. Res. Inst. Math. Sci. Kyoto Univ.
   Ser. A, 3(1967)203-210.
2. On the behavior for large |x| of solutions of parabolic equations
   with unbounded coefficients, Funkcialaj Ekvacioj, 11(1968)169-174.
3. Remarks on the behavior of solutions of second order parabolic
   equations with unbounded coefficients, Funkcialaj Ekvacioj,
   11(1968)197-205.

J. E. Littlewood

1. Mathematical notes (4): On a theorem of Fatou, J. London Math.
   Soc., 2(1927)172-176.

B. A. Mair

1. Fine and parabolic limits for solutions of second-order linear
   parabolic equations on an infinite slab, Trans. Amer. Math. Soc.,
   284(1984)583-599. Erratum, ibid., 291(1985)381.

B. A. Mair and J. C. Taylor

1. Integral representation of positive solutions of the heat equation,
   Théorie du Potentiel (pp. 419-433), Springer, 1984.

P. Mustață

1. Sur la représentation de Poisson des solutions du problème de
   Cauchy pour l'équation de la chaleur, Rev. Roumaine Math. Pures
   Appl., 11(1966)673-691.
2. Un théorème d'unicité de la solution du problème de Cauchy pour
   l'équation linéaire parabolique du second ordre, Ann. Scuola Norm.
   Sup. Pisa Cl. Sci., 21(1967)507-526.

A. Nagel and E. M. Stein

1. On certain maximal functions and approach regions, Advances
   Math., 54(1984)83-106.

L. Nirenberg
1. A strong maximum principle for parabolic equations, <u>Comm.</u>
    <u>Pure Appl. Math.</u>, 6(1953)167-177.
O. A. Oleinik and E. V. Radkevich
1. The method of introducing a parameter in the study of evolutionary
    systems, <u>Uspekhi Mat. Nauk</u>, 33 No. 5 (1978)7-76; <u>Russian Math.</u>
    <u>Surveys,</u> 33 No. 5 (1978)7-84.
H. Pollard
1. One-sided boundedness as a condition for the unique solution of
    certain heat equations, <u>Duke Math. J.</u>, 11(1944)651-653.
R. Redheffer
1. The sharp maximum principle for nonlinear inequalities, <u>Indiana</u>
    <u>Univ. Math. J.</u>, 21(1971)227-248.
S. M. Robinson
1. Some properties of the fundamental solution of the parabolic
    equation, <u>Duke Math. J.</u>, 27(1960)195-220.
P. C. Rosenbloom and D. V. Widder
1. Expansions in terms of heat polynomials and associated functions,
    <u>Trans. Amer. Math. Soc.</u>, 92(1959)220-266.
W. Rudin
1. <u>Real and complex analysis</u>, McGraw-Hill, 1970.
V. L. Shapiro
1. The uniqueness of solutions of the heat equation in an infinite
    strip, <u>Trans. Amer. Math. Soc.</u>, 125(1966)326-361.
J. D. Stafney
1. On uniqueness for certain parabolic partial differential equations,
    <u>Boll. Un. Mat. Italiana Ser. A</u>, 15(1978)237-242.
S. Täcklind
1. Sur les classes quasianalytiques des solutions des équations aux
    dérivées partielles du type parabolique, <u>Nova Acta Regiae Soc.</u>
    <u>Sci. Upsaliensis,</u> 10 No. 3 (1936)1-56.
A. Tychonoff (Tikhonov)
1. Théorèmes d'unicité pour l'équation de la chaleur, <u>Mat. Sb. (Rec.</u>
    <u>Math.</u>), 42(1935)199-216.
W. Walter
1. <u>Differential and integral inequalities</u>, Springer, 1970.
N. A. Watson
1. A theory of subtemperatures in several variables, <u>Proc. London</u>
    <u>Math. Soc.</u> (3), 26(1973)385-417.
2. Classes of subtemperatures on infinite strips, <u>Proc. London</u>
    <u>Math. Soc.</u> (3), 27(1973)723-746.
3. Uniqueness and representation theorems for parabolic equations,
    <u>J. London Math. Soc.</u> (2), 8(1974)311-321.

4. Green functions, potentials, and the Dirichlet problem for the heat equation, Proc. London Math. Soc. (3), 33(1976)251-298. Corrigendum, ibid., 37(1978)32-34.
5. Differentiation of measures and initial values of temperatures, J. London Math. Soc. (2), 16(1977)271-282.
6. Thermal capacity, Proc. London Math. Soc. (3), 37(1978)342-362.
7. The rate of spatial decay of non-negative solutions of linear parabolic equations, Arch. Rational Mech. Anal., 68(1978)121-124.
8. Uniqueness and representation theorems for the inhomogeneous heat equation, J. Math. Anal. Appl., 67(1979)513-524.
9. Positive thermic majorization of temperatures on infinite strips, J. Math. Anal. Appl., 68(1979)477-487.
10. Initial singularities of Gauss-Weierstrass integrals and their relations to Laplace transforms and Hausdorff measures, J. London Math. Soc. (2), 21(1980)336-350.
11. The asymptotic behaviour of temperatures and subtemperatures, Proc. London Math. Soc. (3), 42(1981)501-532.
12. Initial and relative limiting behaviour of temperatures on a strip, J. Austral. Math. Soc. Ser. A, 33(1982)213-228.
13. Solutions of parabolic equations with initial values locally in L^p, J. Math. Anal. Appl., 89(1982)86-94.
14. Boundary measures of solutions of partial differential equations, Mathematika, 29(1982)67-82.
15. Parabolic limits of solutions of weakly coupled parabolic systems, J. Math. Anal. Appl., 95(1983)278-283.
16. The weak maximum principle for parabolic differential inequalities, Rend. Circ. Mat. Palermo (2), 33(1984)421-425.
17. Classes of subtemperatures and uniqueness in the Cauchy problem for the heat equation, J. London Math. Soc. (2), 32(1985)107-115.
18. Growth of solutions of weakly coupled parabolic systems and Laplace's equation, J. Austral. Math. Soc. Ser. A, 41(1986)391-403.
19. On the representation of solutions of the heat equation and weakly coupled parabolic systems, J. London Math. Soc. (2), 34(1986)457-472.

D. V. Widder
1. Positive temperatures on an infinite rod, Trans. Amer. Math. Soc., 55(1944)85-95.
2. The heat equation, Academic Press, 1975.

C. H. Wilcox
1. Positive temperatures with prescribed initial heat distributions, Amer. Math. Monthly, 87(1980)183-186.

G. N. Zolotarev
1. The uniqueness of the solution of the Cauchy problem for systems parabolic in the sense of I. G. Petrovskiǐ, Isv. Vysš. Učebn. Zaved. Mat., 2(1958)118-135.

A. Zygmund
1. On a theorem of Littlewood, Summa Brasil. Math., 2 No. 5 (1949)51-57.

Index

Abundant Vitali covering, 228
  generator of, 232
Adjoint:
  of the heat operator, 43
  of a linear parabolic equation,
    66

Cauchy problem, 27, 86
Chapman-Kolmogorov relation, 31
Class:
  $C_c(\underline{R}^n)$, 37
  $L^p(\mu)$, 182
  R_a, 98
  S_a, 75
  Σ_b, 73
  U_a, 91

Closed lower face, 247

Delta-function property, 5
Difference of signed measures,
    222
  non-negativity of, 223
Dirac δ-measure, 5

Eidel'man's estimates, 26

Fatou theorem, 143
  local, 273
Fourier-Poisson:
  integral, 31
  representation, 31

Fundamental solution:
  of the heat equation, 2
  of a linear parabolic equation, 25
  sharp lower estimate for, 28
  the, 32
  uniqueness of, 95-98

Gauss-Weierstrass integral, 7, 8
Generator, 232
Green's formula:
  for the heat equation, 43
  for a linear parabolic equation,
  66

Hausdorff's maximality theorem,
  253
Heat equation, 2
Heat operator, 43, 247
Hölder:
  condition, 26
  conjugate, 182
  continuous function, 26
Hölder's inequality, converse of,
  273
Huygens property, 31

Initial function, 86

Kolmogorov's equation, 31

$\Lambda(x_0, t_0)$, 247
Lateral surface, 247
Least thermic majorant, 165, 194
Lebesgue:
  class $L^p(\mu)$, 182
  point, 112
  set, 112
Lipschitz condition, 31

Local Fatou theorem, 273
Locally uniform convergence, 54
Lower symmetric derivative, 147

Maximal functions:
  M_Ω, 126
  M_Ω^W, 126
  N, 131
Maximal totally ordered subset,
  253
Maximum principle:
  extended, 39
  on circular cylinders, 33, 66
  strong, 246
  weak, 246
Means:
  M_b, 70, 103
  $M_{b,p}$, 185, 200, 215
Minimality of W, 164
Modulus of continuity, 205

Narrow convergence, 68
Narrow Vitali sense, 156, 223
Negative part, 43
Negative variation, 7, 222
Non-negativity of a difference of
  measures, 223
Normal, 147
  limit, 147

Open parallelogram, 247
Open upper face, 247
Ordered set:
  partially, 253
  totally, 253

Parabolic equation, 25
Parabolic limit, 112
Parabolic measure, 29

Para-cone condition, 128
Parametrix method, 26
Partially ordered set, 253
Poisson:
  integral, 31
  representation, 31
Poisson-Weierstrass integral, 31
Polygonal path, 247
Positive part, 43
Positive variation, 7, 222

Semigroup property, 19, 31
Separability:
  of $C_c(\underline{R}^n)$, 50
  of $L^p(\mu)$, 256
Separable space, 68
Sharp lower estimate for fundamental solution, 28
Signed measure, 7
Singular part, 112
Solution:
  of the heat equation, 1
  of a linear parabolic equation,
  25

Symmetric derivative, 112, 147
  lower, 147
  upper, 147

Temperature, 2
Thermic majorant, 100
Through, 137
Totally ordered set, 253

Uniformly parabolic equation, 25
Union family, 232
Uniqueness class, 86
Upper symmetric derivative, 147

Vague convergence, 68

W, 2
Weak (or weak*) compactness:
  in $L^p(\mu)$, 273
  of measures, 68
Weak (or weak*) convergence, 68, 245
Weierstrass integral, 31